Environmental Justice and Climate Change

Environmental Justice and Climate Change

Assessing Pope Benedict XVI's Ecological Vision for the Catholic Church in the United States

Jame Schaefer and Tobias Winright
Editors

LEXINGTON BOOKS
Lanham • Boulder • New York • Toronto • Plymouth, UK

Published by Lexington Books
A wholly owned subsidiary of Rowman & Littlefield
4501 Forbes Boulevard, Suite 200, Lanham, Maryland 20706
www.rowman.com

10 Thornbury Road, Plymouth PL6 7PP, United Kingdom

British Library Cataloguing in Publication Information Available

Library of Congress Cataloging-in-Publication Data

Environmental justice and climate change : assessing Pope Benedict XVI's ecological vision for the Catholic Church in the United States / edited by Jame Schaefer and Tobias Winright.
p. cm.
Includes bibliographical references and index.
ISBN 978-0-7391-8380-9 (cloth : alk. paper) -- ISBN978-0-7391-8381-6 (electronic)
1. Human ecology--Religious aspects--Catholic Church. 2. Environmental justice--Religious aspects--Christianity. 3. Benedict XVI, Pope, 1927- 4. Climatic changes. 5. Environmentalism--Religious aspects--Catholic Church. 6. Catholic Church--Doctrines. I. Schaefer, Jame, editor of compilation.
BX1795.H82E58 2013
261.8'8--dc23
2013031486

∞™ The paper used in this publication meets the minimum requirements of American National Standard for Information Sciences Permanence of Paper for Printed Library Materials, ANSI/NISO Z39.48-1992.

Printed in the United States of America

Contents

Acknowledgments vii

Preface: If you want to cultivate peace, protect creation, Message on the 2010 World Day of Peace ix
Pope Benedict XVI

Introduction: Celebrating and Advancing Magisterial Discourse on the Ecological Crisis xix
Jame Schaefer

I: Human and Natural Ecology/Human Life and Dignity 1

1 Bonaventure in Benedict: Franciscan Wisdom for Human Ecology 3
Keith Douglass Warner, OFM

2 If You Want Responsibility, Build Relationship: A Personalist Approach to Benedict XVI's Environmental Vision 19
Mary A. Ashley

3 Natural Law and the Natural Environment: Pope Benedict XVI's Vision beyond Utilitarianism and Deontology 43
Michael Baur

II: Solidarity, Justice, Poverty, and the Common Good 59

4 Human, Social, and Natural Ecology: Three Ecologies, One Cosmology, and the Common Good 61
Scott G. Hefelfinger

5 Commodifying Creation?: Pope Benedict XVI's Vision of the Goods of Creation Intended for All 83
Christiana Z. Peppard

6 The Grammar of Creation: Agriculture in the Thought of Pope Benedict XVI 103
Matthew Philipp Whelan

III: The Sacramentality of Creation 125

7 The Way of Wisdom: "Keep hold of instruction; do not let go; guard her, for she is your life" (Prov 3:14) 127
Elizabeth Groppe

8 The World as God's Icon: Creation, Sacramentality, Liturgy 149
Msgr. Kevin W. Irwin

9 Pope Benedict XVI's Cosmic Soteriology and the Advancement of Catechesis on the Environment 173
Jeremiah Vallery

IV: Our Catholic Faith in Action **195**

10 Discernment of the Church and the Dynamics of the Climate Change Convention 197
John T. Brinkman, MM

11 American Lifestyles and Structures of Sin: The Practical Implications of Pope Benedict XVI's Ecological Vision for the American Church 215
David Cloutier

12 American Nature Writing As a Critically-Appropriated Resource for Catholic Ecological Ethics 237
Anselma Dolcich-Ashley

Appendix A: Keynote Address at the Catholic Consultation on Environmental Justice and Climate Change November 7, 2012 257
Bishop Bernard Unabali, Diocese of Bougainville, Papua New Guinea

Appendix B: Homily: Catholic Consultation on Environmental Justice and Climate Change, November 8, 2012 263
Bishop Donald Kettler, Diocese of Fairbanks

Index 267

About the Contributors 275

Acknowledgments

We are especially grateful to the bishops who participated in the consultation at The Catholic University of America, the scholars who presented papers and those who submitted proposals, Dr. William Dinges who facilitated the selection of scholars, and members of the advisory group who vetted the final selections. We are deeply indebted to Cecilia Calvo who represented the U.S. Conference of Catholic Bishops in planning and executing the consultation, Dan Misleh who heads the remarkable efforts of the Catholic Coalition on Climate Change, and Dan DiLeo who served eagerly and resourcefully as the point person for this anthology on behalf of the sponsoring organizations. And, we thank Eric Wrona who contracted for this anthology on behalf of Lexington Books and Ethan Feinstein who ushered it to publication.

Preface

If you want to cultivate peace, protect creation, Message on the 2010 World Day of Peace

Pope Benedict XVI

1. At the beginning of this New Year, I wish to offer heartfelt greetings of peace to all Christian communities, international leaders, and people of good will throughout the world. For this XLIII World Day of Peace I have chosen the theme: *If You Want to Cultivate Peace, Protect Creation*. Respect for creation is of immense consequence, not least because "creation is the beginning and the foundation of all God's works," [1] and its preservation has now become essential for the pacific coexistence of mankind. Man's inhumanity to man has given rise to numerous threats to peace and to authentic and integral human development—wars, international and regional conflicts, acts of terrorism, and violations of human rights. Yet no less troubling are the threats arising from the neglect—if not downright misuse—of the earth and the natural goods that God has given us. For this reason, it is imperative that mankind renew and strengthen "that covenant between human beings and the environment, which should mirror the creative love of God, from whom we come and towards whom we are journeying".[2]
2. In my Encyclical *Caritas in Veritate,* I noted that integral human development is closely linked to the obligations which flow from *man's relationship with the natural environment*. The environment must be seen as God's gift to all people, and the use we make of it entails a shared responsibility for all humanity, especially the poor and future generations. I also observed that whenever nature, and human beings in particular, are seen merely as products of chance or an evolutionary determinism, our overall sense of responsibility wanes.[3] On the other hand, seeing creation as God's gift to humanity helps us understand our vocation and worth as human beings. With the Psalmist, we can exclaim with wonder: "When I look at your heavens, the work of your hands, the moon and the stars which you have established; what is man that you are mind-

ful of him, and the son of man that you care for him?" (Ps 8:4-5). Contemplating the beauty of creation inspires us to recognize the love of the Creator, that Love which "moves the sun and the other stars".[4]

3. Twenty years ago, Pope John Paul II devoted his Message for the World Day of Peace to the theme: *Peace with God the Creator, Peace with All of Creation*. He emphasized our relationship, as God's creatures, with the universe all around us. "In our day," he wrote, "there is a growing awareness that world peace is threatened . . . also by a lack of *due respect for nature*." He added that "*ecological awareness*, rather than being downplayed, needs to be helped to develop and mature, and find fitting expression in concrete programmes and initiatives".[5] Previous Popes had spoken of the relationship between human beings and the environment. In 1971, for example, on the eightieth anniversary of Leo XIII's Encyclical *Rerum Novarum*, Paul VI pointed out that "by an ill-considered exploitation of nature (man) risks destroying it and becoming in his turn the victim of this degradation." He added that "not only is the material environment becoming a permanent menace—pollution and refuse, new illnesses and absolute destructive capacity—but the human framework is no longer under man's control, thus creating an environment for tomorrow which may well be intolerable. This is a wide-ranging social problem which concerns the entire human family".[6]
4. Without entering into the merit of specific technical solutions, the Church is nonetheless concerned, as an "expert in humanity," to call attention to the relationship between the Creator, human beings and the created order. In 1990 John Paul II had spoken of an "ecological crisis" and, in highlighting its primarily ethical character, pointed to the "urgent moral need for a new solidarity".[7] His appeal is all the more pressing today, in the face of signs of a growing crisis which it would be irresponsible not to take seriously. Can we remain indifferent before the problems associated with such realities as climate change, desertification, the deterioration and loss of productivity in vast agricultural areas, the pollution of rivers and aquifers, the loss of biodiversity, the increase of natural catastrophes and the deforestation of equatorial and tropical regions? Can we disregard the growing phenomenon of "environmental refugees," people who are forced by the degradation of their natural habitat to forsake it—and often their possessions as well—in order to face the dangers and uncertainties of forced displacement? Can we remain impassive in the face of actual and potential conflicts involving access to natural resources? All these are issues with a profound impact on the exercise of human rights, such as the right to life, food, health and development.

5. It should be evident that the ecological crisis cannot be viewed in isolation from other related questions, since it is closely linked to the notion of development itself and our understanding of man in his relationship to others and to the rest of creation. Prudence would thus dictate a *profound, long-term review of our model of development,* one which would take into consideration the meaning of the economy and its goals with an eye to correcting its malfunctions and misapplications. The ecological health of the planet calls for this, but it is also demanded by the cultural and moral crisis of humanity whose symptoms have for some time been evident in every part of the world.[8] Humanity needs a *profound cultural renewal;* it needs to *rediscover those values which can serve as the solid basis* for building a brighter future for all. Our present crises—be they economic, food-related, environmental or social—are ultimately also moral crises, and all of them are interrelated. They require us to rethink the path which we are travelling together. Specifically, they call for a lifestyle marked by sobriety and solidarity, with new rules and forms of engagement, one which focuses confidently and courageously on strategies that actually work, while decisively rejecting those that have failed. Only in this way can the current crisis become an opportunity for discernment and new strategic planning.
6. Is it not true that what we call "nature" in a cosmic sense has its origin in "a plan of love and truth"? The world "is not the product of any necessity whatsoever, nor of blind fate or chance . . . . The world proceeds from the free will of God; he wanted to make his creatures share in his being, in his intelligence, and in his goodness".[9] The *Book of Genesis,* in its very first pages, points to the wise design of the cosmos: it comes forth from God's mind and finds its culmination in man and woman, made in the image and likeness of the Creator to "fill the earth" and to "have dominion over" it as "stewards" of God himself (cf. Gen 1:28). The harmony between the Creator, mankind and the created world, as described by Sacred Scripture, was disrupted by the sin of Adam and Eve, by man and woman, who wanted to take the place of God and refused to acknowledge that they were his creatures. As a result, the work of "exercising dominion" over the earth, "tilling it and keeping it," was also disrupted, and conflict arose within and between mankind and the rest of creation (cf. Gen 3:17-19). Human beings let themselves be mastered by selfishness; they misunderstood the meaning of God's command and exploited creation out of a desire to exercise absolute domination over it. But the true meaning of God's original command, as the *Book of Genesis* clearly shows, was not a simple conferral of authority, but rather a summons to responsibility. The wisdom of the ancients had recognized that na-

ture is not at our disposal as "a heap of scattered refuse".[10] Biblical Revelation made us see that nature is a gift of the Creator, who gave it an inbuilt order and enabled man to draw from it the principles needed to "till it and keep it" (cf. Gen. 2:15).[11] Everything that exists belongs to God, who has entrusted it to man, albeit not for his arbitrary use. Once man, instead of acting as God's co-worker, sets himself up in place of God, he ends up provoking a rebellion on the part of nature, "which is more tyrannized than governed by him".[12] Man thus has a duty to exercise responsible stewardship over creation, to care for it and to cultivate it.[13]

7. Sad to say, it is all too evident that large numbers of people in different countries and areas of our planet are experiencing increased hardship because of the negligence or refusal of many others to exercise responsible stewardship over the environment. The Second Vatican Ecumenical Council reminded us that "God has destined the earth and everything it contains for all peoples and nations".[14] The goods of creation belong to humanity as a whole. Yet the current pace of environmental exploitation is seriously endangering the supply of certain natural resources not only for the present generation, but above all for generations yet to come.[15] It is not hard to see that environmental degradation is often due to the lack of far-sighted official policies or to the pursuit of myopic economic interests, which then, tragically, become a serious threat to creation. To combat this phenomenon, economic activity needs to consider the fact that "every economic decision has a moral consequence" [16] and thus show increased respect for the environment. When making use of natural resources, we should be concerned for their protection and consider the cost entailed—environmentally and socially—as an essential part of the overall expenses incurred. The international community and national governments are responsible for sending the right signals in order to combat effectively the misuse of the environment. To protect the environment, and to safeguard natural resources and the climate, there is a need to act in accordance with clearly defined rules, also from the juridical and economic standpoint, while at the same time taking into due account the solidarity we owe to those living in the poorer areas of our world and to future generations.
8. *A greater sense of intergenerational solidarity* is urgently needed. Future generations cannot be saddled with the cost of our use of common environmental resources. "We have inherited from past generations, and we have benefited from the work of our contemporaries; for this reason we have obligations towards all, and we cannot refuse to interest ourselves in those who will come after us, to enlarge the human family. Universal solidarity represents a benefit as well as a duty. *This is a responsibility that present generations*

have towards those of the future, a responsibility that also concerns individual States and the international community".[17] Natural resources should be used in such a way that immediate benefits do not have a negative impact on living creatures, human and not, present and future; that the protection of private property does not conflict with the universal destination of goods;[18] that human activity does not compromise the fruitfulness of the earth, for the benefit of people now and in the future. In addition to a fairer sense of intergenerational solidarity there is also an urgent moral need for a renewed sense of *intragenerational solidarity*, especially in relationships between developing countries and highly industrialized countries: "the international community has an urgent duty to find institutional means of regulating the exploitation of non-renewable resources, involving poor countries in the process, in order to plan together for the future".[19] *The ecological crisis shows the urgency of a solidarity which embraces time and space.* It is important to acknowledge that among the causes of the present ecological crisis is the historical responsibility of the industrialized countries. Yet the less developed countries, and emerging countries in particular, are not exempt from their own responsibilities with regard to creation, for the duty of gradually adopting effective environmental measures and policies is incumbent upon all. This would be accomplished more easily if self-interest played a lesser role in the granting of aid and the sharing of knowledge and cleaner technologies.

9. To be sure, among the basic problems which the international community has to address is that of energy resources and the development of joint and sustainable strategies to satisfy the energy needs of the present and future generations. This means that technologically advanced societies must be prepared to encourage more sober lifestyles, while reducing their energy consumption and improving its efficiency. At the same time there is a need to encourage research into, and utilization of, forms of energy with lower impact on the environment and "a world-wide redistribution of energy resources, so that countries lacking those resources can have access to them".[20] The ecological crisis offers an historic opportunity to develop a common plan of action aimed at orienting the model of global development towards greater respect for creation and for an integral human development inspired by the values proper to charity in truth. I would advocate the adoption of a model of development based on the centrality of the human person, on the promotion and sharing of the common good, on responsibility, on a realization of our need for a changed life-style, and on prudence, the virtue which tells us what needs to be done today in view of what might happen tomorrow.[21]

10. A sustainable comprehensive management of the environment and the resources of the planet demands that human intelligence be directed to technological and scientific research and its practical applications. The "new solidarity" for which John Paul II called in his Message for the 1990 World Day of Peace [22] and the "global solidarity" for which I myself appealed in my Message for the 2009 World Day of Peace [23] are essential attitudes in shaping our efforts to protect creation through a better internationally-coordinated management of the earth's resources, particularly today, when there is an increasingly clear link between combatting environmental degradation and promoting an integral human development. These two realities are inseparable, since "the integral development of individuals necessarily entails a joint effort for the development of humanity as a whole".[24] At present there are a number of scientific developments and innovative approaches which promise to provide satisfactory and balanced solutions to the problem of our relationship to the environment. Encouragement needs to be given, for example, to research into effective ways of exploiting the immense potential of solar energy. Similar attention also needs to be paid to the world-wide problem of water and to the global water cycle system, which is of prime importance for life on earth and whose stability could be seriously jeopardized by climate change. Suitable strategies for rural development centred on small farmers and their families should be explored, as well as the implementation of appropriate policies for the management of forests, for waste disposal and for strengthening the linkage between combatting climate change and overcoming poverty. Ambitious national policies are required, together with a necessary international commitment which will offer important benefits especially in the medium and long term. There is a need, in effect, to move beyond a purely consumerist mentality in order to promote forms of agricultural and industrial production capable of respecting creation and satisfying the primary needs of all. The ecological problem must be dealt with not only because of the chilling prospects of environmental degradation on the horizon; the real motivation must be the quest for authentic world-wide solidarity inspired by the values of charity, justice and the common good. For that matter, as I have stated elsewhere, "technology is never merely technology. It reveals man and his aspirations towards development; it expresses the inner tension that impels him gradually to overcome material limitations. *Technology in this sense is a response to God's command to till and keep the land* (cf. Gen 2:15) that he has entrusted to humanity, and it must serve to reinforce the covenant between human beings and the environment, a covenant that should mirror God's creative love".[25]

11. It is becoming more and more evident that the issue of environmental degradation challenges us to examine our life-style and the prevailing models of consumption and production, which are often unsustainable from a social, environmental and even economic point of view. We can no longer do without a real change of outlook which will result in *new life-styles*, "in which the quest for truth, beauty, goodness and communion with others for the sake of common growth are the factors which determine consumer choices, savings and investments".[26] Education for peace must increasingly begin with far-reaching decisions on the part of individuals, families, communities and states. We are all responsible for the protection and care of the environment. This responsibility knows no boundaries. In accordance with the *principle of subsidiarity* it is important for everyone to be committed at his or her proper level, working to overcome the prevalence of particular interests. A special role in raising awareness and in formation belongs to the different groups present in civil society and to the non-governmental organizations which work with determination and generosity for the spread of ecological responsibility, responsibility which should be ever more deeply anchored in respect for "human ecology." The media also have a responsibility in this regard to offer positive and inspiring models. In a word, concern for the environment calls for a broad global vision of the world; a responsible common effort to move beyond approaches based on selfish nationalistic interests towards a vision constantly open to the needs of all peoples. We cannot remain indifferent to what is happening around us, for the deterioration of any one part of the planet affects us all. Relationships between individuals, social groups and states, like those between human beings and the environment, must be marked by respect and "charity in truth." In this broader context one can only encourage the efforts of the international community to ensure progressive disarmament and a world free of nuclear weapons, whose presence alone threatens the life of the planet and the ongoing integral development of the present generation and of generations yet to come.
12. *The Church has a responsibility towards creation*, and she considers it her duty to exercise that responsibility in public life, in order to protect earth, water and air as gifts of God the Creator meant for everyone, and above all to save mankind from the danger of self-destruction. The degradation of nature is closely linked to the cultural models shaping human coexistence: consequently, "when 'human ecology' is respected within society, environmental ecology also benefits".[27] Young people cannot be asked to respect the environment if they are not helped, within families and society as a whole, to respect themselves. The book of nature is one and indi-

visible; it includes not only the environment but also individual, family and social ethics.[28] Our duties towards the environment flow from our duties towards the person, considered both individually and in relation to others.

13. Hence I readily encourage efforts to promote a greater sense of ecological responsibility which, as I indicated in my Encyclical *Caritas in Veritate,* would safeguard an authentic "human ecology" and thus forcefully reaffirm the inviolability of human life at every stage and in every condition, the dignity of the person and the unique mission of the family, where one is trained in love of neighbour and respect for nature.[29] There is a need to safeguard the human patrimony of society. This patrimony of values originates in and is part of the natural moral law, which is the foundation of respect for the human person and creation.
14. Nor must we forget the very significant fact that many people experience peace and tranquillity, renewal and reinvigoration, when they come into close contact with the beauty and harmony of nature. There exists a certain reciprocity: as we care for creation, we realize that God, through creation, cares for us. On the other hand, a correct understanding of the relationship between man and the environment will not end by absolutizing nature or by considering it more important than the human person. If the Church's magisterium expresses grave misgivings about notions of the environment inspired by ecocentrism and biocentrism, it is because such notions eliminate the difference of identity and worth between the human person and other living things. In the name of a supposedly egalitarian vision of the "dignity" of all living creatures, such notions end up abolishing the distinctiveness and superior role of human beings. They also open the way to a new pantheism tinged with neo-paganism, which would see the source of man's salvation in nature alone, understood in purely naturalistic terms. The Church, for her part, is concerned that the question be approached in a balanced way, with respect for the "grammar" which the Creator has inscribed in his handiwork by giving man the role of a steward and administrator with responsibility over creation, a role which man must certainly not abuse, but also one which he may not abdicate. In the same way, the opposite position, which would absolutize technology and human power, results in a grave assault not only on nature, but also on human dignity itself.[30]
15. *If you want to cultivate peace, protect creation*. The quest for peace by people of good will surely would become easier if all acknowledge the indivisible relationship between God, human beings and the whole of creation. In the light of divine Revelation and in fidelity to the Church's Tradition, Christians have their own contribution to

make. They contemplate the cosmos and its marvels in light of the creative work of the Father and the redemptive work of Christ, who by his death and resurrection has reconciled with God "all things, whether on earth or in heaven" (Col 1:20). Christ, crucified and risen, has bestowed his Spirit of holiness upon mankind, to guide the course of history in anticipation of that day when, with the glorious return of the Saviour, there will be "new heavens and a new earth" (2 Pet 3:13), in which justice and peace will dwell for ever. Protecting the natural environment in order to build a world of peace is thus a duty incumbent upon each and all. It is an urgent challenge, one to be faced with renewed and concerted commitment; it is also a providential opportunity to hand down to coming generations the prospect of a better future for all. May this be clear to world leaders and to those at every level who are concerned for the future of humanity: the protection of creation and peacemaking are profoundly linked! For this reason, I invite all believers to raise a fervent prayer to God, the all-powerful Creator and the Father of mercies, so that all men and women may take to heart the urgent appeal: *If you want to cultivate peace, protect creation.*

From the Vatican, 8 December 2009
Benedictus PP. XVI

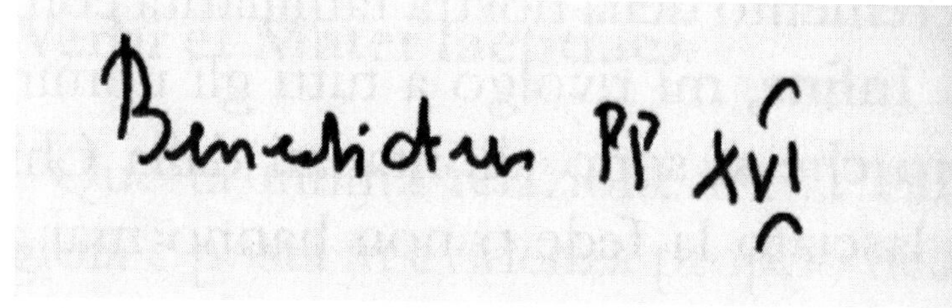

Figure 0.1.

[1] Catechism of the Catholic Church, 198.
[2] Benedict XVI, Message for the 2008 World Day of Peace, 7.
[3] Cf. No.48.
[4] Dante Alighieri, *The Divine Comedy, Paradiso*, XXXIII, 145.
[5] *Message for the 1990 World Day of Peace*, 1.
[6] Apostolic Letter *Octogesima Adveniens*, 21.
[7] *Message for the 1990 World Day of Peace*, 10.
[8] Cf. Benedict XVI, Encyclical Letter *Caritas in Veritate*, 32.
[9] *Catechism of the Catholic Church*, 295.
[10] Heraclitus of Ephesus (c. 535 - c. 475 B.C.), Fragment 22B124, in H. Diels-W. Kranz, Die Fragmente der Vorsokratiker, Weidmann, Berlin,1952, 6th ed.
[11] Cf. Benedict XVI,Encyclical Letter *Caritas in Veritate*, 48.
[12] John Paul II, Encyclical Letter *Centesimus Annus*, 37.

[13] Cf. Benedict XVI, Encyclical Letter *Caritas in Veritate*, 50.

[14] Pastoral Constitution *Gaudium et Spes*, 69.

[15] Cf. John Paul II, Encyclical Letter *Sollicitudo Rei Socialis*, 34.

[16] Benedict XVI, Encyclical Letter *Caritas in Veritate*, 37.

[17] Pontifical Council for Justice and Peace, *Compendium of the Social Doctrine of the Church*, 467; cf. Paul VI, Encyclical Letter *Populorum Progressio*, 17.

[18] Cf. John Paul II, Encyclical Letter *Centesimus Annus*, 30-31, 43.

[19] Benedict XVI, Encyclical Letter *Caritas in Veritate*, 49.

[20] Ibid.

[21] Cf. Saint Thomas Aquinas, S. Th., II-II, q. 49, 5.

[22] Cf. No. 9.

[23] Cf. No. 8.

[24] Paul VI, Encyclical Letter *Populorum Progressio*, 43.

[25] Encyclical Letter *Caritas in Veritate*, 69.

[26] John Paul II, Encyclical Letter *Centesimus Annus*, 36.

[27] Benedict XVI, Encyclical Letter *Caritas in Veritate*, 51.

[28] Cf. ibid., 15, 51.

[29] Cf. ibid., 28, 51, 61; John Paul II, Encyclical Letter *Centesimus Annus*, 38, 39.

[30] Cf. Benedict XVI, Encyclical Letter *Caritas in Veritate*, 70.

NOTES

1. Included in this anthology with permission from Professor D. Giuseppe Costa, S.D.B., Director, Libreria Editrice Vaticana, January 15, 2013. Downloaded from http://www.vatican.va/holy_father/benedict_xvi/messages/peace/documents/hf_ben-xvi_mes_20091208_xliii-world-day-peace_en.html without modifications or changes as required on March 10, 2011. Papal insignia and signature provided by Prof. Costa.

Introduction

Celebrating and Advancing Magisterial Discourse on the Ecological Crisis

Jame Schaefer

Theologians listen when the pope[1] and bishops speak. They have assumed responsibility for preserving and proclaiming the Christian faith for Roman Catholics and how the faithful should live accordingly in the world. Referred to as *the magisterium*, popes and bishops have exercised this teaching authority for centuries. The perils of industrialization and other major social issues during the nineteenth century prompted Pope Leo XIII and subsequent popes to address them through encyclicals and other statements that carry varying degrees of authority (e.g., Gaillardetz 2003). Among the major issues on which popes John Paul II (now Blessed John Paul II) and Benedict XVI (now Pope Emeritus Benedict XVI) have written is the ongoing degradation of Earth that is adversely affecting materially poor and vulnerable people today, projected to severely affect future generations, and threatens the integrity of the biosphere. Throughout their pontificates, they underscored the moral responsibility the faithful have to mitigate these adverse effects, to transform consumptive lifestyles, and to live in harmony with God's creation so its goods are available to sustain all people now and into the future. Bishops throughout the world individually and regionally have issued many statements pertaining to ecological problems within their dioceses (Whittington 2004), emphasizing the urgency with which the faithful must gear their actions toward Earth's ability to sustain human life.

Whereas the magisterium speaks *for* the Catholic Church, theologians speak *from* the Church when addressing the loss of biological diversity, the degradation of ecological systems, threats to the biosphere, and their effects on humans. Theologians bring their specialized fields of inquiry to the service of the Christian faith by reinterpreting the language used so it adequately reflects the faith. They are especially conscious of the need for expressions of faith to be consistent with the deep meanings of the biblical and historical theological tradition, informed by the current scientific understanding of the world, and articulated as profoundly as possible to address ecological problems today. When popes issue encyclicals, messages, and statements pertaining to the human-Earth relationship, theologians listen to their teachings and, as scholars, independently and freely analyze them, engage in research, reflect on their findings, and publish articles, essays, and books that may extend magisterial teachings and point to fruitful ways of addressing issues from a theological perspective. Their efforts are aimed in part to help the magisterium update and articulate the Christian faith more appropriately while drawing their authority from the scholarly academy which demands rigorous analysis, research, and synthesis of their findings. Their work as scholars of the Christian faith parallels the work of the bishops who preserve and promulgate the Christian faith. Though they have special roles to play in relation to the Christian faith, they are both dedicated to it. Occasionally theologians and bishops have collaborated to draft magisterial documents as occurred during the Second Vatican Council.

Leaders of other religious traditions also listen to the magisterium of the Catholic Church, especially the popes, when addressing ecological issues. For example, from the Eastern Orthodox branch of Christianity, the Ecumenical Patriarch Bartholomew, Archbishop of Constantinople and New Rome, has been particularly attentive to ecological problems, has characterized those caused by humans as "sinful" (Stammer 1997), and has exchanged mutual concerns and efforts with John Paul II and Benedict XVI in general and on particular problems occurring in various parts of the world (e.g., Ecumenical Patriarchate 2013; Benedict XVI 2006, 2009a, 2009b; Howden 2002). Similarly, the United States Conference of Catholic Bishops has engaged at least one other Christian denomination in formal dialogue on the natural environment (United Methodist-Catholic Dialogue 2013), and more would be helpful. All Christian denominations and other religions share one planet, a "suffering" Earth (John Paul II 1989), and the cooperative as well as individual efforts of their leaders are essential to stimulate concern about Earth and to guide their faithful toward living in ways that sustain the integrity of our planetary home.

Theologians wish to help the magisterium with this task, and *Ecological Justice and Climate Change: Assessing Pope Benedict XVI's Ecological Vision for the Catholic Church in the United States* provides some examples that are promising. To facilitate this desire, the remainder of this introductory essay highlights the overall contributions that Benedict XVI has made to ecological teachings. Explained subsequently is the process for planning the November 2012 consultation that yielded the essays in this anthology and key features of the consultation. An overview of the contents of this volume follows with emphasis on the papers given and the contributions of two bishops who serve in opposite parts of two hemispheres. We close with hope that Pope Francis and the Catholic bishops of the United States will advance magisterial teachings on the human-Earth relationship and will find in theologians' efforts some fruitful directions.

POPE BENEDICT XVI'S LEGACY

Benedict XVI played a significant role in advancing concern about the ongoing threats to the integrity of Earth. He built upon the seminal efforts of his predecessor, John Paul II, who issued the first pontifical statement dedicated to the ecological crisis on the 1990 World Day of Peace and identified it as a moral problem for which the faithful are responsible. He left for Benedict XVI a legacy of statements, messages, and parts of encyclicals that seed ideas for future teachings. Benedict XVI embraced and advanced his predecessor's teachings. As essays in this anthology indicate, he repeatedly lamented the exploitation and destruction of God's creation—God's garden, God's gift. He urged the faithful to share in the responsibility of caring for this gift so its goods are available to all people, and he gave special attention to the materially poor and vulnerable who are often thwarted by economic, social, and/or political circumstances from sustaining their lives. He warned against overconsumption by some while others could not secure the necessities of life for themselves, their families, and their communities. He underscored the connection between ecological degradation and socio-economic inequities in his continual quest to promote justice among peoples and nations of the world in light of the growing chasm between the wealth of some and poverty of others. Though he lauded some technologies that have been helpful for enhancing human life and well-being, he expressed concern about the adverse effects that fossil fuels and other technologies are having on people today, are projected to impact future generations, and threaten the integrity of Earth. Several teachings were directed explicitly toward the perils of human-forced climate change. And, in his message on the 2010 World Day of Peace which prefaced this anthology, he repeated a theme that had become common in his teachings—the need to address ecological degradation in order to realize peace in the world.

Among Benedict XVI's most developed contributions to magisterial teachings on the ecological crisis is his understanding of the sacramental character of the world, which several essayists explore variously and richly in this anthology. In encyclicals, messages, and statements, he retrieved this biblical and traditional theological understanding that God's gift of the creation–the Earth and the totality of its vast expanding surroundings–is revelatory of God, that we can know something about God in and through the world, and that we can learn from studying the world how we should be functioning. Because Benedict XVI's ecological writings have been prolific during the short period of his pontificate, two collections of and commentaries on them have been published thus far. One is *The Environment* (Benedict 2012), and the other is *Ten Commandments for the Environment: Pope Benedict XVI Speaks out for Creation and Justice* (Benedict XVI and Woodeene Koenig-Bricker 2009). Many articles on his writings have appeared in journals and in *Confronting the Climate*

Crisis: Catholic Theological Perspectives (Schaefer 2011). We are delighted that this anthology adds another. His efforts have gone beyond writing to practical initiatives including the installation of solar panels to power the lighting, heating, and cooling of a portion of Vatican City, authorizing the Vatican's bank to purchase carbon credits by funding a Hungarian forest that would make Vatican City the only country fully carbon neutral, and using a new hybrid Popemobile. His ecological commitments overall have resulted in some people referring to him as "the green pope" (Stone 2013).

THE 2012 CONSULTATION ON BENEDICT XVI'S TEACHINGS

Appreciating Benedict XVI's prophetic ecological endeavors, representatives of the United States Conference of Catholic Bishops (USCCB), the Catholic Coalition on Climate Change (Coalition), and the Institute for Policy Research and Catholic Studies at The Catholic University of America (CUA) recognized the need for careful reflection on Benedict XVI's ecological vision and its implications for the Catholic Church in the United States. They met in May 2011 to discuss the possibility of holding a consultation during which Catholic scholars could present papers and engage in discussion with bishops regarding the implications of Benedict XVI's teachings. Previous consultations had been held in response to the U.S. Catholic bishops call in 1991 to "*theologians, scripture scholars, and ethicists to help explore, deepen, and advance the insights of our Catholic tradition and its relation to the environment and other religious perspectives on these matters*" in their pastoral document *Renewing the Earth: An Invitation to Reflection and Action on Environment in Light of Catholic Social Teaching*. These consultations provided an opportunity to consider environmental issues more deeply from a faith perspective, to ask questions of one another, and to recognize some challenges ahead. The consultation on "Ecology and Catholic Theology: Contribution and Challenge" that was held at Mount Angel Abbey in Portland, Oregon during the summer of 1995 provided the papers and material that were eventually published as *"And God Saw that It was Good:" Catholic Theology and the Environment*, edited by Drew Christiansen, S.J. and Walter E. Grazer (1996). Following this precedent, a decision was made to hold a consultation at The Catholic University of America to be attended by several interested bishops, selected scholars, staff of the sponsoring organizations, and leaders of national Catholic organizations.

Early in the planning process, the USCCB, Coalition, and CUA staff formed an advisory group of Catholic scholars. They proved invaluable in assisting the sponsoring organizations to refine the categories of scholarship for the consultation, draft a call for papers, sort and prioritize the papers that best fit the themes and goals of the consultation, and shape the structure and flow of the consultation. Forty-four proposals were received from scholars, William D. Dinges, Ph.D., Ordinary Professor of Religious Studies in the School of Theology and Religious Studies at the Catholic University of America, narrowed the pool to twenty-five, and twelve were subsequently selected by the following scholars who joined Dr. Dinges in this effort: Sr. Ilia Delio, OSF., Ph.D., Research Fellow, Woodstock Theological Center at Georgetown University; Sr. Mary Ann Hinsdale, IHM., Ph.D., Associate Professor of Theology, Boston College; Jame Schaefer, Ph.D., Associate Professor of Theology, Marquette University; Lucia Silecchia, Esq., Professor of Law, Columbus School of Law, The Catholic University of America; Br. Keith Warner, OFM, Ph.D., Associate Adjunct Lecturer, Santa Clara University; and Tobias Winright, Ph.D., Associate Professor of Theological Ethics, St. Louis University. Several phone conferences were held between September 2011 and August 2012 to finalize and group the selections.

To accommodate the bishops who expressed interest in the consultation, November 8-10, 2012 was chosen to coincide with the semi-annual meeting of the U.S. Catholic bishops in Baltimore, Maryland. Participating in the consultation were Bishop Frank Dewane (Venice, Florida), Bishop Donald Kettler (Fairbanks, Alaska), Bishop John Ricard (Emeritus of Pensacola-Tallahassee, Florida), Bishop William Skylstad (Emeritus of Spokane, Washington), and Bishop Jaime Soto (Sacramento, California). Bishop Bernard Unabali of Bougainville, Papua New Guinea was invited to share his experience in assisting with the relocation of some of the world's first "climate refugees," the Carteret Islanders.

The bishops-scholars consultation proved to be a joyous and stimulating event. From Thursday afternoon through Noon on Saturday, the participants gathered to listen and respond to one another. We prayed together. We opened to God's grace for strength to deal with the vexing reality of human-forced climate change and its justice implications. We participated in a Eucharistic liturgy at which Bishop Kettler delivered a thought-provoking homily that closes this anthology. We shared concerns and problems particular to our missions. And, we expressed our mutual desire to collaborate in the future. The culminating session was both celebratory and poignant. Under the rubric of "Meaning, Messages, and Messengers," the panelists found in the consultation an opportunity to explore promising magisterial teachings, especially by Benedict XVI, and theological reflections based on research by scholars who are striving to address human-forced climate change and other ecological problems. Other meaningful outcomes of the consultation included an opportunity to express mutual appreciation for the varied ways in which the bishops, scholars, and staff of the sponsoring organizations have been addressing the climate crisis, to identify directions in which to move forward concurrently to mitigate the adverse effects that are especially devastating for the most materially poor and vulnerable people in the world, and to provide a forum for emerging scholars who are eager to address ecological issues.

Surfacing throughout the bishops-scholars consultation was the message that there are many riches in the Catholic theological tradition for retrieving, updating by contemporary findings (the natural sciences, social sciences, and humanities), and creatively working to address human-forced climate change. This realization needs to be celebrated, developed by scholars, and made more readily available to the faithful for their stimulation and application at all levels of social, economic, and political endeavor. Because scholars are required by their profession to produce works for the academy, making them available at palatable levels for other adults and youths may require collaboration with skillful speakers like the Coalition's Catholic Climate Ambassadors, writers, and media experts.

The messengers are many and varied—bishops, pastors, priests, catechetical leaders, Catholic agencies and organizations, media, scholars, and institutions of higher learning. The bishops are primary messengers, and their voices may be especially effective and timely when issuing reflections on ecological issues on appropriate memorial days (e.g., the feast of St. Francis of Assisi—October 4, Earth Day—April 22, and Climate Change Day—December 1), encouraging reflection by priests and incorporation into their homilies, and assuring that seminaries in their dioceses cover magisterial teachings, creation theology, and theological discourse on the human responsibility to God for functioning in ways that aim to assure the sustainability of Earth. Pastors of parishes are messengers by working with priests and parishioners to inform and raise their awareness of magisterial teachings and theological discourse on the climate crisis and other ecological issues that may be pertinent to their locales. Priests are messengers by delivering homilies on magisterial teachings and theological insights on the human-Earth relationship and the obligations to function compatibly with other species within shared ecological systems as constituents of the larger biosphere. Priests are also messengers when encountering parishioners at the confessional, guiding them in seeking reconciliation with God and God's creation for actions that cause ecological degradation, and helping them commit to acting more prudently, justly, and moderately in the world. Communities of professed religious are messengers by striving to live sustainably as an example to others, investing wisely in Earth-sustainable technologies and funds, and developing ecologically compatible initiatives that reflect their missions. Catechetical leaders of adult and youth education and activities are messengers by developing projects aimed at facilitating reflection and action to mitigate the phenomenon of climate change and other ecological problems. The Catholic Coalition on Climate Change, the National Catholic Rural Life Conference, Catholic Relief Services, and other organizations and agencies play vital roles as messengers who educate, act on ecological issues from a faith perspective, and serve the most poor and vulnerable who are adversely affected by climate and other ecological crises. The Catholic media carry messages to all who are interested and can choose to highlight magisterial statements and pertinent endeavors that focus on ecological problems. Catholic institutions at formative school, college, and university learning can be messengers by providing opportunities for students to learn about ecological issues from an interdisciplinary perspective and to recognize their moral obligations to mitigate adverse effects. Catholic scholars are messengers by continuing to research, publish, and engage students in academic endeavors that deepen their understanding of ecological issues and address them from theological perspectives informed by other disciplines. Other messengers within the Catholic community may exist and may surface, notably young people who have left the Catholic Church and discover the capability of the magisterial teachings and the Catholic theological tradition to address ecological issues. Hopefully, their voices will be raised through words and actions.

OVERVIEW OF THE ANTHOLOGY

This anthology opened with Benedict XVI's Message on the 2010 World Day of Peace. Delivered twenty years after John Paul II issued the first papal statement dedicated to the ecological crisis, Benedict XVI's *If You Want to Cultivate Peace, Protect Creation* establishes the imperative to bring about peace in the world by protecting and assuring equitable access to the water and air as gifts from God intended for all. He applies many of his prior constructs to the task, including human ecology, integral human development, and intergenerational solidarity, and he calls upon all people of good will to assure that the vulnerable and poor in our midst have access to these gifts and that they are available to future generations. The quest for peace will become easier, he insists (#14), "if all acknowledge the indivisible relationship between God, human beings and the whole of creation."

The papers presented by the scholars center around four themes, the first of which focus on Benedict XVI's teachings on *human and natural ecology* and *human life and dignity*. In "Bonaventure in Benedict: Franciscan Wisdom for Human Ecology," Br. Keith Douglas Warner, OFM retraces Benedict XVI's intellectual footsteps and explains how St. Bonaventure's wisdom theology helps interpret the pontiff's understanding of "human ecology." Warner identifies several distinct features of Bonaventure's creation theology that are found in Benedict XVI's teaching, especially Bonaventure's understanding of God's creation as the "book of nature" through which the faithful can better understand God, and concludes convincingly that the pontiff's use of Bonaventure's thinking has the potential for engaging Catholics to care for Earth. Mary A. Ashley's essay, "If You Want Responsibility, Build Relationship: A Personalist Approach to Benedict XVI's Environmental Vision," advocates a "personalist" theology as found in the pontiff's teachings and in the 2004 *Compendium on the Social Doctrine of the Church*. In contrast with the dominant approaches to ecology found among some Protestants and especially by secular modernism, Ashley believes that a Catholic personalist perspective distinctively understands human existence as centered on a divine love that tends to move "outward" to encompass our more-than-human world. In "Natural Law and the Natural Environment: Pope Benedict XVI's Vision Beyond Utilitarianism and Deontology," Michael Baur finds the pontiff's teachings deeply embedded in the Catholic "natural law" tradition and identifies three metaphysical premises (the convertibility of being and goodness, the convertibility of being and order, and the uniquely intellectual nature of the human being) that resonate with but are distinguishable from some insights of contemporary environmental philosophy. Baur concludes that Benedict XVI provides an intellectually defensible and helpful alternative to deontological and utilitarian approaches to addressing environmental concerns.

The next set of three essays explore *solidarity, justice, poverty, and the common good* in Benedict XVI's teachings vis-à-vis ecology. In "Human, Social, and Natural Ecology: Three Ecologies, One Cosmology, and the Common Good," Scott Hefelfinger examines the pope's use of "human ecology" which situates human and natural ecology within the framework of the common good as conceptualized by Thomas Aquinas. Hefelfinger argues that taking this approach opens a way beyond anthropocentrism to a cosmocentrism that redounds to the good of all creatures in ways that are appropriate to their natures. This perspective also offers timely and creative suggestions for the Catholic Church in the United States as it struggles to come to terms with today's ecological challenges. Christiana Z. Peppard builds upon the premise that humans are forcing changes on Earth systems in "Commodifying Creation? Benedict XVI's Vision of the Goods of Creation Intended for All." Focusing on the significance of Catholic social teaching in relation to the central concepts of the universal destination of the goods of creation and fundamental human rights, she gives special attention to air and water to show that papal encyclicals in recent decades have prophetically specified that some goods are so vital that they transcend market value. These goods constitute right to life issues that Catholics should take seriously and require changes to current economic systems and practices. In "The Grammar of Creation: Agriculture in the Thought of Pope Benedict XVI," Matthew P. Whelan studies Benedict XVI's treatment of agriculture, focusing primarily on his encyclical *Caritas in Veritate* and his addresses to the Food and Agriculture Organization of the United Nations on the occasion of World Food Day. Among the central questions Whelan probes are the need to pay attention to the grammar of creation, how to read it appropriately, and how the practice of agriculture would look accordingly. To attend to the grammar of creation, Whelan argues, is to attend to what Benedict calls "the astonishing experience of gift," and making good use of the common gifts of creation is essential to the practice of agriculture. Whelan concludes by tracing the continuity between Benedict XVI's approach to agriculture with natural scientists, including soil scientist Albert Howard, plant geneticist Wes Jackson, and entomologist Miguel Altieri.

The *sacramental character of the world* is a topic that Benedict XVI explored early in his scholarly career and advanced as Cardinal Josef Ratzinger and subsequently during his pontificate. Though three essays focus specifically on this topic, it also surfaces in several other essays in this anthology because it is so foundational to his theological discourse. In "The Way of Wisdom: 'Keep hold of instruction; do not let go; guard her, for she is your life' (Prov 3:14)," Elizabeth Groppe points to Benedict XVI's teaching that the faithful should view the world as an expression of God's loving plan (wisdom) to be valued intrinsically and used responsibly and respectfully. She surveys similar thinking in the wisdom literature of the Bible and the Christian theological tradition, notes how it was challenged by modern science, and suggests possibilities for recovering the wisdom tradition today. Especially important for mitigating the climate crisis is the need to study the functioning of living and inanimate entities in relation to one another and to strive to mimic their interactions. Monsignor Kevin Irwin's essay on "The World as God's Icon: Creation, Sacramentality, Liturgy" follows in which he explains today's environmental crisis as both a challenge and an opportunity for the Catholic Church to respond pastorally from its rich theological tradition. By framing major Catholic beliefs (creation theology, the incarnation, the principles of sacramentality and mediation, participation, the common good, and beauty) within the context of the liturgy and the sacraments, appealing to the adage *lex orandi, lex credendi,* and examining magisterial teachings, especially Benedict XVI's, the world can be recognized as God's icon where the divine is experienced and revealed. Irwin finds this insight instrumental in increasing believers' awareness of and appreciation for the interdependence of the entire creation. In "Pope Benedict XVI's Cosmic Soteriology and the Advancement of Catechesis on the Environment," Jeremiah Vallery considers the catechetical approach to the ecological crisis as a social justice issue insufficient to convince the faithful to care for Earth and finds a more effective basis in Benedict XVI's cosmic soteriology that was influenced by Teilhard de Chardin. Appropriating passages from the pontiff's works and constructing catechesis around his three pillars of the cosmic extent of Christ's sacrificial death and resurrection, liturgical worship, and the common eschatological destiny of humans and Earth, Vallery demonstrates how these essential aspects of the

Catholic faith relate to the natural environment and can serve to motivate the faithful to protect God's creation.

The final three essays focus on *the Catholic faith in action*. Maryknoll Missionary John T. Brinkman approaches his topic, "Discernment of the Church and the Dynamics of the Climate Change Convention," from the perspective of the history of religions and the realization that all humans are drawn by their faiths to acknowledge their common inheritance of Earth and their common destiny in resolving the climate crisis. Benedict XVI contributes to this realization, Brinkman insists, by intimately connecting economic and social equity concerns with ecological integrity. Emphasizing this nexus within the ongoing international discourse on climate change should stimulate attention to the spiritual and moral issues underlying the need to take action to mitigate the devastating effects on the most vulnerable people. In "American Lifestyles and Structures of Sin: The Practical Implications of Pope Benedict XVI's Ecological Vision for the American Church," David Cloutier focuses on Benedict XVI's teachings in his 2009 encyclical, *Caritas in Veritate*. As other scholars have noted, Benedict XVI ties climate change to other moral issues and directs the faithful to rethink their lives in two specific ways–reduction of energy use and rejection of hedonistic lifestyles. Cloutier argues that the suburban ideal constitutes a structure of sin that must be challenged on both personal and social levels. The desire for luxury, especially in transportation and housing, must be recognized as sinful in order to prompt a conversion to more sustainable lifestyles and a desire for the Kingdom of God. The last of these informative and stimulating essays is Anselma T. Dolcich-Ashley's "American Nature Writing as a Critically-Appropriated Resource for Catholic Ecological Ethics." Concerned with what prompts a change of mind and heart to make a commitment to take action on environmental crises, she turns to classic American nature writers for some insight. She finds in selected writings of John Muir, Aldo Leopold, and Annie Dillard the capacity to put readers into a place of personal encounter with non-human nature, to enable us to examine prior assumptions, and to transform both our rational understanding and emotional responses. This distinctive genre of American nature writing also provides interdisciplinary cultural resources for Americans to develop positive responses to the gratuitousness of God's creation, to identify sinful action and experience contrition, and to contemplate God's concurrent transcendence of and immanence in the world.

Closing this anthology are two poignant entries. The first is the text of the presentation by the Most Reverend Bernard Unabali, Bishop of Bougainville, Papua New Guinea, who shares a remarkable story of the Carteret Islanders who are refugees of the climate crisis. He begins with the historical background of the Carteret Islands, describes the Islanders, their traditions, and their geographical setting, and explains the devastating effects that human-forced climate change has had on them. Depicted in the film *Sun Come Up* (2011), the rising sea levels exacerbated by more frequent storms and high tides have forced the relocation of all Islanders to the mainland. Bishop Unabali assisted in this effort by appealing to the people of his diocese who live on the mainland to accept their moral obligation to help their brothers and sisters in Christ. The Islanders responded to this generous offer, some were relocated to Bougainville, and others began independently to move from their ancestral grounds while trying to maintain their cultural identity. Because the people of his diocese were so deeply motivated by their faith to act on the dire circumstances of the Carteret Islanders, Bishop Unabali is hopeful that this gift of faith will continue to be embraced and passed to future generations.

The second appendix consists of the homily given by the Most Rev. Donald J. Kettler at the Eucharistic Mass celebrated by the Most Rev. William S. Skylstad, Bishop Emeritus of Spokane and Honorary Chairman of the Catholic Coalition on Climate Change. In his homily, Bishop Kettler shared two major concerns he is addressing in his diocese. One is the plight of the Eskimos in the Yup'ik village of Newtok where rising ocean waters and a later autumn freeze are washing away the protective coastland, threatening to inundate the village within ten to fifteen years, and forcing them to abandon the subsistence lifestyle of hunting, fishing, and gathering that was established by their ancestors thousands of years ago. The difficulties these Indigenous people in the North Western Hemisphere are facing parallels the immense problems the Carteret Islanders are experiencing in the South Eastern Hemisphere, thereby manifesting the vulnerability of some of the most materially poor people in the world to the adverse effects of the climate crisis. Bishop Kettler's related concern is the ongoing exploitation of non-renewable sources of energy in Alaska that is harming the lives of the local people and the ecological systems in which they function. He is hopeful that Catholics will become "bridge builders" by drawing upon magisterial teachings, grappling with ethical questions, fostering understanding among people, and providing guidance on a path forward to respect human and non-human life and to safeguard the goods of Earth for future generations.

ADVANCING MAGISTERIAL TEACHINGS—HOPE IN POPE FRANCIS AND THE BISHOPS

Pope Francis has stimulated our hope that he will advance magisterial teachings beyond his predecessors and provide an effective voice for the Catholic Church that is needed today as the effects of human-forced climate change on people, other species, ecological systems, and the biosphere are recognized. Why do we have hope? As many have noted, he chose as his namesake St. Francis of Assisi, the patron saint of people who promote ecology, the epitome of humility and simplicity, and the lover of humans and other animals. Pope Francis adopted a simple lifestyle, thereby providing an example for Catholics to follow Christ more closely, stand in solidarity with the materially poor and vulnerable, and minimize our consumption of the goods of Earth. Pope Francis presented himself humbly to the world by choosing to live in community with other professed religious and handling mundane tasks for himself by himself. And, during his inaugural Mass at St. Peter's Square on the Feast of St. Joseph, Pope Francis (2013) proffered the foster father of Jesus as the exemplar of a "protector" for all, including "all creation." The newly installed pontiff explained:

> The vocation of being a 'protector,' however, is not just something involving us Christians alone; it also has a prior dimension which is simply human, involving everyone. It means protecting all creation, the beauty of the created world, as the Book of Genesis tells us and as Saint Francis of Assisi showed us. It means respecting each of God's creatures and respecting the environment in which we live. It means protecting people, showing loving concern for each and every person, especially children, the elderly, those in need, who are often the last we think about. It means caring for one another in our families: husbands and wives first protect one another, and then, as parents, they care for their children, and children themselves, in time, protect their parents. It means building sincere friendships in which we protect one another in trust, respect, and goodness. In the end, everything has been entrusted to our protection, and all of us are responsible for it. Be protectors of God's gifts!

He continued:

> I would like to ask all those who have positions of responsibility in economic, political and social life, and all men and women of goodwill: let us be 'protectors' of creation, protectors of God's plan inscribed in nature, protectors of one another and of the environment. Let us not allow omens of destruction and death to accompany the advance of this world!

Thus, we pin our hopes on these promising signs of the direction Pope Francis may take to address climate change, the loss of biological diversity, and other issues that threaten the sustainability of Earth. We anticipate that he will deepen and demonstrate the indisputable connection between ecological degradation and the ongoing plight of materially poor and vulnerable people now and in the future as Benedict XVI underscored repeatedly throughout his pontificate. We anticipate that Pope Francis will help focus the faithful on the Franciscan and Ignatian sense of God's presence in and through the world and the outpouring of God's grace to us to live respectfully and responsibly in relation to one another and all creatures of Earth as mutual members of the Earth community. And, we anticipate from his example of humility that he will lead us into a more realistic theological anthropology in which we humbly recognize our interrelatedness and interconnectedness with other species and abiota from whom and with whom we evolved into existence as God's ongoing creation and upon whom we are utterly dependent for sustaining our temporal bodies as we yearn for everlasting life in the presence of God.

Can we also hope for the first encyclical dedicated to the climate crisis and related ecological issues of which there are many? For calling attention to the ongoing loss of biological diversity and other human interferences with Earth's processes and systems that must be abated? For identifying with leaders of other world religions a shared basis for addressing ecological degradation (e.g., a sense of the sacred in and through the world) and pointing to ways in which the faithful should live in relation to other species, ecological systems, and the biosphere? We are open to being surprised by this thoughtful, sincere, humble, and inspiring person who has become the 267th pope of the Church.

Finally, we have hope for the Catholic Church in the United States. Because our country is one of the leading contributors to the present climate crisis through its consumption of fossil fuels that emit greenhouse gases (U.S. EPA 2013) and because Catholics constitute approximately 24 percent of the U.S. population (Pew Forum 2013), the Church has a moral obligation to respond as a composite of laity and hierarchy. We hope that the bishops will lead the faithful in awakening to the reality of the climate crisis, the unjust plight of the vulnerable and materially poor who are most adversely affected now and into the future, and the threats to the sustainability of Earth. We also hope that the bishops will join theologians in tempering a strictly instrumental valuation of the goods of Earth with an understanding of the intrinsic value of other species, the air, the land, the waters, and ecological systems in and for themselves. If called upon by the bishops, scholars are willing and ready to help, thereby continuing the cooperation that was evident during the 2012 Catholic Consultation on Environmental Justice and Climate Change: Assessing Pope Benedict XVI's Ecological Vision for the Catholic Church in the United States.[2]

NOTES

1. The pope also serves as the Bishop of Rome.
2. Many thanks to Dan DiLeo, Project Manager for the Catholic Coalition on Climate Change, and Cecilia Calvo, Project Coordinator for the Environmental Justice Program and Climate Change of the U.S. Conference of Catholic Bishops, for providing background information leading up to the consultation. I am also grateful to Dan Misleh, Founding Executive Director of the Catholic Coalition on Climate Change, and Tobias Winright, Associate Professor of Theological Ethics at St. Louis University, for their suggestions when this introductory essay was nearing completion.

REFERENCES

Benedict XVI. 2006. "Letter of His Holiness Benedict XVI to His Holiness Bartholomew I, Ecumenical Patriarch, on the Occasion of the Sixth Symposium on 'Religion, Science, and the Environment.'" The Vatican, July 6. Accessed June 16, 2013. http://www.vatican.va/holy_father/benedict_xvi/letters/2006/documents/hf_ben-xvi_let_20060706_bartolomeo-i_en.html.

———. 2009a. "Letter of Pope Benedict XVI to Ecumenical Patriarch Bartholomew on the Occasion of the Eighth International Symposium on Religion, Science, and the Environment." The Vatican, October 12. Accessed June 16, 2013. http://www.patriarchate.org/events/usvisit2009/addresses-rse/Pope-Benedict-RSE-letter#sthash.kTsyZmCy.dpuf.

———. 2009b. "Papal Letter to Mississippi River Symposium." The Vatican. October 22. Accessed June 16, 2013. http://www.zenit.org/en/articles/papal-letter-to-mississippi-river-symposium.

———. 2009c. "If You Want to Cultivate Peace, Protect Creation." Message of His Holiness Pope Benedict XVI for the Celebration of the World Day of Peace. The Vatican, December 8. Accessed June 19, 2012. http://www.vatican.va/holy_father/benedict_xvi/messages/peace/documents/hf_ben-xvi_mes_20091208_xliii-world-day-peace_en.html.

———. 2012a. *The Environment*. Huntington IN: Our Sunday Visitor.

———. 2012b. "Address of His Holiness Pope Benedict XVI to the Members of the Diplomatic Corps Accredited to the Holy See." Vatican City, January 9. Accessed June 16, 2013. http://www.vatican.va/holy_father/benedict_xvi/speeches/2012/january/documents/hf_ben-xvi_spe_20120109_diplomatic-corps_en.html.

Benedict XVI and Woodeene Koenig-Bricker. 2009. *Ten Commandments for the Environment: Pope Speaks Out for Creation and Justice*. Notre Dame IN: Ave Maria Press.

Christiansen, S.J., Drew and Walter E. Grazer, eds. 1996. *And God Saw that It was Good:* Catholic Theology and the Environment. Washington DC: United States Conference of Catholic Bishops.

Ecumenical Patriarchate of Constantinople. 2013. "Orthodoxy and the Environment." Accessed June 16. http://www.patriarchate.org/environment.

Francis. 2013. "Homily of Pope Francis." Mass, Imposition of the Pallium and Bestowal of the Fisherman's Ring for the Beginning of the Petrine Ministry of the Bishop of Rome. St. Peter's Square, Vatican City, March 19. Accessed June 27. http://www.vatican.va/holy_father/francesco/homilies/2013/documents/papa-francesco_20130319_omelia-inizio-pontificato_en.html.

Gaillardetz, Richard R. 2003. *By What Authority? A Primer on Scripture, the Magisterium, and the Sense of the Faithful*. Collegeville MN: Liturgical Press.

Howden, Daniel. 2002. "The Green Patriarch - Bartholomew I." BBC News World Edition, June 12. Accessed June 16, 2013. http://news.bbc.co.uk/2/hi/europe/2040567.stm.

John Paul II. 1989. "Peace with God the Creator, Peace with All of Creation." Message of His Holiness Pope John Paul II for the Celebration of the World Day of Peace, 1 January 2010. The Vatican, December 8. Accessed June 16, 2013. http://www.vatican.va/holy_father/john_paul_ii/messages/peace/documents/hf_jp-ii_mes_19891208_xxiii-world-day-for-peace_en.html.

———. 2003. "Pope's Message to Patriarch of Constantinople on the Environment." Zenit, June 11. Accessed June 16, 2013. http://www.zenit.org/en/articles/pope-s-message-to-patriarch-of-constantinople-on-the-environment.

Myers, Norman. 2001. "Environmental Refugees: A Growing Phenomenon of the 21st Century." *Proceedings of the Royal Society of London*. Accessed June 30, 2013. http://www.nicholas.duke.edu/people/faculty/myers/myers2001.pdf.

———, and Jennifer Kent. 1995. *Environmental Exodus: An Emergent Crisis in the Global Arena*. The Climate Institute. June. Accessed June 30, 2013. http://www.climate.org/PDF/Environmental%20Exodus.pdf.

Pew Forum. 2013. "The Global Catholic Population." Accessed June 28. http://www.pewforum.org/Christian/Catholic/The-Global-Catholic-Population.aspx

Pontifical Academy of Sciences. 2011. *Fate of Mountain Glaciers in the Anthropocene: A Report by the Working Group*. Vatican: Pontifical Academy of Sciences.

Pullella, Philip. 2007. "Pope Leads Eco-friendly Youth Rally." *Reuters*, September 2. Accessed September 3, 2011. http://www.reuters.com/article/idUSL0111597220070901. Also see video from http://www.reuters.com/news/video?videoId=65392.

Schaefer, Jame. 2011. "Solidarity, Subsidiarity, and Preference for the Poor." In *Confronting the Climate Crisis: Catholic Theological Perspectives*, 389-425. Milwaukee WI: Marquette University Press.

Seewald, Peter. 2010. *Light of the World: The Pope, the Church and the Signs of the Times*. San Francisco: Ignatius Press.

Stammer, Larry B. 1997. "Harming the Environment is Sinful." *Los Angeles Times*. November 9. Accessed June 16, 2013. http://articles.latimes.com/1997/nov/09/news/mn-51974.

Stone, Daniel. 2013. "How Green Was the 'Green Pope'?" *National Geographic News*, February 28. Accessed June 27. http://news.nationalgeographic.com/news/2013/02/130228-environmental-pope-green-efficiency-vatican-city/

Sun Come Up. 2011. "Academy Award Nominee, Best Documentary." New Day Films. http://www.suncomeup.com/film/Home.html

United Methodist-Catholic Dialogue. 2012. "Heaven and Earth Are Full of Your Glory: A United Methodist and Roman Catholic Statement on the Eucharist and Ecology." Accessed June 16, 2013. http://www.usccb.org/beliefs-and-teachings/ecumenical-and-interreligious/ecumenical/methodist/upload/Heaven-and-Earth-are-Full-of-Your-Glory-Methodist-Catholic-Dialogue-Agreed-Statement-Round-Seven.pdf;http://www.vatican.va/roman_curia/pontifical_councils/chrstuni/meth-council-docs/rc_pc_chrstuni_doc_20060125_bolen-methodists_en.html.

United States Conference of Catholic Bishops. 1991. *Renewing the Earth: An Invitation to Reflection & Action on the Environment in Light of Catholic Social Teaching*. Washington DC: USCCB.

United States Environmental Protection Agency. 2013. "Global Greenhouse Gas Emissions Data." Accessed June 29. http://www.epa.gov/climatechange/ghgemissions/global.html.

Whittington, Heather. 2004. "The Catholic Church on Ecological Degradation." Interdisciplinary Minor in Environmental Ethics, Marquette University. Accessed June 16, 2013. http://www.inee.mu.edu/CatholicChurchonEnvironmentalDegradation.htm.

I

Human and Natural Ecology/Human Life and Dignity

ONE

Bonaventure in Benedict

Franciscan Wisdom for Human Ecology

Keith Douglass Warner, OFM

John Paul II introduced the expression "human ecology" into Catholic teaching with *Centesimus Annus* (1991, #38).[1] This expression has appeared in other Vatican documents, and it was used extensively by Pope Benedict XVI in several high profile addresses (e.g., 2009, 2010a). However, Catholics in the United States, both leaders and laity, have generally ignored this term. It is scarcely used. Why is this? I do not intend to answer this question directly. Instead I will draw from the thought of St. Bonaventure, a thirteenth century Franciscan theologian and doctor of the church, to propose an explanation for why Pope Benedict deployed this expression. The answer lies, in part, in Josef Ratzinger's intellectual journey and his understanding of how to access wisdom from the Catholic tradition for its application to the needs of the contemporary world. The future Pope Benedict (Ratzinger 1971) engaged Bonaventure as a "dialogue partner" when he researched and wrote his *habilitationsschrift*.[2] Pope Benedict continued to express great affection for Bonaventure and his Franciscan wisdom. By retracing Pope Benedict's intellectual footsteps we can better understand the potential of the expression "human ecology" to engage Catholics to care for God's creation. This methodology of retrieving wisdom from our Catholic tradition and presenting it in a form meaningful to our contemporary society holds promise for Catholic engagement with climate protection, but perhaps also for preaching the Gospel message in an era of new evangelization.

Here is a brief outline of the argument in this chapter. For eight centuries, the Franciscan tradition has drawn from St. Francis' passionate love of creation to develop an intellectual tradition that expresses a positive view of the created order. Francis is held up as the premier model of Christian care for creation, and was named patron saint of "those who promote ecology" by Pope John Paul II in 1979 (see Warner 2011b). Francis was not what we would call a theologian, but his spiritual and intuitive understanding of God expressing love through creation was brought into scholastic thought by early Franciscans, most notably Bonaventure (1217-1274). Bonaventure conveyed Franciscan insights through his philosophy and theology. For example, Francis found Christ present in all creation, singing affectionately of animals and elements in familial terms. Bonaventure elaborated Francis' insights to express a radically Christocentric vision of all created reality.

Roughly 700 years after the life of Bonaventure, Josef Ratzinger became one of the most important Catholic intellectuals of the twentieth century. Aiden Nichols, OP (2007) lists 99 books and 374 journal articles published by Ratzinger between 1954 and 2004. Through his *habilitationsschrift*, "The Theology of History in Bonaventure," Ratzinger engaged Bonaventure's wisdom theology that explains how God reveals God's self in creation and time. Ratzinger's writings echo some elements of Bonaventure's religious philosophy. Indeed, as Nichols has observed, Ratzinger follows in the footsteps of philosopher-theologians in the Catholic wisdom tradition, as exemplified by Augustine and Bonaventure. As will be discussed below, Bonaventure's influence on Ratzinger was less in terms of content and more through his thought structure.

Bonaventure articulated a symbol-laden, integrated understanding of God's self-revelation through all reality. For Bonaventure, since God is relational, all created reality is necessarily relational. Bonaventure proposed a Christian metaphysics of the good which is a metaphysics of relationality (Delio 1999, 2007). Francis was one important influence on Bonaventure, and Bonaventure was one important influence on Pope Benedict. Thus, we might consider how Franciscan wisdom might help us understand Pope Benedict's understanding of humanity's relationship to creation in salvation history.

In this light, the expression "human ecology" communicates with contemporary resonance a classic understanding of creation within the Catholic tradition. This expression has the potential to speak from the Catholic tradition to contemporary human society, which is more influenced by technoscientific materialism than Christian faith. Indeed, we may consider this expression to have a surplus of meaning, for it invites the Catholic community to consider how it should be interpreted (Ricoeur 1976). By recognizing the wisdom tradition that underlies Pope Benedict's use of this expression, the US Catholic Church may be able to use it to appeal more broadly for climate protection initiatives.

With the argument now outlined, this essay will develop it in the following order. It begins with an introduction to Bonaventure's Franciscan wisdom. It then presents his fundamental philosophical insights that convey his understanding of the importance of creation to salvation history. The essay then returns to consider evidence for Bonaventure's influence on Pope Benedict and concludes by proposing several steps for applying this framework more fruitfully in the US Catholic Church.

BONAVENTURE AND FRANCISCAN LOVE OF WISDOM

Francis of Assisi cannot be considered a philosopher or theologian in any formal sense, but recent scholarship has characterized his popular, affective, embodied, and intuitive approach to incarnational spirituality. He can be well described as a "vernacular theologian" (Monti 2001; for background on the term "vernacular theology," see McGinn 1998). Francis exerted a profound influence on the trajectory of medieval Catholic spirituality, and through the Franciscan movement, on theology, philosophy, spirituality and art, social ethics and the practice of social engagement.

Recent scholarship demonstrates how Francis' spirituality inspired the more learned and sophisticated thought of his subsequent followers. Our contemporary, popular understanding of Francis as a model of sentimental spirituality distorts his witness and the significance of the Franciscan movement he launched. Efforts by Franciscans to return to the spirit of their founders, prompted by Vatican II, have yielded an understanding of Francis' Christian discipleship that is more radical and potent. Scholarship is now articulating the influence Francis had on this movement, and through his followers, upon church and society. Francis launched a lay reform movement that emphasized devotion to the Incarnation, Eucharistic adoration, an inclusive, familial spirituality, and practical expressions of compassion within society. The Franciscan movement grew quickly to be one of the most influential currents in medieval Catholicism.

As the direct and indirect influence of Francis' charismatic witness on medieval Catholic thought has come into clearer focus, contemporary scholars have defined one expression of this as the Franciscan intellectual tradition (Osborne 2003, 2008). There were 5,000 Franciscan Friars by the end of Francis' life, and 30,000 within a generation. Franciscan and Dominican Friars arose concurrently with the rise of medieval universities and contributed much to their development in the Middle Ages. Franciscans made substantial contributions to the fields of philosophy, theology, natural sciences, and socio-economic ethics in medieval Europe and beyond. Contributions to this development can be termed the Franciscan intellectual tradition–this is broader than theology and philosophy–which is a branch of the Catholic intellectual tradition. The very idea that a Franciscan intellectual tradition existed in the historical past may be news to some.[3] Indeed, without appropriate academic tools, it may be difficult to discern such a phenomenon since the most important contributions by Franciscan thinkers were so fully incorporated into the broader Catholic tradition that their origins in the Franciscan movement are indiscernible.

Francis' vernacular theology inspired the development of academic institutions that were themselves undergoing transformation under the influence of new discoveries and the retrieval of ancient knowledge from the Greeks. The Franciscan approach complements the other two major intellectual traditions within Catholicism: the Augustinian and Thomistic (or Dominican). All three conduct theology within the Catholic tradition, yet in their diverse interpretive approaches, they provide a broader array of theological resources.

No components of a tradition interpret themselves. To access wisdom from any historical religious tradition requires a retrieval methodology (Schaefer 2009). Recent scholars have undertaken the retrieval of the Franciscan intellectual tradition, including its theology, cosmology, and social philosophy. This retrieval project articulates these features of Franciscan thought with contemporary forms of knowledge and social concerns. The word "tradition" comes from the Latin *tradere,* meaning to transmit or deliver. This suggests that traditions are not static treasures to be defended, but rather living memories and values, and ways of knowing and being that are shared from one generation down to the next. This requires a method that addresses the following questions:

1. What components of our tradition do we wish to select for retrieval?
2. How do we appropriately interpret these elements in light of present need?
3. How can these components be appropriately combined with contemporary forms of knowledge?

These are the tasks for the retrieval of the Franciscan intellectual tradition, and as a methodology, retrieval holds out broad potential for engaging and sharing the riches of the Catholic tradition (Warner 2011a). This essay cannot fully address these questions, but it can explore their applicability in proposing the contemporary relevance of Bonaventure's thought for Catholic care for creation.

Bonaventure is arguably the greatest example of a Franciscan intellectual leader. In his work we find the intuition and spirit of Francis translated into formal philosophy and systematic theology. When Bonaventure became a Franciscan, he was attracted to the spirituality of Francis as expressed by his followers. As a scholar and university teacher, Bonaventure might seem to have little in common with the poor man of Assisi, but they share a radically Christocentric spirituality, a belief that God is revealed through creation, and an understanding that all creation is essentially good and relational in character.

Bonaventure is a remarkable intellectual figure in his own right, yet he cannot be properly understood apart from his identity as a Franciscan Friar. He joined the friars at an early age, spent more than 15 years studying, teaching and living as a Friar Minor at the University of Paris, and was the seventh minister general of the Franciscan order. He held that office for 17 years, steering the order through internal divisions and external attacks that threatened its integrity and survival. He wrote one of the most remarkable and influential works in medieval Christian spirituality, *Itinerarium Mentis in Deum* (which can be best translated as *The Journey of the Soul into God*).[4] Bonaventure, like any great thinker, can be read from multiple perspectives. Scholarly analysis of the influences on Bonaventure suggests they be ranked thus: Scripture, Augustine, Francis, Aristotle, Pseudo-Dionysius, Hugh of St. Victor, and Anselm (Bougerol 1964). Bonaventure's thought has enjoyed a resurgence of interest, in part because it engaged and transformed patristic philosophical and theological concepts, and synthesized a more conducive framework for dialogue with contemporary sciences including ecology and evolution.

This essay will highlight his contributions as a Franciscan religious philosopher, drawing on the ancient understanding of philosophy as love of wisdom. Bonaventure, following Augustine, sought to express an integrative and practical wisdom. Integrative wisdom holds that the purpose of (Christian) philosophy is to unify reality in the human mind and its understanding (Hayes 1994). Neither a fragmentary approach to knowledge and knowing, nor a dualistic worldview, could be compatible with philosophy in this tradition. Franciscan philosophy is practical, meaning that knowledge never exists for the sake of possession, but rather to grow in love of God and help everyone follow in the footprints of Jesus Christ (Boehner 2005; Hayes 1997). Bonaventure conveyed his understanding of wisdom as holistic and integrative: "There are some dimensions of wisdom that relate to our intellect, others that relate to our desires, and others that are to be lived out. Therefore, wisdom ought to take possession of the entire person, that is with respect to the intellect, the affective life, and the person's action" (Hayes 1999, 334). This Franciscan Catholic understanding of philosophy as integral and practical wisdom diverges fundamentally from many contemporary forms of academic philosophy.

Bonaventure's thought integrated several intellectual and spiritual currents. He achieved a remarkable synthesis of speculative theology and affective spirituality, of Eastern and Western Christian spirituality, of abstract symbolism and practical embodied experience. A leading expert in Bonaventure's thought, Ewert Cousins (1978) described Bonaventure as achieving for spirituality what Thomas of Aquinas did for theology and what Dante did for medieval culture. Cousins described Bonaventure as platonizing Franciscanism and Franciscanizing neoplatonism, by which he meant that Bonaventure wove together into a coherent, seamless system Scripture, Patristic thought, scholastic philosophy and the passion of Francis' Incarnational spirituality. Bonaventure's thought is fundamentally synthetic, meaning that he synthesized a coherent and complete whole out of these distinct influences. The elements of his thought were historical or traditional, but his synthesis created a fresh approach to Christian wisdom for the people of his era.

For the purpose of this chapter, the historical influences on and theological content of Bonaventure's thought are less important than the implications of his thought structure for understanding Franciscan wisdom. Bonaventure can be approached through our modern lenses of theology and religious philosophy; however, the integral character of his approach to wisdom does not fit comfortably within the compartments of these modern academic disciplines. The noted scholar Etienne Gilson (1938) claimed that Bonaventure's thought is of such remarkable harmony and unity that one sees the whole or nothing at all. His thought structure seems peculiar to those with a modern, technoscientific worldview, which is highly compartmentalized, fragmented and reductionistic. However, Bonaventure's approach to wisdom may be the very balm we most need.

Bonaventure's integral thought structure can be represented by a circle, suggesting completeness or wholeness. In God all life originates, finds expression in the time and space of the created order, and discovers its ultimate destiny in return to God. The Trinity is the template for this circular movement, and Bonaventure understands God to be an infinite primordial mystery of self-communicative love. God the Father is the origin, sustainer, and consummation of all created reality. Bonaventure (1970, 10) describes the action of the Incarnation of Jesus Christ as central to the manifestation of God's love: "Such is the metaphysical Center that leads us back and this is the sum total of our metaphysics: concerned with emanation, exemplarity, and consummation, that is, illumination through spiritual radiations and return to the Supreme Being. And in this you will be a true metaphysician." Everything flows from the Father (emanation) and ultimately returns to Him (consummation). Exemplarity expresses the "conviction that all of created reality is grounded in the divine archetypal reality and manifest the mystery of the divine in the created realm to some limited degree" (Hayes 2003b, 95). The concept of exemplarity is essential to understanding Bonaventure's thought (Bowman 1975). Exemplarity reflects Augustine's presentation of "the divine idea." This strongly influenced Bonaventure who stated that "things" have a two-fold existence, to be themselves and to be in the divine mind as exemplars (Cullen 2006). The concepts of emanation, exemplarity and consummation are inter-related, dynamic components operating through every dimension of Bonaventure's thought system. Together they function as a conceptual framework for expressing God's love, God's activity in the world, the Incarnation, and the wisdom expressed through creation (Hayes 1976). Bonaventure's thought structure, in the spirit of St. Francis, thus provided a paradigm for understanding the relationship of Jesus Christ to all created reality.

BONAVENTURE'S METAPHORS AND TRINITARIAN METAPHYSICS

Bonaventure understood creation to have an essential role in salvation history. He did not express what we would recognize today as an environmental ethic but, rather, a (medieval Franciscan) Catholic worldview that could not conceive of creation independently of the relationship between God and humanity. Creation is not a distinct topic for him, but instead one of several themes that integrate his many writings. He conveyed his theological wisdom of creation through a system of religious metaphors, a metaphysics rooted in the Trinity, and a vision for human beings in salvation history.

Bonaventure deployed a rich system of symbols to convey how God reveals God's self through creation. Cousins (1978) proposed a gothic cathedral as the best organizing metaphor to convey the integral character of Bonaventure's symbol system to modern people. The superstructure of a cathedral represents the designed cosmos that discloses the mind of its creator. The entire interior of the cathedral is designed to focus attention on the centrality of the Crucified Jesus; this is the purpose of creation and the cathedral. The massive doors, doorways, and stained glass windows recount salvation history. The light that enters the cathedral illuminates those inside with wisdom, grace and love.

The following are examples of Bonaventure's religious metaphors of creation and how he used them to communicate God's plan for the cosmos.

1. Word, speech and book: Bonaventure understood reality as two books, "one written within, namely the eternal Art and Wisdom of God; and the other written without, namely, the perceptible world" (Hayes 2003a, 255). In his *Collation on the Six Days*, Bonaventure wrote: "the entire world is, as it were, a kind of book in which the Creator can be known in terms of power, wisdom and goodness which shine through in creatures" (Hayes 1999, 63). Creation is thus an external Word of God, the "speech" of God, expressed in finite time and history. Creation discloses God and God's love, and these metaphors help us recognize this deeper purpose. Components of creation and creation as a whole serve as symbols of the divine (Hayes 2003a). Creation, like Jesus, comes from God, helps us to perceive and understand God, and leads us back to God.

2. Mirror: the book of creation is akin to a mirror, for as one studies creation, one learns more about God, and as one comes to understand God, one recognizes God's love in creation. This points to the importance of the natural sciences to help understand the character of God and God's artistry revealed through creation and its purpose.
3. Circle: Bonaventure used the circle many times in his works as a symbol of the Trinity itself, and to describe the movement of divine life into, through and back from creation. Salvation history is played out in a dynamic setting.
4. Water and river: Bonaventure used a fountain as a theological image for God's love. The Father is the "fountain fullness" from which the river of reality flows, both within the mystery of God's self and outside the divinity in the form of creation. Bonaventure frequently described creation as a river that flows from that spring of God, spreading across the land to purify it and make it fertile, and flowing back to its origin (Delio 2001).
5. Song: the Latin *carmen* can be translated as song or poem. Bonaventure compared the universe to a beautifully composed song, another image that he borrowed from Augustine. Yet, he developed this metaphor further, insisting that full appreciation of a song requires grasping the entire melody–not only the individual notes, but also their inter-relation with pitch, rhythm and tone. His use of this image points to the necessity of understanding the individual components of creation as well as their integral whole. Only by grasping the whole can the harmonious structure of creation become clear.
6. Light and window: light is a metaphor for divine reality and divine life. Bonaventure understood every creature (not only humans) to have within itself a shining forth of divine life. The entire material cosmos is a window to the divine, and its rich diversity of creatures reflects the depth and richness of God (Hayes 2001).

God created the cosmos and created it with purpose, intention. Two Latin keywords used by Bonaventure illustrate God's intention for creation: *manifestare* and *participare*. "The cosmos manifests the mystery of God in the nondivine. And creation is called into being so as to participate in ever deeper levels in the mystery of the divine life" (Hayes 1997, 53). Creation's purpose is to communicate to us who God is, and how we humans are to respond. The created order is to help us perceive, understand and love God. In technical terms, this is a semiotic metaphysics (Cullen 2006). Creation is capable of bearing that communication, and we humans are capable of understanding that message. Question: could contemporary Catholics believe that creation communicates something of God to human beings?

To draw from these positive metaphors for creation would be helpful today. However, this "use" of Bonaventure's metaphorical language to assert creation's goodness is a superficial reading of what he sought to convey. There is a very tight relationship between creation and the God as Trinity. Indeed, the Trinity is the foundation of Bonaventure's entire theological program, and the Trinity for Bonaventure is relational. Bonaventure drew from the fourth-century Cappadocian Patristic tradition that understood Trinity to be a community of divine persons, not Trinity as the unity of God's substance per Augustine.[5] Bonaventure's theology of the Trinity holds together divine essence and divine relationality. They must always be presented in a common frame. They cannot be properly examined or presented independently. Relationship or relationality is the basis for a Trinitarian theology of God, not vice versa. For Bonaventure, God is in essence communicable. "The main point of Bonaventure's Trinitarian theology is that the very nature of God is relational, and that it is only in and through meditation on this basic relational nature of God that one can formulate the Trinity of Father, Son and Spirit" (Osborne 2011, 119). In other words, before we can even use the term "Trinity" we have to recognize the relational dimension of God.

Being (*Summum Ens*) itself is good and relational. Bonaventure's approach to metaphysics reflects his Trinitarian theology: God is good and God is relational; therefore, being is good and relational; therefore, all reality is good and relational. Trinitarian life finds expression in the structure of the universe, from macrocosm to microcosm. "The reality of Christ pertains to the very structure of reality: as Word, to the reality of God; as incarnate Word, to the reality of the universe created by God" (Hayes 1994, 72). This has significant implications for the relationship of scientific inquiry and theological inquiry. From a Franciscan perspective, a priori, these cannot essentially contradict each other. Question: could contemporary Catholics believe that the deeper, metaphysical structures of created reality reflect something of the character of God?

Bonaventure understood humanity to have a special relationship with the cosmos and other forms of life, and to have a critical, distinct role in salvation history. In the *Breviloquium* (1963, 77), he wrote:

> Hence, it is undubitably true that we human beings are the end of all existing things. All material things are made to serve man, and to enkindle in him the fire of love and praise for the Maker of the universe through whose providence all is governed. Therefore the fabric of [God's] sensitive body is like a house made for man by the supreme Architect to serve until such time as he may come to the house not made by human hands . . . in the heavens.

Although we are, according to Bonaventure, bent over by sin and not able to fully perceive how creation calls us into union with its creator, he nonetheless has a positive understanding of humanity and its role in the created order. The book of creation has been rendered opaque by our sin. Just as God created creation for fecundity and harmony, the vocation of human beings is to foster these. We are intended for a central role in God's plan for creation, for as Zachary Hayes, OFM (1994, 68) observes,

> The material world stands most properly at the service of humanity when it enables human beings to realize the end of God's creative activity by awakening in them the conscious appreciation, love, and praise for the Giver of the gift of created existence. It is thus that humanity gives a conscious, loving voice to what otherwise would remain a mute creation.

In other words, the destiny of humanity and the material cosmos are utterly entwined.

Here again the image of the circle conveys the movement of humanity and creation through time. As Hayes (1994, 79) further explains,

> The circle is a symbolic expression of the conviction that creation is the movement of finite being from nothing into historical existence and ultimately to that fullness of personal life in union with God that Christians understand to be salvation. Creation and salvation, though distinguishable conceptually, are inseparably related. In its fullest sense, salvation is the actualization of the deepest potential that lies at the heart of created reality by reason of the creative love of God. . . . The visible creation is the objective expression of the Word that lies at the center of the divinity. . . . The theology of the return of creation to God is, in essence, the theology of history. If we move this line of thought but a step further, it appears that eschatology is the attempt to articulate that point at which the curve of creation and history bends back on its point of origin. It is the doctrine of ultimate fulfillment of the created order. . . . The return of creation to God, then, is in essence a journey through history, and the fulfillment of the universe is inseparably related to that of humanity.

Bonaventure understood consummation—and how creation and humanity travel the path of salvation history together—as bringing all created reality into ultimate communion with God. Question: could contemporary Catholics believe that human beings have an essential role in bringing to consummation all creation's fulfillment in God?

POPE BENEDICT'S JOURNEY INTO THE MIND OF BONAVENTURE

Benedict XVI spoke in very positive ways about his appreciation for the Franciscan tradition. He wrote his doctoral dissertation on the Church as People and House of God in Augustine, and framed this in terms of Christianity's philosophical dialogue with late Roman culture. He wanted to continue his research into the dialogue between theology, philosophy and culture by investigating the interrelation of "revelation–history–metaphysics." He focused his *habilitationsschrift* to address questions of the relationship between salvation history and metaphysics in the context of the retrieval of Aristotle's corpus in the thirteenth century and the currents of religious thought about the character of time.[6] Bonaventure's work was more amenable to this line of inquiry than Aquinas's, since Ratzinger (1989, xii) wanted to dialogue with medieval conceptualizations of revelation through time. This was of considerable academic concern in the thirteenth century: what is time, how is God operating within time, and how is time as humans know it related to salvation history? Bonaventure engaged this question within the context of the debates of his era. The content and structure of Ratzinger's thesis, and the critics of his argument and conclusions, are not directly relevant to this present article.[7] This section addresses two questions: Did Bonaventure's thought and its structure influence Josef Ratzinger? And if so, what Bonaventurian influence on Benedict's conceptualization of human ecology might we discern?

As Pope Benedict XVI, he did not speak of any direct influence of Bonaventure on his philosophy or theology of creation. However, he did give three audiences, or informal presentations, on Bonaventure and the Franciscans on successive weeks in 2010:

> Today I would like to talk about St Bonaventure of Bagnoregio. I confide to you that in broaching this subject I feel a certain nostalgia, for I am thinking back to my research as a young scholar on this author who was particularly dear to me. My knowledge of him had quite an impact on my formation (Benedict XVI 2010b).
>
> . . . [A]mong St Bonaventure's various merits was the ability to interpret authentically and faithfully St Francis of Assisi, whom he venerated and studied with deep love (Benedict XVI 2010c).
>
> I would like to study with you some other aspects of the doctrine of St Bonaventure of Bagnoregio. He is an eminent theologian who deserves to be set beside another great thinker, a contemporary of his, St. Thomas Aquinas (Benedict XVI 2010d).

Pope Benedict spoke quite eloquently of the philosophical and theological foundations that can support an authentic dialogue between faith and reason, theology and science.

At a presentation to scientists, he (Benedict XVI 2008) deployed several metaphors and images that had been used by Bonaventure:

> The imagery of nature as a book has its roots in Christianity and has been held dear by many scientists . . . It is a book whose history, whose evolution, whose "writing" and meaning, we "read" according to the different approaches of the sciences, while all the time presupposing the foundational presence of the author who has wished to reveal himself therein. This image also helps us to understand that the world, far from originating out of chaos, resembles an ordered book; it is a cos-

> mos. . . . We may not at first be able to see the harmony both of the whole and of the relations of the individual parts, or their relationship to the whole. Yet, there always remains a broad range of intelligible events, and the process is rational in that it reveals an order of evident correspondences and undeniable finalities . . . And thanks to the natural sciences we have greatly increased our understanding of the uniqueness of humanity's place in the cosmos.

This address is significant in three ways. First, it deploys some of the same metaphors for creation and its attributes (book, reading, harmony) as Bonaventure. Next, it conveys a metaphysics consistent with that of Bonaventure, for creation, by disclosing God's purpose for all reality, is communicative. In other words, a semiotic metaphysics provides the possibility of creation bearing God's message. Third, it articulates an integral relation of the parts and whole, and in this sense, it echoes Bonaventure's wisdom theology.

CONCLUSION: FRANCISCAN WISDOM OPENS UP MEANINGS OF HUMAN ECOLOGY

Pope Benedict had a very positive understanding of science, as the quote above demonstrates. Elsewhere he and his predecessor spoke of the importance of a constructive dialogue between theology and science (Russell et al. 1990). Both Pope John Paul and Benedict, however, used the modern word "ecology" to communicate a philosophical understanding of creation–including humanity–that is rooted deep within the Catholic tradition.

Ecology is the scientific study of the relationships that living organisms have with each other and their environment. It undermines the reductionistic tendencies in current technoscientific thought common in our culture. Strictly speaking, it is scientific discipline, a field of the life sciences. However, ecology has been called the "subversive science" because it asserts that all of nature is related. Ecology has contributed a revolutionary concept, the ecosystem, even though it points to an ancient insight. An ecosystem is "a functional system of complementary relations between living organisms and their environment, delimited by arbitrarily chosen boundaries, which in space and time appear to maintain a steady yet dynamic equilibrium. An ecosystem thus has physical parts with particular relationships–the *structure* of the system–that together take part in dynamic processes–the *function* of the system" (Gliessman 1998, 17; emphasis original). The term "appear" is important because, in fact, no functional systems in nature actually maintain equilibrium indefinitely.

The modifier "human" in the expression "human ecology" is necessary to communicate that human beings are a part of nature or creation, a fact that seems, curiously, to be frequently forgotten by many scientists, and in some cases religious leaders. The evidence presented in this essay suggests that Pope Benedict uses "human ecology," very much in the spirit of Bonaventure's thought, to communicate the essentially relational character of God *and* of all metaphysical reality. God is essentially relational, and therefore God's cosmos is too. Here the metaphor of the circle indicates that essential reality. Perhaps Bonaventure's circle metaphor can help us engage the essential relationally character of life proposed by the ecosystem concept.

This chapter has traced elements in Bonaventure's wisdom theology that Pope Benedict deployed, even though he did not cite his work. Here we might consider that Bonaventure's thought may have influenced Josef Ratzinger in a broader sense, as many elements of his religious philosophy of creation were broadly shared across much of Catholic history. The Catholic tradition has long held that understanding creation helps us understand God. This is expressed by the Old Testament, and more fully developed by Augustine. Bonaventure merely built upon this tradition (Hayes 2001). By understanding modern scientific ecology, we can better read the book of nature. Pope Benedict's teachings challenge us to engage contemporary concerns and ways of thinking, while simultaneously drawing on the full array of philosophical, intellectual and theological resources in our big, broad, dynamic Catholic tradition.

To return to the original question: US Catholics, of all kinds, have ignored "human ecology" as an expression. It is scarcely used. Why is this? The answer lies in part due to our ignorance of the wisdom in our Catholic tradition. The presentation of creation by Bonaventure, and Pope Benedict, may appear novel to contemporary Christian streams of thought, but in reality, it is deeply traditional. For us today, Bonaventure's work challenges us to consider the breadth, diversity, and wisdom in our Catholic tradition. As a doctor of the church, he offers us many valuable teachings about creation and its role in salvation history–teachings that have been lost or forgotten. We are inheritors to an impressive, coherent, attractive understanding of God's love expressed to us through creation. This discussion suggests the following tasks for the US Catholic Church:

1. Recover the breadth of Catholic philosophy as wisdom tradition grounded in the Incarnation of Jesus Christ and the Holy Trinity. To take full advantage of the expression "human ecology" we should ground this concept in our Catholic philosophical tradition, including cosmology and metaphysics. This is a more robust position from which to dialogue with contemporary culture and science.
2. Propose human ecology as a broad, positive framework to guide us in taking our appropriate place in and as a part of creation,

drawing from our tradition of Trinitarian theology, common good, and shared vocation of humanity.

3. Draw from the Franciscan intellectual tradition that integrates affective inquiry and social engagement. Knowledge alone is not adequate to guide the human to a balanced relationship with creation, nor to the sense of religious purpose God intends for all created reality (Warner 2012). Bonaventure's philosophy can inform our efforts to integrate love and praxis with knowledge, and to enhance the integrity of our witness to God's love for all creation and our human duties to respond in kind.

Pope Benedict's use of "human ecology" thus challenges us to simultaneously engage contemporary concerns and ways of thinking, but on the full array of philosophical, intellectual and theological resources in our big, broad, dynamic Catholic tradition. I am not suggesting that the St. Francis Pledge be replaced by the St. Bonaventure Pledge; however, I do think that we who attempt to interpret and extend Pope Benedict's teaching about human ecology and climate protection can draw wisdom and inspiration from Bonaventure's wisdom.

NOTES

1. The generous support of the Markkula Center for Applied Ethics at Santa Clara University made participation in the conference and this publication possible.

2. The *habilitationsschrift* is a post-doctoral research project (usually a scholarly book) used by European academic institutions to evaluate the candidacy for a senior professorship. Ratzinger completed his in 1959, and it was first published in English in 1971.

3. Much scholarship of the Franciscan movement emphasized tension and discontinuity between the primitive and intellectual expressions of the Franciscan movement; more recent work emphasizes continuity through institutionalization. Two examples of the former are Desbonnets (1988) and Landini (1968). Two examples of the latter are Cousins (1981) and Blastic (1998).

4. The best edition is Bonaventure (2002). The notes on translation and references to other components of Bonaventure's corpus are particularly helpful. Another fine version can be found in Cousins (1978).

5. The literature on Bonaventure's understanding of Trinity is extensive. Some examples include: Hayes (1994), Delio (2001), and Cullen (2006). Bonaventure's theology of the Trinity was also influenced by John Damascene, Pseudo-Dionysius, and Richard of St. Victor. See Hayes (1992).

6. For a personal narrative of his motivations for and struggles with his *habilitationsschrift*, see pages 105-114 in Ratzinger (1998).

7. For a review of the critical responses to Ratzinger's discussion of Bonaventure's apocalypticism, see Cullen (2006, 177-186).

REFERENCES

Benedict XVI. 2008. "Address of His Holiness Benedict XVI to Members of the Pontifical Academy of Sciences on the Occasion of Their Plenary Assembly." October 31. Accessed May 21, 2013. http://catinfor.com/en/vatican_en/?p=3269.

———. 2009. *Caritas in Veritate*. Accessed May 27, 2013. http://www.vatican.va/holy_father/benedict_xvi/encyclicals/documents/hf_ben-xvi_enc_20090629_caritas-in-veritate_en.html.

———. 2010a. "If You Want to Cultivate Peace, Protect Creation." Message for the World Day of Peace. January 1. Accessed May 27, 2013. http://www.vatican.va/holy_father/benedict_xvi/messages/peace/documents/hf_ben-xvi_mes_20091208_xliii-world-day-peace_en.html.

———. 2010b. "St. Bonaventure." General Audience. March 3. Accessed 28 September 2012. http://www.vatican.va/holy_father/benedit_xvi/audiences/2010/documents/hf_ben-xvi_aud_20100303_en.html.

———. 2010c. "St. Bonaventure (2)." General Audience. March 10. Accessed September 28, 2012. http://www.vatican.va/holy_father/benedict_xvi/audiences/2010documents/hf_ben-xvi_aud_20100310_en.html.

———. 2010d. "St. Bonaventure (3)." General Audience. March 17. Accessed September 28, 2012. http://www.vatican.va/holy-father/benedict_xvi/audiences/2010/documents/hf_ben-xvi_aud_20100317_en.html.

Blastic, Michael, OFM. 1998. "'It Pleases Me That You Should Teach Sacred Theology': Franciscans Doing Theology." *Franciscan Studies* 55:1-25.

Boehner, Philotheus, OFM. 2005. "The Spirit of Franciscan Philosophy." *Greyfriars Review* 19 (3):237-56 (republished from *Franciscan Studies* 2: 217-37, with Latin texts translated by Robert Karris OFM).

Bonaventure. 1963. *The Breviloquium*. Translated by Jose De Vinck. Patterson, New Jersey: St Anthony Guild Press.

———. 1970. *Collations on the Six Days*. Patterson, New Jersey: St. Anthony Guild Press.

———. 2002. *Itinerarium Mentis in Deum*. Translated by Philotheus Boehner OFM and Zachary Hayes, OFM. St. Bonavnenture NY: Franciscan Institute, Saint Bonaventure University.

Bougerol, Jacques G. 1964. *Introduction to the Works of Bonaventure*. Patterson NJ: St. Anthony Guild Press.

Bowman, Leonard J. 1975. "The Cosmic Exemplarism of Bonaventure." *The Journal of Religion* 55(2):181-98.

Cousins, Ewert. 1978. *Bonaventure: The Soul's Journey Into God, The Tree of Life, and The Life of St. Francis*. New York: Paulist Press.

———. 1981. "Francis of Assisi and Bonaventure: Mysticism and Theological Interpretation." In *The Other Side of God*, edited by Peter L. Berger, 74-103. New York: Anchor Press.

Cullen, Christopher M., SJ. 2006. *Bonaventure*. New York: Oxford University Press.

Delio, Ilia, OSF. 1999. "Bonaventure's Metaphysics of the Good." *Theological Studies* 60:228-46.

———. 2001. *Simply Bonaventure: An Introduction to His Life, Thought and Writings*. Hyde Park, New York: New City Press.

———. 2007. "Theology, Metaphysics and the Centrality of Christ." *Theological Studies* 68:254-73.

Desbonnets, Theophile. 1988. *From Intuition to Institution*. Chicago: Franciscan Herald Press.

Gilson, Etienne. 1938. *The Philosophy of Saint Bonaventure*. New York: Sheed and Ward.

Gliessman, Stephen R. 1998. *Agroecology: Ecological Processes in Sustainable Agriculture*. Chelsea, Michigan: Sleeping Bear Press.

Hayes, Zachary, OFM. 1976. "Incarnation and Creation in St. Bonaventure." In *Studies Honoring Ignatius Brady, Friar Minor*, edited by Romano Stephen Almagno and Conrad L. Harkins, 309-29. St. Bonaventure, New York: The Franciscan Institute Press.
———. 1992. *The Hidden Center: Spirituality and Speculative Christology in St. Bonaventure*. St. Bonaventure NY: Franciscan Institute Press.
———. 1994. "Bonaventure: Mystery of the Triune God." In *The History of Franciscan Theology*, edited by Kenan B. Osborne, OFM, 34-126. St. Bonaventure, New York: Franciscan Institute Press.
———. 1997. "Franciscan Tradition as Wisdom Tradition." *Spirit and Life: A Journal of Contemporary Franciscanism* 7:27-40.
———. 1999. *Bonaventure: Mystical Writings*. New York: Crossroad Publishing Company.
———. 2001. *The Gift of Being: A Theology of Creation*. Collegeville MN: The Liturgical Press.
———. 2003a. "The Cosmos, a Symbol of the Divine." In *Franciscan Theology of the Environment: An Introductory Reader*, edited by Dawn M. Nothwehr, OSF, 249-67. Quincy, Illinois: Franciscan Press.
———. 2003b. "Is Creation a Window to the Divine? A Bonaventurian Response." In *Franciscans and Creation: What is Our Responsibility?* edited by Elise Saggau, OSF, 91-100. St. Bonaventure, NY: Franciscan Institute Publications.
John Paul II. 1991. *Centesimus Annus*. Accessed May 27, 2013. http://www.vatican.va/holy_father/john_paul_ii/encyclicals/documents/hf_jp-ii_enc_01051991_centesimus-annus_en.html.
Landini, Lawrence C. 1968. *The Causes of the Clericalization of the Order of Friars Minor 1209-1260 in the Light of Early Franciscan Sources*. Roma: Pontificia Universitas Gregoriana.
McGinn, Bernard. 1998. *The Flowering of Mysticism*. New York: Herder and Herder.
Monti, Dominic, OFM. 2001. "Francis as Vernacular Theologian." In *The Franciscan Intellectual Tradition*, edited by Elise Saggau, OSF, 21-42. St. Bonaventure, New York: Franciscan Institute Press.
Nichols, Aidan, OP. 2007. *The Thought of Pope Benedict XVI: An Introduction to the Theology of Joseph Ratzinger*. London: Burns & Oates Ltd.
Osborne, Kenan B., OFM. 2003. *The Franciscan Intellectual Tradition: Tracing Its Origins and Identifying Its Central Components*. St. Bonaventure NY: Franciscan Institute.
———. 2008. "The Franciscan Intellectual Tradition: What Is It? Why Is It Important?" *Association of Franciscan Colleges and Universities Journal* 5(1):1-26.
———. 2011. "The Trinity in Bonaventure." In *The Cambridge Companion to the Trinity*, edited by Peter C. Phan, 108-127. New York: Cambridge University Press.
Ratzinger, Joseph. 1971. *The Theology of History in Bonaventure*. Translated by Zachary Hayes, OFM. Chicago: Franciscan Herald Press.
———. 1998. *Milestones: Memoirs, 1927-1977*. San Francisco: Ignatius Press.
Ricoeur, Paul. 1976. *Interpretation Theory: Discourse and the Surplus of Meaning*. Ft. Worth, Texas: Texas Christian University Press.
Russell, R.J., W.R. Stoeger, Pope John Paul II, and G.V. Coyne, SJ. 1990. *John Paul II on Science and Religion: Reflections on the New View from Rome*. Rome, Italy: Vatican Observatory Publications.
Schaefer, Jame. 2009. *Theological Foundations for Environmental Ethics: Reconstructing Patristic and Medieval Concepts*. Washington DC: Georgetown University Press.
Warner, Keith Douglass, OFM. 2011a. "Franciscan Environmental Ethics: Imagining Creation as a Community of Care." *Journal of the Society of Christian Ethics* 31(1):143-160.
———. 2011b. "Retrieving St. Francis: Tradition and Innovation for Our Ecological Vocation." In *Green Discipleship: Catholic Theological Ethics and the Environment*, edited by Tobias Winright, 114-127. Winona MN: Anselm Academic.
———. 2012. *Knowledge for Love: Franciscan Science as the Pursuit of Wisdom*. St. Bonaventure NY: Franciscan Institute Press.

TWO

If You Want Responsibility, Build Relationship

A Personalist Approach to Benedict XVI's Environmental Vision

Mary A. Ashley

I think we can agree that, at present, Catholic environmental action is spotty at best. But even though Catholics—at least *as* Catholics—seem to be hanging back, we could be in the very forefront of the American environmental movement. Here's why: Although the first decades of the movement reflected a strongly holist character, a personalist environmentalism is actually both more basic and more durable. And the Catholic Church, because of its strongly personalist theology, and given its relative influence in American life, is the institution best able to promote a personalist environmentalism. I will argue, then, that Catholicism stands ready to offer a personalist environmentalism grounded in love as its own unique gift to the environmental movement.

This argument will be advanced in five steps. First, I will clarify what I mean by a personalist environmentalism and distinguish it from the holist alternative. Second, I will explain that, while environmental holism has its advantages, the personalist option is both more fundamental and more likely to sustain widespread commitment in the critical decades ahead. Third, I will reference Pope Benedict XVI and the 2004 *Compendium* on social doctrine to outline a quintessentially Catholic personalism that, because it begins with God's presence in our hearts, tends naturally to spread outward so as to encompass our "more-than-human world"

(Abram 1996). In order to highlight the distinctiveness of Catholicism's personalist environmentalism, I will contrast it with the alternatives offered by secular modernity and most forms of Protestantism. Fourth, I will reference Church teaching on "integral human development" to show that love's primacy over "use" is the interpretive key that enables us to understand Benedict's oft-repeated teaching that our own "human ecology" underlies our efforts to support "environmental ecology." Finally, I will offer "Kinship Care" as a practical starting point for a Catholic and personalist environmentalism grounded in love.

ENVIRONMENTAL HOLISM AND ITS PERSONALIST ALTERNATIVE

When talking about the wider world that surrounds us, taking a "holist" approach has been customary. Our approach is holist when we assume that Big Picture view of Earth as one interconnected and "whole" system that is often termed "the environment." The iconic image of a sunlit "Blue Marble" floating in space is one familiar expression of environmental holism. In the first decades of the environmental movement, the holist approach got almost all of the press. That is changing as more people become aware of the "personalist" alternative.[1]

While holism attempts to stand apart from the Earth so as to perceive a unified system, a personalist environmentalism begins with the individual human being in the midst of his or her ongoing experience. It starts, in other words, with the first-person viewpoint that each of us holds as we attempt to understand and respond to all we encounter in the course of an ordinary day. Accordingly, a personalist environmentalism attends to our relationships, and especially our direct encounters, with the life around us. This category, of course, includes other people but can also encompass living creatures of every kind. Looking through a personalist lens, we recognize the relational possibilities in the circumstances close at hand. A backyard, for example, becomes an opportunity to plant native flora so as to support bees and other native pollinators. The necessity of replacing a computer becomes an opportunity to ensure that it is recycled in the safest way possible, so that no danger is posed to the worker who will take it apart. Or we begin to see our own food choices as influencing the quality of life in rural areas—as that life is experienced by both people and animals.

The holist approach is indispensable for at least two reasons. First, because it assumes a standpoint external to any ecological system that it examines, it brings a degree of detachment to the way that we observe nature. We can thus use a holist approach to gain information that is relatively objective and so more amenable to precise analysis. Second, when we adopt the holist concern to restore ecological functioning, we are confronted with our duty to respect Earth's fragility. A holist environmentalism will counsel humility and urge caution.

I suggest, however, that unless one is in a role that requires a scientific kind of detachment, a choice for the personalist starting point is likely to yield the greater long-term benefit. Now what could underlie such a claim, given that holism has important players, like ecological science and a humble respect for the Earth, on its team? I offer two reasons.

ADVANTAGES OF THE PERSONALIST APPROACH

The first reason that personalist environmentalism is generally superior to environmental holism is that the former engages the problem of environmental degradation at its source in the human person. From a first-person viewpoint, the tenacious root of our crisis resides deep within "my" self and in every human being. I refer here to that basic moral stance that we hold toward the world's nonhuman entities. These include both individual organisms and larger ecological systems like, say, the wetland at the edge of town. Put in the starkest terms, this stance will fall somewhere on a continuum between an abusive exploitation and a loving and responsible care.

How does it happen that "I" or anyone else intends either to exploit or to love some nonhuman being? Given our pluralistic context, it isn't likely that some authority could compel a decision to move in the direction of care; and love, by definition, can never be forced. Each community, rather, will generate ways to cultivate its members' affections. Our Catholic community strives to offer us just this sort of formation. It seeks, in particular, to lead us to that encounter with God in Christ which can open our spirits and awaken our love.

In addition to this, a personalist environmentalism offers an intrinsically satisfying and so more sustainable solution to our ongoing crisis than holistic environmentalism. I ground this assertion in the human's relational nature as demonstrated by social science and affirmed by Church teaching. Benedict (2009b, #53) states that "[a]s a spiritual being, the human creature is defined through interpersonal relations. The more authentically he or she lives these relations, the more his or her own personal identity matures. It is not by isolation that man establishes his worth, but by placing himself in relation with others and with God. Hence these relations take on fundamental importance." The *Compendium of the Social Doctrine of the Church* which Benedict (2005c, #27) has explicitly endorsed is unequivocal on this matter. It states that "[s]ocial activity constitutes [our] very nature," such that we "can only grow and realize [our] vocation in relation with others" (Pontifical Council for Justice and Peace 2004, #149; hereafter Pontifical Council). Furthermore, because we only flourish *through* our relationships, we come to care deeply about our particular "others" and to desire that they also survive and thrive (Myers 2007, 52).

Just as our relational capacity extends to babies and other nonverbal human beings, it extends to other animals. All we need do is watch an animal move in pursuit of some goal, or express some feeling that corresponds with its present experience, or develop over time. Notably, it is these perceptions themselves, and not conscious thinking, that engender our recognition of the animal as a true social other who merits our care (Stern 2000, 26-32, 69-123; Myers 2007, 7-8, 16, 50, 63, 65-88). Although the research has yet to be done, this kind of cross-species connection may also extend to plants that we understand to be growing. It may extend as well to those larger entities, like the wetland mentioned above, that we perceive as coherent unities and that appear to be either moving toward greater ecological integrity, or to be unraveling in that same regard.[2]

Positive interaction with our fellow creatures, therefore, when no threat is present and we give our appreciative attention, leads naturally to concern for those creatures. Our Catholic faith can play an important role here. Just as with our intra-human interactions, our basic moral stance—whether it leans toward care or exploitation—will tend to reinforce itself. A choice for "care," for example, will likely evoke a closer observation, and in so attending, we are more likely to apprehend that creature's own viewpoint, and begin to understand its needs. This is, in other words, a virtuous cycle in which our choice to enact God's love will tend to move us toward that kind of encounter that will in itself evoke our more generous response. A propensity to exploit, on the other hand, discourages recognition of the creature's needs and causes us to minimize what is necessary for the creature's own survival. This, in turn, makes us become indifferent to another creature's viewpoint and disengage from its reality. Alienated from the creature's good, we then lack compunction against its further exploitation.

A personalist environmentalism engages us at the level of our affinity for life itself. And it is this engagement *per se* that can encourage our commitment. When such engagement is lacking, our efforts feel both frustrating and fragile (Selznick 1992, 183-206). For this reason, most of us require frequent encounters with nature to sustain commitment. The personalist alternative, in other words, puts Catholicism's strongly relational understanding of the human person at the center of every effort to generate and sustain the political will necessary to reverse current trends. In essence, if we are to preserve Earth's systems over the long haul, we must take our environmentalism to heart. Indeed, I maintain that the distinctiveness of our social teaching, as coupled with Catholicism's relative influence in U.S society, makes the American Church uniquely suited to promote the personalist option. I will support this claim by first offering a few statistics, and then outlining the core logic of the Catholic worldview as compared to its secular and Protestant counterparts.

A CATHOLIC PERSONALISM THAT EXTENDS TO OUR MORE-THAN-HUMAN WORLD

Demographic studies indicate that Protestantism (51 percent) is the only religion with which Americans identify more than Catholicism (24 percent). After Catholicism, the next largest category with which Americans self-identify is "secular unaffiliated" (6 percent) (Pew Forum 2013).[3] Although I am not aware of research that shows just how "secular" these "unaffiliateds" really are, there is little controversy regarding the moral logic that underlies much of secular culture. Although philosophers such as J. Baird Callicott (1992), Martha C. Nussbaum (2006), and Paul W. Taylor (1986) articulate the sort of cross-species ethics that could appeal to both religious and "secular" audiences—in Taylor's case, for example, by grounding the organism's worth in its efforts to value itself through pursuit of what it needs to thrive—the diverse groups that make up American society have yet to agree on a basis for according intrinsic value to either human or nonhuman beings.

Despite this lack of consensus around intrinsic value, however, all people acknowledge the reality of human wants and needs, or "interests." Hence, public discourse regarding the more-than-human world tends to default to an essentially economic concern with its potential usefulness as a material resource (Evernden 1993, 3-34). Secular culture, in other words, can tend to see the world in terms of two Hobbesian categories: 1) Humans as fundamentally self-interested and thus asocial; and 2) everything else that might be useful in furthering those human interests. As Thomas Hobbes (1985) makes clear, a strongly self-interested human will seek to control other humans as well. In sum, because secular culture lacks a shared basis for value, it can tend to frame the world as divided into the users and the used. Consequently, we end up with the logic of domination—the logic of master and slave.

As Christians, however, we believe in a loving God who made every creature to be "good" (Pontifical Council 2004, #133, 451). We also understand that the first couple's sin somehow damaged humanity's goodness. Catholics disagree with most Protestants, however, as to the locus at which that damage occurred. While Protestants believe that we retain some of God's goodness, they tend to understand the damage to be so deep that human sinfulness is nearly absolute. Catholics, in contrast, hold that our God-given capacities, while unequivocally wounded by sin, remain nevertheless essentially sound (Gustafson 1978, 6-12; Pontifical Council 2004, #140, 142, 453).[4]

Now this Catholic interpretation of original sin could have framed humans as merely adequate to assist with God's purposes on Earth. In actuality, however, Catholic teaching is at pains to lift up human possibility. As the *Compendium* declares, humans "have received an incomparable dignity from God himself," and in fact reflect God's *"living image"*

(Pontifical Council 2004, #105, emphasis in original). This conception of the human person, furthermore, is centered on love, and thus is profoundly moral (Benedict 2005c, #1). Indeed, Benedict (2005c, #1, 31) begins his first encyclical with the notion that God has "inscribed" love into our "very nature" as our truest vocation. Here Benedict (2009b, #3, 6, 30) references not merely sentimental love, but rather the sort of love that, because it has incorporated the truths of faith and of reason, is also intelligent and just.

As we have seen, secular modernity is centered on the self-interested human as striving to dominate every useful thing. Most forms of Protestantism, because they assume a human who has lost almost all original goodness, are centered on a God who is strongly transcendent. Accordingly, God's grace flows "downward," if you will, to the believer and then outward as love for the neighbor. Catholicism, however, locates God's goodness—actually locates God in a sense—within the human person and every creature. This is the strongly immanent God of Catholicism's characteristic "sacramental imagination."

Here it is important to unfold this Catholic emphasis on God's immanence, because it holds immense implications for Catholic environmental response. As per Benedict's (2007c, #3) felicitous phrase, God is "goodness in person." As good, God is love itself and so the source of all love (2009b, #1, 5; 2007c, #27). As personal, God draws near and loves us "first," inviting us to offer our own love in response (2009b, #1; cf. 1 John 4:19). In fact, it is this personal encounter with God that orders human life. As Benedict (2007c, #27) declares, "Life in its true sense is a relationship. And life in its totality is a relationship with him who is the source of life."

Benedict (2007c, #28) goes on to explain that this relationship with God is established "through communion with Jesus." Importantly, Benedict (2005b; Pontifical Council 2004, #454), like St. Paul, understands Christ as "firstborn of all creation," in the sense of exercising a cosmic leadership (Col 1:15). Christ is cosmic because "all things were created through him and for him," as St. Paul asserts; and He leads by "creat[ing] anew" the harmonious relationships "that sin destroyed." Consequently, "communion with Jesus Christ draws us into his 'being for all,'" such that "[u]nion with Christ is also union with all those to whom he gives himself" (2007c, #28; 2009b, #14). To follow Jesus, then, is to take on Christ's cosmic compassion.

That is a mighty tall order, isn't it? Who do we think we are to even entertain the possibility of renewal on a cosmic scale? And yet the Church assures us that God has made us for this task. More specifically, God has made us to be both generous and free. "For the great majority of people," Benedict (2007c, #46) teaches, "there remains in the depths of their being an ultimate interior openness to truth, to love, to God." And when we align our will with this "interior openness," such that God can further "permeate" our being, we also "open" in generosity to our neighbors (#45-47). Since God, furthermore, has given us the capacity to reflect on our choices, we are free to transcend our selfish impulses and become ever more open-hearted. Accordingly, humans are, quite simply, "made for gift," as Benedict (2009b, #34) puts it. In giving of ourselves, therefore, we image God, even "converse" with God, and in fact become nothing less than a "partner" in God's plan for renewal (2009b, #34; Pontifical Council 2004, #128, 452).

At this point, we have followed our Catholic personalism from its source in a loving God who lives within us, out through a conception of Christ as universally compassionate, to a human being both able and free to channel God's love to a suffering world. Given this, the question now becomes how the contemporary Magisterium understands this human love for a more-than-human world? To begin, it primarily sees this love as integral to Catholic identity. Every Catholic lives "in the presence of"—and so in relation to—"all the other creatures" (Pontifical Council 2004, #113). And because God's love "by its very nature must be shared with others," it tends to expand outward until, as Benedict (2005c, #18, quoting 1 Corinthians 15:28) proclaims, "God is 'all in all'." Furthermore, we need not choose between loving God and loving creatures. Benedict (2010d) assures us that "love for Christ generates love for others and also for all God's creatures." Our God, as love itself, is never in competition with God's Creation. As such, we can approach loving creatures via three steps which I term "receive," "respect," and "respond." We begin with our own receptivity. The *Compendium* (Pontifical Council 2004, #130) characterizes the human as uniquely "open to all created beings," "open to the fullness of being, to the unlimited horizon of being." This is exalted language, to be sure, but we "open" when we approach nature with appreciation and let it evoke our wonder (#487).

Second, we recognize the creature's goodness, and respect that good. Given that "respect for creation" has become something of a cliché, I want to acknowledge the *Compendium*'s (Pontifical Council 2004, #113) careful treatment of this step. It tells us that in order to "respect" our fellow creatures we must first "recognize them for what they are." This raises the question of exactly "what" they might be, in the sense of just what sort of meaning we ought to assign to their lives. The *Compendium* centers its answer on one word, "good," which is underpinned by simple humility. Humans have barely begun to comprehend otherkind—either as members of larger wholes like species and ecosystems, or as specific individuals, each with its own history and unique point of view (Taylor 1986, 122-23). Recall also that all value, including our own, derives from God. It follows that the Magisterium would want us to evaluate the creature's significance from God's standpoint, not our own. From the standpoint of divine reality, the creature is, essentially and unequivocally, "good": "good" because God loves it, and "good" in that it reveals God's love. There are therefore at least these two senses in which nonhuman creatures are good, and each holds an important implication.

God's love for a creature accords that being a value which is independent of our human wants and needs. For this reason, we can never regard it as only "a mere object to be manipulated and exploited" (Pontifical Council 2004, #463). To the contrary, the *Compendium* (#464) reminds us that "Christian culture has always recognized the creatures that surround man as gifts of God to be nurtured and safeguarded." Benedict (2007) himself affirms a certain horizontal dimension to our relations with nonhuman beings, speaking in terms of "the freedom and equality of all God's creatures." One year later, in an address to diocesan clergy, Benedict (2008d) was emphatic on this point, declaring that "the task of 'subduing' ["the earth"] was never intended to enslave it but rather as a task of being guardians of Creation and developing its gifts."

The second sense of the creature's goodness, by virtue of which it reveals God's love, is of course its sacramental aspect. Here, the *Compendium* (Pontifical Council 2004, #113) asserts that the goodness of each creature, understood from God's point of view, equates precisely to its "truth." Although I cannot here parse the full significance of this statement, which derives from an influential argument originated by St. Augustine, one key implication is that each creature reveals a degree of goodness and a partial truth which opens onto God as goodness and truth itself.[5] If we approach the creature with "*gratitude*" to its Creator, and allow it to evoke our "*appreciation,*" we can expect to encounter nothing less than the "*mystery of God who created and sustains it*" (#487). As Benedict (2010a) describes, it is love that underlies this capacity to see God. Consequently, when we grow in love, this ability grows as well:

> Our perception of the world depends essentially on the presence within us of God's Spirit. It is a sort of "resonance": those whose hearts are empty only perceive flat images lacking in depth. On the other hand, the more we are inhabited by God the more we are sensitive to his presence in our surroundings: in all creatures and especially in other human beings.

The *Compendium* (Pontifical Council 2004, #113) frames this process as a grand adventure, asserting that "the challenge" implied in discovering creaturely goodness "should lift [us] up as on wings."

We are still left, however, with the question of just how our recognition of creaturely goodness should inform our "respect" for it. The *Compendium* (Pontifical Council 2004, #473) offers only the broadest of brush strokes here as well, but again, it is easy to read this as deliberate. Rather than resolve every detail, its treatment remains open-ended so as to encourage our own engagement with the possibilities. We are given only two guidelines, the first of which is that we need not remain entirely "hands-off" of the nonhuman natural world. "Nature" is not sacred in itself, such that it must be left alone. This accords with Benedict's (2009b, #48) own understanding that we may use nature "responsibly to satisfy our legitimate needs." The second guideline comes in the form of an image. After asserting our duty to nurture and safeguard living creatures, the *Compendium* (Pontifical Council 2004, #464) recommends a return to the Benedictine and Franciscan notion of a "kinship" with our "creaturely environment." It specifically identifies such "kinship" as having the potential to foster "an attitude of respect for every reality of the surrounding world."

When this recommendation is reflected upon in light of climate change and other examples of environmental degradation, it would seem that few have acted on the Magisterium's suggestion that we let a cross-species kind of "kinship" inform Catholic environmental response. One reason might be that the kinship image sits rather uneasily within environmental holism, with its focus on how various components function to support some whole. The image of kinship is entirely consistent, however, with the personalist approach, given its emphasis on loving relationship, encounter, interaction, and care. When "I" step into the kinship frame, I commit myself to upholding the well-being of each member of my creaturely family—so far as possible, given my personal resources and my obligations to those other humans who are both close and not-so-close to me. In striving to fulfill this commitment, I extend God's love "outward" as far as possible. Ilia Delio et al. (2008, 48) explain how St. Francis of Assisi understood love to follow such a concentric ordering, incorporating:

> [F]irst, a grasp of God's love in his own life; second, a realization that he was loved by God and thus called to love himself; third, a love of the human person outside himself signified by the leper, followed by a love of creatures. To put this progression of love in another way, once Francis experienced the love of God in his own life, he could begin to experience that love in other persons, in all creatures large and small and in the diverse things of creation.

Delio et al. (2008, 48) contend that such love "forged Francis into an 'ecological' brother." Benedict (2010d), who himself praises St. Francis' "sense of universal brotherhood," imagines such love as attaining to a cross-species "communion." In anticipation of World Youth Day, Benedict (2008b) acknowledged the despair of "young people" who witness "great damage done to the natural environment through human greed and [who] struggle to find ways to live in greater harmony with nature and with one another." At the inception of the event, he developed this thought in greater detail:

> My dear friends, God's creation is one and it is good. The concerns for non-violence, sustainable development, justice and peace, and care for our environment are of vital importance for humanity. Our world has

> grown weary of greed, exploitation and division, of the tedium of false idols and piecemeal responses, and the pain of false promises. Our hearts and minds are yearning for a vision of life where love endures, where gifts are shared, where unity is built and where identity is found in respectful communion."

Evolutionary biology maintains that other organisms are our kin in a literal sense. Here, the *Compendium* and Benedict use the kinship image to interweave the discourse of biology—as concerned with life in all its diversity—with the discourse of family love. The effect is to invite us out of a way of life that condones our unrestricted "use" of nonhuman creatures and into an alternative that sets no limit on our relational possibilities, so long as these are governed by love. As the metaphorical core of an extensive framework of meaning, the kinship image cannot serve to ground any detailed system of rules. Rather, the accordance to other creatures the respect due to kin tells us that the essence of our response, as the Magisterium's third and final step in loving creatures, is simply "responsible care."

The *Compendium* (Pontifical Council 2004, #451, 112) describes how, just as in the beginning, when "[t]he Lord entrusted all of creation to [our] responsibility, charging [man and woman] to care for its harmony and development," humans remain "in relationship with others above all as those to whom the lives of others have been entrusted." Importantly, it is precisely this vocation to an essentially boundless kind of responsibility that underlies our privileged position in relation to nonhuman nature. Put differently, our privilege in regard to the other creatures rests on God's trust—on God's hope, if you will—that we will care for those creatures and strengthen the web that connects us all. Put even more plainly, because God is love, we humans, as made in God's own "image," actually image God most clearly when we love God's human and nonhuman creation.

Let us now see how this Catholic personalism informs our contemporary environmental crisis. Recall that American secular culture, because it accepts no ultimate basis of value, tends to frame the human as a user of other beings. In this view, environmental degradation isn't necessarily wrong; it is just not smart given the negative impact on our own quality of life. For this reason, secular society tends to focus on how it can maintain conventional ways of life—without paying too high a price (Evernden 1999, 11). And although some very important alternative voices have arisen within Protestant environmental ethics—such as L. Shannon Jung (1993), Sallie McFague (1993; 2008), Larry Rasmussen (1996) and George E. Tinker (2008)—the mainstream view sees humans themselves, because profoundly impaired by the Fall, as constituting the source of the problem. In the words of Protestant ethicist James M. Gustafson (2007, 95), the best that we can do is to figure out "[t]he patterns of interdependence of things on each other, which are in theological terms the patterns of divine governance" and then essentially do whatever is necessary to get out of God's way. Protestant ethicist Roger H. Crook (2007, 290, 297-98) similarly asserts that "[t]he ecological problem is that of sustaining a growing population with finite, limited resources" and endorses human "responsibility" for what is "not ours but God's." Mainline Protestantism thus holds that God evicted humanity from our original home in God's garden, and that we must remain at a safe distance given the depth of our brokenness.

The Catholic Church, in contrast, holds that we are already at home in God's world. Because we carry God's presence in our hearts, we can always feel at home in both our human family and among our various nonhuman kin (Pontifical Council 2004, #464). Benedict (2006b) insists that "Jesus makes us feel at home on this earth, sanctified by his presence. He asks us to make it a home in which all are welcome." The problem from the Catholic perspective is that we don't grasp the full extent of the love that God would have us both give and receive. As Benedict (2005) laments in his inaugural homily, we wander, "lost in the desert" when we could be "build[ing] God's garden for all to live in."

If confusion is the problem, the solution lies in our "conversion" to the creative possibilities inherent in our prodigious capacity for ethical relationship (Pontifical Council 2004, #135-37). We need to receive, respect and respond so that God's presence in creatures can nurture our spirits, and the divine love within can be channeled outward to reach the life around us. Benedict (2009b, #1-2) regards love as the "power," the "principal driving force"—the indispensible engine, if you will—behind all enduring social change.[6] In his words, "[w]e contribute to a better world only by personally doing good now, with full commitment and wherever we have the opportunity, independently of partisan programmes" (2005c, #31). Insisting that "[t]he Christian programme" is most authentically "'a heart which sees' where love is needed and acts accordingly," Benedict (2005c, #31; 2009b, #2) envisions love as the principle of both "micro-" and "macro-relationships." It is "at the level of the heart," then, where "mobilization" for real change must occur (#20).

LOVE LINKS "HUMAN ECOLOGY" WITH "ENVIRONMENTAL ECOLOGY"

We can understand the primacy of love to be the underlying principle that elucidates Benedict's conviction that a "human ecology" underlies "environmental ecology." Pope John Paul II (1991, #39) originated the phrase "human ecology" in *Centesimus Annus* to refer to the intrafamilial matrix in which the human being "receives his first formative ideas about truth and goodness, and learns what it means to love and to be loved, and thus what it actually means to be a person." It is within a healthy human ecology that humans "become aware of their dignity" and "develop their potentialities."

Since *Centesimus Annus,* John Paul and various curial officials have explored additional facets of human ecology. Benedict's (2009b, #51) emphasis, however, has been on human ecology as the basis for environmental ecology. He asserts that:

> The deterioration of nature is in fact closely connected to the culture that shapes human coexistence: *when "human ecology" is respected within society, environmental ecology also benefits.* Just as human virtues are interrelated, such that the weakening of one places others at risk, so the ecological system is based on respect for a plan that affects both the health of society and its good relationship with nature.

In order to protect nature, it is not enough to intervene with economic incentives or deterrents; not even an apposite education is sufficient. These are important steps, but *the decisive issue is the overall moral tenor of society.*

In and of itself, this statement seems a bit obscure, prompting us to ask exactly what Benedict means by "culture," "plan," and "overall moral tenor." Similarly, the sections immediately preceding and following this statement are also abstruse. Interestingly, Benedict does not—in either *Caritas in Veritate* or any other document—provide a straightforward explanation of precisely how human ecology generates environmental ecology.

Despite this lack of explicit elucidation of what is meant by "culture," "plan," and "overall moral tenor," a closer look at a few key documents can clarify their relation.[7] In order to do this, I will proceed via three steps. First, I will place this question within its larger context, which is not human ecology *per se*, but an "integral" sort of personal development that both begins and ends in a loving God. Second, I will use a personalist lens to parse this context, demonstrating its close fit with the Catholic and personalist progression of Creator, person, society and broader creation sketched above. I will then argue that we can reasonably infer that "love," as the principle which orders every antecedent element in this context, is also the principle that links human with environmental ecology.

"Integral human development," as supported by a culture of "transcendent humanism," is Benedict's primary focus in *Caritas in Veritate* (2009). Given that this encyclical reflects on Pope Paul VI's own treatment of economic development in *Populorum Progressio* (1967), I will here reference Paul as well. In *Caritas in Veritate*, Benedict argues that economic discourse tends to conflate two distinct models of culture which, although similarly involving human development of some sort, differ greatly in their view of human nature and human relationships. According to the first model, which I will call materialist humanism, the human acknowledges only the material aspect of reality and either rejects or is indifferent to God and the transcendental dimension of human life. This model encourages everyone to see the pursuit of material wealth as an end in itself, regardless of whether they seek only basic necessities or are pursuing the greatest possible profit. Therefore, while a materialist humanism appears to liberate, it actually "stifl[es]" and "imprisons," such that people "no longer gather together in friendship but out of self-interest" (Paul VI 1967, #18, 19). In Benedict's (2009b, #11) words, lacking faith in God and "[w]ithout the perspective of eternal life, human progress in [such a] world is denied breathing-space. Enclosed within history, it runs the risk of being reduced to the mere accumulation of wealth."

Paul VI (1967, #16), however, articulates a "transcendent humanism" that, because it incorporates humans' spiritual dimension, is far superior to the materialist worldview. In this second cultural model, the human person is "a 'unity of body and soul,' born of God's creative love and destined for eternal life" and so "endowed with transcendent meaning and aspirations" (Benedict 2009b, #48). Accordingly, any conception of development "must include not just material growth but also spiritual growth" (#76). Indeed, Benedict (#29) lifts up God's role as "*the guarantor of man's true development*" who both establishes humans' "transcendent dignity and feeds their innate yearning to 'be more.'"

Benedict (2009b, #53) explains how transcendent humanism's incorporation of faith in a loving God effects a pervasive revision of cultural meanings. As made by such a God, humans are "spiritual being[s]" who grow through loving relationships. Human life is most fundamentally a "vocation" to engage in such relationships so as to "grow in humanity" (2009b, #16; Paul VI 1967, #15, 16). Development, therefore, is properly concerned with those conditions necessary to support each person's "integral" growth, rather than narrow economic expansion and the accumulation of temporal goods (Paul VI 1967, #21). In other words, whereas a materialist humanism holds material wealth to be its primary measure of value, a "transcendent" and "integral" humanism understands value in terms of the human person fully engaged in loving relationship. As Benedict (2009b, #25, referencing *Gaudium et Spes*, #63) declares, "the primary capital to be safeguarded and valued is . . . the human person in his or her integrity: 'Man is the source, the focus and the aim of all economic and social life.'"

Benedict's (2006a) conception of an "integral" kind of development as encompassed by a transcendent humanism incorporates each of the more-central elements of the personalist progression. Its starting point is simply faith in a loving God, which "anchors every human effort to build a civilization of love." This faith, in turn, underwrites our commitment to "the innate dignity of the human person made in God's image," a dignity which, as we've seen, rests on our special vocation to an essentially boundless kind of care. It is at this point in the progression that Benedict brings in human ecology, stating that "children and young people are by nature receptive, generous, idealistic and open to transcendence. They need above all else to be exposed to love and to develop in a healthy human ecology, where they can come to realize that they have not been cast into the world by chance, but through a gift that is part of God's loving plan." Following John Paul, Benedict sees the family as playing a critical role in forming humans able to image God's love. Benedict (2009a) speaks of the Catholic family, for example, as offering "a place of faith and of loving concern for the true and enduring good of each of its members," so as to foster "that human ecology which our world so urgently needs: a milieu in which children learn to love and cherish others." "In God's plan for the family," he explains,

> the love of husband and wife bears fruit in new life, and finds daily expression in the loving efforts of parents to ensure an integral human and spiritual formation for their children. In the family each person, whether the smallest child or the oldest relative, is valued for himself or herself, and not seen simply as a means to some other end. Here we begin to glimpse something of the essential role of the family as the first building-block of a well-ordered and welcoming society.

For Benedict, then, the family's significance rests on its unique ability to offer members the experience of being loved for who they are in themselves, regardless of what they can do for someone else. Familial love, in other words, as the fulcrum of the experience of being loved for oneself, is the quintessence of a healthy human ecology.

Within a healthy family life, we learn to understand God as love and ourselves as lovable. In addition, healthy families underwrite, as per the text quoted immediately above, "a well-ordered and welcoming society" in the sense of that broader "human ecology" where we practice our potential for responsible care on a larger scale (Benedict 2009a). Given that this potential extends to all God's creatures, this application of our personalist lens seems to hold at least two implications. First, it indicates that the elevation of love over "use"—as that principle which governs the more-central elements of Catholicism's personalist theological progression—ought to characterize our practice of environmental ecology as well. Thus, we can understand these more-central elements—faith in God's love, and a commitment to human dignity and a robust human ecology—to support a love for nonhuman creatures as that final element which represents the outermost "circle," the still-mysterious frontier if you will, of Christian love.

The second implication of a personalist approach to human ecology is the more practical notion that the foundational spiritual and moral capacities developed within healthy families—and all human ecologies—will also facilitate our love for otherkind. This is the meaning of Benedict's

(2009a, #51) contention, expressed in that perplexing excerpt quoted near the beginning of this section, that "[t]he deterioration of nature is closely connected to the culture that shapes human coexistence: *when 'human ecology' is respected within society, environmental ecology also benefits*." We can now understand how Benedict uses those terms that previously appeared so enigmatic. "Culture" refers to the opposition between materialist and transcendent humanism. "Plan," as used in the more complete quote above, signifies God's loving plan for all creation. And Benedict's further insistence that only a shift in the "*the overall moral tenor of society*" will suffice to protect nature fits well with the possibility that a greater love for both human- and otherkind might indeed prove more effective than "economic incentives or deterrents," or even a particularly "apposite," in the sense of a relevant and timely, sort of environmental education.

Although the suggestion that we turn to familial love to inform our Catholic environmental action might appear startling, it coheres well with Benedict's (2008a, #2; 2009b, #50, 69; 2010b, #10) characterization of the family as "*the prototype of every social order*" and his assertion, thrice repeated, that the "covenant between human beings and the environment should mirror the creative love of God." Even as an imaginational stretch, however, it is a message that merits energetic promulgation, especially given that the Catholic Church's elevation of human being is easily misconstrued. There is a temptation to interpret statements that raise up the human person, such as the above characterization of the human as "the focus and the aim of all economic and social life," in a way that puts human complexity above our capacity to love. If we think only of our own self-interest—narrowly conceived—we can easily convince ourselves that since God loves human persons because we are relatively complex beings with an amazing capacity to think about and act on the larger world, we therefore can and ought to take advantage of those creatures who are relatively simple.

As Benedict suggests, however, this attitude can cause humans to "enslave" creation. To understand our complexity as the basis of our privilege forces us to negotiate two antithetical ways of life: We are God's children seeking the good of our human brothers and sisters, but also slave-masters attempting to dominate every other sort of being. In contrast, the *Compendium*'s (Pontifical Council 2004, #451) chapter on "Safeguarding the Environment" begins by reminding us that our lived experience of liberation from Pharoah's Egypt is the very "foundation" of our faith. It goes on to emphasize God's choice to treat us as partners, and suggests that we extend this same respect to "the world" (#452). There is a clear implication here that the God we worship wouldn't make a slave—of any species. The alternative to enslaving creation, then, is to acknowledge our relative complexity, but elevate "love" above it so as to govern how we understand and express that complexity. In this way, environmental action can begin and end with human well-being, but with our well-being as those creatures to whom God has entrusted the responsibility to care for all the rest.

KINSHIP CARE LETS US LOVE GOD'S MORE-THAN-HUMAN WORLD

Chapter eight of the Letter to the Romans says that the whole of Creation has been groaning in travail because of the bondage to which it has been subjected, awaiting the revelation of God's sons: it will feel liberated when creatures, men and women who are children of God, treat it according to God's perspective. I believe that we can establish exactly this as a reality today. Creation is groaning—we perceive it, we almost hear it—and awaits human beings who will preserve it in accordance with God. The brutal consumption of Creation begins where God is not, where matter is henceforth only material for us, where we ourselves are the ultimate demand, where the whole is merely our property and we consume it for ourselves alone. And the wasting of creation begins when we no longer recognize any need superior to our own, but see only ourselves (Benedict 2008d).

In this reflection, Benedict declares his confidence in our ability to treat creation "according to God's perspective." But is Benedict's vision a practical possibility? To recast this question in terms of the personalist outlook: "I" am open to accepting a greater responsibility for those creatures that I affect in the course of my everyday life, but how can I do so given that their needs vastly exceed my personal resources? While a detailed description of a Catholic environmental praxis cannot be attempted here, it is possible to articulate a basic policy, which I term "Kinship Care."[8] In brief, "I" follow Kinship Care when I aim to enact a relative equality while also respecting my concentric obligations. We can understand Kinship Care in terms of three fairly simple ideas: relative equality, concentric fidelity, and responsible care.

"Relative equality" becomes possible when I recognize that every creature has some needs that are more vital and some that are more peripheral to its well-being. A need is more vital when it is more important —more central, if you will—to life and health. I enact a relative equality with another creature when I accord equal importance to those needs that are equally vital to both the other creature and to me. Such a comparison is rough, of course, but nevertheless easy to grasp. For example, if both my dog Max and I need exercise, I enact a relative equality with Max when I make sure that we both get the exercise we need. Or, I enact a relative equality with sharks when I decline to order shark fin soup out of recognition that a shark's fin is probably more important to the shark's health than it is to my own.

Although there are of course a great many situations where we could choose such a horizontal relation, just the fact that humans, too, are earthlings with bodily needs tells us that it is not possible to enact a relative equality with every creature. "Concentric fidelity," however, gives us one basic way to rank those situations. Concentric fidelity assumes that, while God is not in competition with any element of God's creation, it is usually necessary for me to attain at least some minimal level of well-being before I am able to be of service to my human neighbor. We all understand why the flight attendant tells us to "put your own oxygen mask on first." It also assumes that those humans whom we have helped to learn love within the context of a healthy human ecology will be the humans best-prepared to love the wider world. We can understand concentric fidelity, then, to adhere loosely to St. Francis' own schema mentioned above, which is itself aligned with the Catholic Church's progressive and radial ordering of God, human person, human other and broader creation.

Yet even within this framework, one might ask how "I" can possibly combine relative equality and concentric fidelity into any sort of environmental ethic, even if it is a personalist environmental ethic that is by definition incomplete? Wouldn't all my time and energy go toward meeting human needs? Certainly relative equality and concentric fidelity seem to pull us in opposite directions. Relative equality expresses our longing for a most-expansive "respectful communion," or felt solidarity, with our fellow creatures. Concentric fidelity acknowledges that we are constantly negotiating all sorts of relationships, each with its own cooperative and competitive aspect, such that prudent distinctions must be made. But although these resist combination in the abstract, we can and indeed already do integrate them in the course of daily life. Parents, for example, and all those who hold responsibility within a family, accomplish this when they carefully manage resources so as to meet both their own needs and the changing needs of each dependent family member while living in a socially responsible manner, e.g., using non-toxic cleaning products, purchasing the most-humane sort of eggs, keeping tires pumped up so as to cut down on MPG, volunteering at the local food bank, cutting apart those six-pack connectors that tend to end up around the necks of wild animals, taking used motor oil to the recycling center, contacting local politicians to advocate for more and wider bike lanes, harvesting organic tomatoes from the balcony or backyard, and gently escorting spiders out the door. In everyday life, then, we tend to respond to various needs as we encounter them, and we usually encounter them in a complicated tangle. As such, it is neither possible nor desirable to enact concentric care in a rigid or systematic way.

"Responsible care" integrates the first two elements by balancing "my" sense of the most important needs against my most pressing obligations. It does this by allowing me to "use" otherkind, but only when such use is made necessary by some obligation that I have to my own well-being, to the well-being of particular members of my human family, or to my human ecology generally speaking. We enact responsible care, therefore, when we refrain from inflicting any unnecessary harm on a nonhuman creature. Certainly, we also image God when we respond to a creature in a way that is straightforwardly benevolent, such as when we rescue a deer that we just happen to see fall through ice and founder in a frozen pond. An obligation to avoid unjustifiable harm, however, actually requires more of us than beneficence alone. It probably bears on a great many more of our choices. And it is a prerequisite for any act of care—as expressed, for example, in a physician's promise to "first, do no harm." As applied to nonhuman creatures, a commitment to resist any inclination to harm them unnecessarily is actually both more loving and more binding, and thus potentially more effective.

I offer Kinship Care as a practical starting point for a personalist environmentalism consistent with the key elements in Benedict's own thought. Like Benedict, Kinship Care assumes a strongly relational human being whose faith in God generates a compassion that is universal in its scope. And while it acknowledges the necessity of some "use" of the nonhuman natural world, it justifies this use on the basis of love in its alternative aspect as a fidelity to those closest to us. Kinship Care's integration of these two aspects of love yields a duty to avoid harming them unnecessarily, in accordance with Benedict's (2009b, #48) view that while we may never regard nonhuman creatures as "object[s] to be manipulated and exploited," we may nonetheless use them, as the *Compendium* (Pontifical Council 2004, #463) puts it, "responsibly to satisfy our legitimate needs."

Thus, for most of us, most of the time, a commitment to love our creaturely family will engage us in the ongoing challenge of protecting them from unjustifiable harm. It is important to note, however, that while approaching the world from "God's perspective" may give us greater sensitivity to creaturely goodness, we will also gain a keener sense of creaturely suffering and death. I don't want to minimize the fact that the cross-species kind of justice broadly outlined by Kinship Care won't eliminate our need to use other creatures. Nevertheless, if we assume, after Benedict, that the unnecessary use of nonhuman creatures represents a kind of enslavement of those creatures, then Kinship Care doesn't sanction any such enslavement. It supports, rather, a way of life that avoids domination so as to affect a practical solidarity with all created being.

CONCLUSION

I have argued that Catholicism has a unique gift to offer the environmental movement: a personalist environmentalism grounded in God's love. I began by defining a personalist environmentalism and explaining how it is both more fundamental and more sustainable than its better-known holist alternative.[9] I then outlined Catholicism's personalist moral progression which understands God's love to radiate outward through caring relationships, and described how Benedict and the *Compendium* urge us to receive, respect and respond so as to extend this love to all creation. Next, I examined Catholic social teaching on integral human development to demonstrate that it is reasonable to infer that Benedict understands the experience of being loved for oneself to both generate a healthy human ecology and prepare us to engage in a whole-hearted environmentalism. Finally, I offered Kinship Care as one way of living out the implications of Benedict's thought. As Catholics, we have guidance sufficient to enable our next steps, which would appear to involve the formation of hearts which "see" the full range of circumstances "where love is needed." From formation we can move to mobilization, and then to cultivation of our earthly home, so as to transform it into Benedict's vision of "God's garden for all."

NOTES

1. Although some use "personalist" to reference the notion that human beings possess a greater value, or higher status, relative to other living beings, I here use the term in its more dynamic sense as that "point of view" that inheres in the individual person as he or she moves, perceives and interacts in everyday life, i.e., that inheres in Maurice Merleau-Ponty's "phenomenal body" (2002). Although the environmental movement has yet to incorporate "personalist" into its vocabulary, a quick review of general-interest periodicals published in the last decade indicates an escalating trend toward stories that describe particular people involved in a "hands-on" and ongoing interaction with a small-scale kind of nature. Examples of such projects include community gardens, the "daylighting" of underground waterways, the rehabilitation of injured wildlife, raising backyard chickens or bees, and home visits that enable elderly or disabled persons to keep their pets.

2. At least three conditions would be necessary to enable a human to experience a relationship with a living entity that is not an animal, such as a plant or ecosystem. These include that entity's development or deterioration over time, the human's attention to those changes, as well as comprehension of what those changes signify regarding the entity's continued survival and flourishing.

3. I follow the Pew Forum's own categories here, such that my figure for "secular unaffiliateds" excludes the 1.6 percent of Americans who identify as "atheist" and the 2.4 percent who identify as "agnostic."

4. Unfortunately, space does not permit a more detailed description of Protestant views on original sin. I follow Gustafson's now-classic, yet rather generic, treatment of this topic. Episcopalians and many Methodists are closer to Catholics in this regard than to Baptists, Calvinists, and Lutherans.

5. Augustine lifts up the correspondence between goodness and being in *Confessions* VII.12. Aquinas' *Summa Theologiae* extensively develops Augustine's idea and argues for an equivalence between truth and goodness-as-perfected being. As numerous texts in the *Summa* inform this conception, I recommend Jean Porter's discussion (1990, 34-43).

6. While Benedict holds personalist and political action to be equally legitimate routes to the common good, he argues that authentic social action has an irreducibly personalist aspect. His starting point is support for human dignity, which he understands to require further commitments to both personal well-being and personal freedom. When attempting to advance well-being, we should never be satisfied with the achievement of a supposed good that merely appears desirable in the abstract, but must always seek the actual well-being of affected individuals (*Caritas in Veritate*, #7). Benedict also emphasizes the necessity of preserving freedom in the form of a sufficient scope for personal moral choice. For this reason, he criticizes the assumption that social action ought to focus entirely on structural reform. Social movements which bypass individual perspectives so as to focus entirely on reforming structures are unlikely to accord adequate protection to personal moral agency. Such movements, because they fail to attend adequately to moral choice, can also tend to a dangerous naiveté in regard to human sin (2005c, #28; 2007c, #16-25; 2009b, #7).

7. In addition to the sources cited parenthetically, Benedict mentions "human ecology" in his 2007 and 2010 World Day of Peace messages, General Audience (Aug 26, 2009), Address to the Diplomatic Corps (Jan 11, 2010), and Post-Synodal Apostolic Exhortation *Africae Munus* (Nov 19, 2011). His comments, however, are brief and add little to my discussion here. In two of these sources—the 2007 World Day of Peace message and the General Audience (Aug 26, 2009)—Benedict does reference the connection between "human ecology" and "integral human development."

8. Multiple sources inform my conception of Kinship Care. Most prominent among these are David Ross' "What Makes Right Acts Right?" (2002), Taylor (1986), Peter Wenz's "The Concentric Circle Theory" (1988), Ferré (1994), Lakoff and Johnson (1999, 310-311), and Wennberg's section on "The Logic of the Line" (2003, 11-14). I take the specific notion that non-maleficence is a more stringent duty than beneficence from Ross (2002) and Taylor (1986, 172-73, 197).

9. I direct those seeking a more comprehensive treatment of personalist ethics, including personalist environmental ethics, to Peterson (2009). In a sense, Peterson's book as a whole addresses the burning question of how small-scale actions, as clearly insufficient in themselves, connect to larger-scale structural change; but see especially, "Toward an Immanently Utopian Political Ethics." Excellent theological and Roman Catholic treatments of personalist environmentalism can be found in Edwards (2006) and Delio et al. (2008).

REFERENCES

Abram, David. 1996. *The Spell of the Sensuous: Perception and Language in a More-than-Human World.* New York: Vintage Books.

Augustine of Hippo. 1961. *Confessions.* Translated by R. S. Pine-Coffin. New York: Viking Penguin.

Benedict XVI. 2005a. Mass, Imposition of the Pallium and Conferral of the Fisherman's Ring for the Beginning of the Petrine Ministry of the Bishop of Rome, St. Peter's Square, Rome. April 29. Accessed March 15, 2013. http://www.vatican.va/holy_father/benedict_xvi/homilies/2005/documents/hf_ben-xvi_hom_20050424_inizio-pontificato_en.html.

———. 2005b. General Audience. Rome. September 7. Accessed March 15, 2013. www.vatican.va/holy_father/ benedict_xvi/ audiences/2005/documents/hf_ben-xvi_aud_20050907_en.html - 10k - 2005-09-14

———. 2005c. *Deus Caritas Est.* Encyclical Letter on Christian Love. Accessed March 15, 2013. http://www.vatican.va/holy_father/benedict_xvi/encyclicals/documents/_20051225_deus-caritas-est_en.html .

———. 2006a. Letter to the Participants in the Twelfth Plenary Assembly of the Pontifical Academy of Social Sciences. Accessed March 15, 2013. http:///www.vatican.va/holy_father/john_paul_ii/messages/peace/documents/hf_jp-ii_mes_08121999_xxxiii-world-day-for-peace_en.html.

———. 2006b. Angelus. Rome. December 24. Accessed March 15, 2013. http://www.vatican.va/holy_father/benedict_xvi/ angelus/ 2006/documents/ hf_ben-xvi_ang_20061224_en.html - 7k - 2007-01-03.

———. 2006c. Angelus. Rome. December 24. Accessed March 15, 2013. http://www.vatican.va/holy_father/benedict_xvi/ angelus/ 2006/documents/ hf_ben-xvi_ang_20061224_en.html - 7k - 2007-01-03.

———. 2007a. Message for the Celebration of the World Day of Peace. Accessed March 15, 2013. http://www.vatican.va/holy_father/benedict_xvi/messages/peace/documents/hf_ben-xvi_mes_20061208_xl-world-day-peace_en.html.

———. 2007b. Homily. September 9. Accessed March 15, 2013. http://www.vatican.va/holy_father/benedict_xvi/homilies/2007/documents/hf_ben-xvi_hom_20070909_wien_en.html.

———. 2007c. *Spe salvi.* Encyclical Letter on Christian Hope. Accessed March 15, 2013. http://www.vatican.va/ holy_father/benedict_xvi/ encyclicals/documents/hf_ben-xvi_enc_20071130_spe-salvi_en.html.

———. 2008a. Message for the Celebration of the World Day of Peace. Accessed March 15, 2013. http://www.vatican.va/holy_father/benedict_xvi/messages/peace/documents/hf_ben-xvi_mes_20071208_xli-world-day-peace_en.html.

———. 2008b. Message of His Holiness Benedict XVI to the Beloved People of Australia and to the Young Pilgrims Taking Part in World Youth Day 2008. Accessed March 15, 2013. http://www.vatican.va/holy_father/benedict_xvi/messages/pont-messages/2008/documents/hf_ben-xvi_mes_20080704_australia_en.html.

———. 2008c. Address Welcoming Celebration by the Young People, World Youth Day. Barangaroo, Sydney Harbour. Accessed March 15, 2013. http://www.vatican.va/holy_father/benedict_xvi/speeches/2008/july/documents/hf_ben-xvi_spe_20080717_barangaroo_en.html.

———. 2008d. Meeting with the Clergy of the Diocese of Bolzano-Bressanone. Accessed March 15, 2013. http://www.vatican.va/holy_father/benedict_xvi/speeches/2008/august/documents/hf_ben-xvi_spe_20080806_clero-bressanone_en.html.

———. 2009a. Homily. Mount of Precipice-Nazareth. May 14. Accessed March 15, 2013. http://www.vatican.va/holy_father/benedict_xvi/homilies/2009/documents/hf_ben-xvi_hom_20090514_precipizio_en.html.

———. 2009b. *Caritas in Veritate.* Encyclical Letter on Integral Human Development in Charity and Truth. Accessed March 15, 2013. http://www.vatican.va/holy_father/benedict _xvi/ encyclicals/documents/ hf_ben-xvi_enc_20090629_caritas-in-veritate_en.html.

———. 2009c. General Audience. August 26. Accessed March 15, 2013. http://www.vatican.va/holy_father/benedict_xvi/audiences/2009/documents/hf_ben-xvi_aud_20090826_en.html.

———. 2010a. Homily, Solemnity of Mary, Mother of God. Vatican Basilica. Accessed March 15, 2013. http://www.vatican.va/holy_father/benedict_xvi/homilies/2010/documents/hf_ben- xvi_hom_20100101_world-day-peace_en.html.

———. 2010b. Message for the Celebration of the World Day of Peace. Accessed March 15, 2013. http://www.vatican.va/holy_father/benedict_xvi/messages/peace/documents/ hf_ben-xvi_mes_20091208_xliii-world-day-peace_en.html.

———. 2010c. Address to the Members of the Diplomatic Corps for the Traditional Exchange of New Year Greetings. http://www.vatican.va/holy_father/benedict_xvi/speeches/2010/january/documents/hf_ben-xvi_spe_20100111_diplomatic-corps_en.html.

———. 2010d. General Audience. Rome. January 27. Accessed March 15, 2013. http://www.vatican.va/holy father/benedict xvi/audiences/2010/documents/hf ben-xvi aud 20100127 en.html - 21k - 2010-02-03.

———. 2011. Post-synodal Apostolic Exhortation Africae munus on the Church in Africa. November 19. Accessed March 15, 2013. http://www.vatican.va/holy_father/benedict_xvi/apost_exhortations/documents/hf_ben-xvi_exh_20111119_africae-munus_en.html.

Callicott, J. Baird. 1992. "Animal Liberation and Environmental Ethics: Back Together Again." In *The Animal Rights/Environmental Ethics Debate,* edited by Eugene C. Hargrove, 249-261. Albany: State University of New York Press.

Crook, Roger H. 2007. *An Introduction to Christian Ethics.* Upper Saddle River NJ: Prentice Hall.

Delio, Ilia., Keith D. Warner, and Pamela Wood. 2008. *Care for Creation: A Franciscan Spirituality of the Earth.* Cincinnati OH: St. Anthony Messenger Press.

Edwards, Denis. 2006. *Ecology at the Heart of Faith: The Change of Heart that Leads to a New Way of Living on Earth.* Maryknoll NY: Orbis Books .

Evernden, Neil. 1999. *The Natural Alien: Humankind and Environment*. Toronto: University of Toronto Press.

Ferré, Frederick. 1994. "Personalistic Organicism: Paradox or Paradigm?" In *Philosophy and the Natural Environment* Eds Robin Attfield and Andrew Belsey, 59-73.

Gaudium et Spes. 1965. Pastoral Constitution on the Church in the Modern World. Accessed March 15, 2013. http://www.vatican.va/archive/hist_councils/ii_vatican_council/documents/vat-ii_const_19651207_gaudium-et-spes_en.html.

Gustafson, James M. 1978. *Protestant and Roman Catholic Ethics: Prospects for Rapprochement.* Chicago: University of Chicago Press.

———. 2007. *Moral Discernment in the Christian Life: Essays in Theological Ethics.* Edited and translated by Theo A. Boer and Paul E. Capetz. Louisville: Westminster John Knox Press.

Hobbes, Thomas. 1985. *Leviathan.* Edited and translated by C. B. Macpherson. London: Penguin Classics.

John Paul II. 1990. Message for the Celebration of the World Day of Peace. Accessed March 15, 2013. http://www.vatican.va/holy_father/john_paul_ii/messages/peace/documents/hf_jp-ii_mes_08121999_xxxiii-world-day-for-peace_en.html.

———. 1991. *Centesimus Annus.* Encyclical letter. Accessed March 15, 2013. http://www.vatican.va/holy_father/john_paul_ii/encyclicals/documents/hf_jp-ii_enc_01051991_centesimus-annus_en.html.

Jung, L. Shannon. 1993. *We are Home: A Spirituality of the Environment.* Mahwah NJ: Paulist Press.

Lakoff, George, and Mark Johnson. 1999. *Philosophy in the Flesh: The Embodied Mind and Its Challenge to Western Thought.* New York: Basic Books.

McFague, Sallie. 1993. *The Body of God: An Ecological Theology.* Minneapolis: Fortress Press.

———. 2008. *A New Climate for Theology: God, the World, and Global Warming.* Minneapolis: Fortress Press.

Merleau-Ponty, M. 1945/2002. *Phenomenology of Perception.* Translated by Colin Smith. London: Routledge Classics

Myers, Gene (Olin E.). 2003. "No Longer the Lonely Species: A Post-Mead Perspective on Animals and Sociology." *International Journal of Sociology and Social Policy* 23 (3): 46-68.

———. 2007. *The Significance of Children and Animals: Social Development and Our Connections to Other Species.* 2d, rev. ed. West Lafayette, IN: Purdue University Press.

Nussbaum, Martha C. 2006. *Frontiers of Justice: Disability, Nationality, Species Membership.* London: The Belknap Press.

Paul VI. 1967. *Populorum Progressio.* Encyclical Letter on the Development of Peoples. Accessed March 15, 2013. http://www.vatican.va/holy_father/paul_vi/encyclicals/documents/ hf_p-vi_enc_26031967_populorum_en.html.

Peterson, Anna L. 2009. *Everyday Ethics and Social Change: The Education of Desire*. New York: Columbia University Press

Pew Forums. 2013. February 5. Accessed March 15, 2013. http://religions.pewforum.org/reports/.

Pontifical Council for Justice and Peace. 2004. *Compendium of the Social Doctrine of the Church.* Accessed March 15, 2013. http://www.vatican.va/roman_curia/pontifical_councils/justpeace/documents/rc_pc_justpeace_doc_20060526_compendio-dott-soc_en.html.

Porter, Jean. 1990. *The Recovery of Virtue: The Relevance of Aquinas for Christian Ethics.* Louisville: Westminster/John Knox Press.

Rasmussen, Larry. 1996. *Earth Community: Earth Ethics.* Maryknoll NY: Orbis Books.

Ross, David. 2002. *The Right and the Good.* Edited by Philip Stratton-Lake. Oxford: Clarendon Press.

Selznick, Philip. 1992. *The Moral Commonwealth: Social Theory and the Promise of Community.* Berkeley: University of California Press.

Stern, Daniel S. 2000. *The Interpersonal World of the Infant: A View from Psychoanalysis and Developmental Psychology.* New York: Basic Books.

Taylor, Paul W. 1986. *Respect for Nature: A Theory of Environmental Ethics.* Princeton: Princeton University Press.

Tinker, George E. 2008. *American Indian Liberation: A Theology of Sovereignty.* Maryknoll NY: Orbis Books.

Wennberg, Robert N. 2003. *God, Humans, and Animals: An Invitation to Enlarge Our Moral Universe.* Grand Rapids: William B. Eerdmans Publishing Company.

Wenz, Peter S. 1988. *Environmental Justice.* Albany: State University of New York Press.

THREE

Natural Law and the Natural Environment

Pope Benedict XVI's Vision beyond Utilitarianism and Deontology

Michael Baur

In his 2009 encyclical letter *Caritas in Veritate*, Pope Benedict XVI (2009b, #53), citing an encyclical letter of Pope Paul VI (1967, #85), observes that "the world is in trouble because of the lack of thinking." To be sure, the world is in trouble for reasons unconnected to the lack of thinking; but, even while that is true, the lack of thinking does partly account for why the world is in trouble. As Benedict XVI (2009b, #53) goes on to say:

> [A] new trajectory of thinking is needed in order to arrive at a better understanding of the implications of our being one family; interaction among peoples of the world calls us to embark upon this new trajectory, so that integration can signify solidarity rather than marginalization. Thinking of this kind requires a *deeper critical evaluation of the category of relation*. This is a task that cannot be undertaken by the social sciences alone, insofar as the contribution of disciplines such as metaphysics and theology is needed if man's transcendent dignity is to be properly understood.

For Benedict XVI, a deeper, theological and metaphysical evaluation of the category of "relation" is necessary if we are to achieve a proper understanding of the human being's "transcendent dignity." For some contemporary thinkers, this position might seem to be hopelessly paradoxical or even incoherent. After all, many contemporary thinkers are apt to believe that the human creature can have "transcendent dignity" only if the being and goodness of the human creature is not conditioned by or dependent upon any relation or relatedness to anything else, including the natural environment.

As this chapter seeks to show, the apparent paradox in Benedict XVI's statement will begin to disappear if one resists the rather understandable temptation to interpret his thought by relying on presuppositions borrowed from contemporary ethical theories. More specifically, this chapter aims to show that Benedict XVI's teachings—embedded as they are within a rich tradition of Catholic "natural law" thinking—are importantly distinguishable from contemporary utilitarian and deontological views. Furthermore, this chapter seeks to demonstrate that Benedict XVI's "natural law" account offers an intellectually defensible alternative to contemporary modes of environmental thinking.

As part of his "natural law" account, Benedict XVI endorses three important yet easily-overlooked metaphysical premises. These three premises have to do with: a) the convertibility of being and goodness; b) the convertibility of being and order; and c) the uniquely intellectual nature of the human being. According to the first premise, every instance of being—precisely insofar as it is an instance of being—is also an instance of goodness (thus wherever there is being, there is also goodness). According to the second premise, every instance of intelligible order—precisely insofar as it is an instance of intelligible order—is also an instance of being (thus wherever there is intelligible order, there is also being and therefore also goodness). According to the third premise, the human being—by virtue of her/his unique intellectual nature—is uniquely capable of reflecting (through acts of understanding) the immanent orderliness and goodness that belongs to any being within the natural world. When one understands what is implied by these three premises built in to Benedict XVI's "natural law" account, it becomes possible to appreciate how his account offers a vision that captures some of the fundamental insights of contemporary environmental thinking without falling prey to some of its problems and shortcomings.

BEING, GOODNESS, AND ORDER

Many contemporary thinkers working in the area of environmental ethics will readily agree that human beings have a moral obligation to care for the (non-human) natural environment. There is a great deal of disagreement, however, regarding just *why* human beings have such a moral obligation. Furthermore, modern arguments offered for why human beings ought to care for the natural environment have not been altogether satisfactory. For example, contemporary utilitarian thinkers such as Peter Singer (1979) have argued that there is a moral obligation to care for the environment since such care will promote "the greatest happiness for the greatest number"; that is, such care will in the long run increase the overall amount of pleasure and decrease the overall amount of pain for beings that are capable of experiencing pleasure and pain. A key problem with the utilitarian, sentience-based position in favor of care for the environment is that this position offers no basis for caring about non-sentient living beings (e.g., plants) or non-living beings (e.g., stalactites and stalagmites) if care for such beings cannot be linked in some way to the pleasures and pains that might be experienced by sentient beings. If the annihilation of some non-sentient natural form (e.g., some crystal formation on a remote part of earth) does not decrease the net amount of happiness or pleasure to be experienced by sentient beings, then—on the utilitarian account—there is no moral reason for humans to refrain from annihilating the non-sentient natural form.

Some contemporary deontological thinkers have sought to overcome the limitations of sentience-based, utilitarian arguments by arguing that non-sentient beings, including even non-living beings, are capable of possessing "intrinsic worth" or "inherent worth." According to these deontological accounts, it is the "intrinsic worth" of all beings (including even non-living beings) that grounds the human obligation to care for the natural environment, even when such care will not in any way enhance the pleasures or diminish the pains that might be experienced by sentient beings (Brennan 1988). A key problem with the deontological position regarding care for the environment is that it fails to account for how some beings—even beings that are said to possess "intrinsic worth"—might be used (or even used up) for the purposes of satisfying legitimate human ends.

Against the backdrop of contemporary utilitarian and deontological theories, Benedict XVI (2007c, 338) holds that it is important to achieve conceptual clarity and rigor regarding our metaphysical commitments. If we fail to ask and answer deeper metaphysical questions, we are apt to misunderstand our own nature as human beings, and as a result also misunderstand the norms that ought to govern our relationship to the rest of the created order. The metaphysical questions that need to be asked and answered should concern us, not only in connection with "abstract philosophical considerations," but also for the sake of addressing "the concrete situation of our society." Unfortunately, according to Benedict (2009b, #31), modern intellectual endeavors tend to be characterized by a kind of fragmentation that has had deleterious consequences for moral thought and action. This fragmentation has made it difficult in our contemporary context for "faith, theology, metaphysics, and science" to "come together in a collaborative effort in the service of humanity." For Benedict XVI, one should not lose sight of the fact that metaphysical superficiality and theoretical fragmentation can have deeply problematic practical as well as theoretical consequences:

> The excessive segmentation of knowledge, the rejection of metaphysics by the human sciences, the difficulties encountered by dialogue between science and theology are damaging not only to the development of knowledge, but also to the development of peoples, because these things make it harder to see the integral good of man in its various dimensions.

The "broadening [of] our concept of reason and its application" is indispensable if we are to succeed in adequately weighing all the elements involved in the question of development and in the solution of socio-economic problems.

In light of contemporary challenges, Benedict XVI offers what promises to be an intellectually defensible and practically viable way of thinking about the natural environment and about human obligations with respect to that environment. Benedict's theorizing about human morality and about the natural environment is indebted in large measure to his understanding and appropriation of classical thinkers—including especially St. Augustine (354-430) and St. Thomas Aquinas (1225-1274)—within the Church's "natural law" tradition. Benedict XVI (2008, #13) holds that, in spite of the fragmentation and confusion which frequently characterize modern metaphysical and moral thought, it is possible to achieve a reasonable degree of conceptual clarity and consensus. We can achieve such clarity and consensus, he argues, if we turn to "natural law" which can serve as a "common moral law" in the midst of contemporary dissonance and disagreement. According to Benedict, human beings are

> capable of discovering, at least in its essential lines, *this common moral law* which, over and above cultural differences, enables human beings to come to a common understanding regarding the most important aspects of good and evil, justice and injustice. It is essential to go back to this fundamental law, committing our finest intellectual energies to this quest, and not letting ourselves be discouraged by mistakes and misunderstandings. Values grounded in the natural law are indeed present, albeit in a fragmentary and not always consistent way, in international accords, in universally recognized forms of authority, in the principles of humanitarian law incorporated in the legislation of individual States or the statutes of international bodies.

The "natural law" tradition to which Benedict XVI explicitly appeals is exceedingly rich, and thus resists any simplistic, superficial characterization. Nevertheless, it is possible to identify at least three central theoretical commitments that belong to this tradition and which will prove to be especially illuminating as we seek to appreciate his thought regarding human beings and their relation to the natural environment.

First of all, Benedict XVI affirms what has come to be known in philosophical and theological circles as the "convertibility" of being and goodness. To say that being and goodness are "convertible" is to say that every instance of being, precisely insofar as it is an instance of being, is also an instance of goodness; and every instance of goodness, precisely insofar as it is an instance of goodness, is also an instance of being. Even though the meaning of the term "being" may not be the same as the meaning of the term "goodness," it nevertheless remains the case that any proper referent of the term "being" is also always a proper referent of the term "goodness." Thus while "being" and "goodness" differ in meaning, they do not differ in reference. It follows from this that every being, even if the being is non-living and non-sentient, is good in some respect, and thus has some degree of value or worth that is not dependent on its instrumental value or worth for some other being.

Benedict XVI has expressed an especially deep and long-lasting affinity and appreciation for the thought of St. Augustine; indeed, Benedict (as Joseph Ratzinger) wrote his doctoral thesis on Augustine (Ratzinger 1954). And thus perhaps fittingly, it was St. Augustine who most famously expressed the traditional Catholic teaching regarding the convertibility of being and goodness. If being and goodness are convertible, argues St. Augustine, then we must conclude that all beings—no matter how seemingly base or ignoble—are good. Furthermore, since every being—insofar as it is a being—is good, it also follows for Augustine (1961, 4.12; also for Benedict XVI) what we call "evil" is not any kind of being in its own right, but is rather a defect or privation in some existing being:

> All of nature, therefore, is good, since the Creator of all nature is supremely good. But nature is not supremely and immutably good as is the Creator of it. Thus the good in created things can be diminished and augmented. For good to be diminished is evil; still, however much it is diminished, something must remain of its original nature as long as it exists at all. For no matter what kind or however insignificant a thing may be, the good which is its "nature" cannot be destroyed without the thing itself being destroyed. There is good reason, therefore, to praise an uncorrupted thing, and if it were indeed an incorruptible thing which could not be destroyed, it would doubtless be all the more worthy of praise. When, however, a thing is corrupted, its corruption is an evil because it is, by just so much, a privation of the good. Where there is no privation of the good, there is no evil. Where there is evil, there is a corresponding diminution of the good. As long, then, as a thing is being corrupted, there is good in it of which it is being deprived . . . . If, however, the corruption comes to be total and entire, there is no good left either, because it is no longer an entity at all. Wherefore corruption cannot consume the good without also consuming the thing itself . . . . Whenever a thing is consumed by corruption, not even the corruption remains, for it is nothing in itself, having no subsistent being in which to exist.

By virtue of his affirmation of the traditional Catholic teaching regarding the convertibility of being and goodness, Benedict XVI can hold that every being—no matter how lowly—is good in itself. This allows Benedict XVI to accept one of the key teachings of deontological environmental ethics, namely that there is "intrinsic worth" or "inherent worth" in every part of nature, and not just in those parts of nature which belong to or serve the interests of sentient and/or living beings. But there are still further implications to Benedict XVI's rich and metaphysically-informed environmental vision. Benedict XVI also accepts—secondly—what might be called the convertibility of being and order. To say that being and order are "convertible" is to say that every instance of being, precisely insofar as it is an instance of being, is also an instance of order; and every instance of order, precisely insofar as it is an instance of order, is also an instance of being. Because every instance of being is also an instance of goodness, it follows (according to the metaphysical view endorsed by Benedict XVI) that every instance of order is also an instance of goodness. If there were no order whatsoever, then there would also be no goodness and thus no being. Once again, it is St. Augustine who most famously elucidated this traditional Catholic view regarding the convertibility of being and order, and also the convertibility of order and goodness.

Augustine readily acknowledges that some instances of order may be better or worse than others; but an instance of order is said to be bad, not just insofar as it is an instance of order, but rather insofar as it is an instance of order that is lacking a higher or more fitting kind of order that ought to exist but does not. What might be fitting, orderly, and thus good in one context (i.e., within the context of one ordering) can turn out to be unfitting, disorderly, and thus bad in some other context (i.e., within the context of some other ordering). As Augustine (1953, ch. 23) explains:

> [A] form is called bad either in comparison with something more handsome or more beautiful, this form being less, that greater, not in size but in comeliness; or because it is out of harmony with the thing to which it is applied, so that it seems alien and unsuitable. It is as if a man should walk forth into a public place naked, which nakedness does not offend if seen in a bath. Likewise also order is called bad when order itself is maintained in an inferior degree. Hence not order, but rather disorder, is bad; since either the ordering is less than it should be, or not as it should be. Yet where there is any measure, any form, any order, there is some good and some nature; but where there is no measure, no form, no order, there is no good, no nature.

The crucial point here is that it is only through some kind of order (or form) that any natural kind and thus any natural thing can exist in the first place; and correspondingly, it is only through some kind of order (or form) that any natural goodness can exist. If all order or form were taken away, then all being and thus all goodness would also disappear.

ORDER, LAW, AND NATURAL LAW

Like Augustine before him, Thomas Aquinas accepted a metaphysically informed account of the convertibility of being and goodness, as well as the convertibility of being and order. Despite this agreement, Aquinas went beyond Augustine in order to develop a systematic account of law and natural law, which was to exercise a deep and abiding influence on subsequent Catholic thought, including the thought of Benedict XVI.

Following Augustine, Aquinas observes that when one apprehends the ordering among parts in a thing, one also apprehends that which gives being and goodness to the thing. For Aquinas, the intelligible ordering of parts (or form) within a thing gives being to the thing; but this very same ordering (or form) also gives goodness to the thing. After all, says Aquinas (1981, 2|2.109.2), "good consists in order," and evil consists in a lack of due order (1|2.75.1.1). For Aquinas, then, wherever there is intelligible order of a certain kind, there is also being and goodness of a certain kind. The being and goodness about which we are speaking can be the goodness of some individual substance such as a plant or animal; or else it can be the being and goodness of some "composition" such as a team, an army, a political community, or even the entire created universe (1|2.17.4).

A crucial implication of the Augustinian-Thomistic view regarding the convertibility of being and order is the view that order is not something super-added to being; it is not the case that beings first exist apart from all order, and then (subsequently and externally) have order imposed upon them. On the contrary, a being can be a being in the first place only if there is some kind of order or ordering that makes it what it is. Stated more fully: a being could not exist and act as the particular kind of being that it is, if there were no (internal) ordering among its parts, and if there were no (external) ordering or context within which the being acted and expressed its true nature. The crucial claim here is that order is not something that has to be imposed upon beings violently or in a manner that contravenes their nature; quite on the contrary, order is nothing other than the patterning or proportionality which enables beings to exist and to act as the kinds of beings that they are in the first place.

For Augustine as well as for Aquinas, law or lawfulness is nothing other than a particular kind of order or ordering. It is a kind of order or ordering in accordance with which one being is said to belong to a community and contribute to the good of that community, even as it contributes to its own good. Thus for Augustine and Aquinas (and as we shall see, for Benedict XVI), a being's placement within a larger whole or within a larger community is not something that is imposed upon the being externally or violently; instead, a creature's placement within a larger whole or community (including the larger whole or community which is the entire created order) is the condition under which the creature becomes most fully and most properly the kind of being that it is in the first place.

This Augustinian-Thomistic understanding of order and lawfulness will strike many contemporary readers as rather counter-intuitive. This is because many contemporary readers, influenced by modern positivistic accounts of law such as that articulated by John Austin (1998), have grown accustomed to thinking of law as nothing other than an externally-imposed command or set of commands backed by threats or force. Against all such positivist accounts, Augustine and Aquinas offer a "natural law" account according to which the law as such guides individual beings to the common good of a larger whole or community, but not by means of externally-imposed orders or commands; it does so rather by means of principles whose operative force is *internal* to those individual beings that are subject to the law.

Along these lines, Aquinas (1981, 1|2.93.5) argues that human beings cannot "legislate" or "make law" for non-rational creatures. Even though human beings may exercise a great deal of control over non-rational creatures, humans are—strictly speaking —unable to make law for non-rational creatures. When human beings exercise control over non-rational creatures, they do not (and cannot) lay down or legislate any principles that might become principles belonging "internally" or "naturally" to those non-rational creatures themselves. Thus, for example, when a farmer plows a field by controlling the actions of oxen, the actions taking place are not the actions of the oxen, but always only the actions of the farmer himself. The farmer merely uses the oxen as a means or as an instrument to accomplish what remains always only the farmer's own end and never becomes the end of the oxen themselves (cf. Baur 2012). When the farmer exercises control over the oxen, he does not make law for the oxen; for to make law for the oxen would be to prescribe a principle of action or motion which would belong to the oxen as an internal or "natural" principle of the oxen's own actions and motions. By contrast, some human beings can make law for other human beings, since some human beings can prescribe principles of action that—precisely because they are understood and adopted by these other human beings—can become the internal or internalized principles of those other human beings' own actions. It is possible, of course, for some positive laws to be externally and violently imposed on the human beings who are made subject to such laws. But the more it is the case that positive law is imposed externally or violently on humans, the more it is the case that such law lacks the character of lawfulness and instead takes on the character of tyranny. It is for this reason that Aquinas (1981, 1|2.96.4), following Augustine (1964, 1.5), says that "unjust law is no law at all." For an "unjust law" is one that is lacking in some due order or proportionality. To the extent that an existing law lacks some due order or proportionality, it is less capable of becoming reasonably adopted as the internal or internalized principle of the actions of those human beings who are subject to it. But if an existing law is not adopted and internalized by those who are subject to it, then it must in some measure be imposed externally and violently—and thus it has the character of tyranny rather than lawfulness.

Some important implications can be developed from the Augustinian-Thomistic view that law or lawfulness, properly speaking, operates as a principle which is internal to those individual beings which are subject to the law. First of all, individual beings that are subject to law are not made to act in accordance with the law because of some external or violent force that must be imposed on them. Instead, they are made to act in accordance with the law (and thus to act for the sake of the common good served by the law) on the basis of principles that are internal to their own being as individuals. This is why Aquinas (1981, 1|2.91.6) argues that sensuous inclinations and instincts in animals have the character of law or lawfulness. By acting on the basis of their own sensuous inclinations (for example, by copulating, and producing and caring for their own offspring), individual animals act in a way that is natural to them; they act in a way that accords with their very own desires and inclinations. And yet, by acting in accordance with their own inner strivings and inclinations, they also act so as to benefit the entire natural community (the species) to which they belong. The law or lawfulness that guides individual animals to act for the sake of the common good of the species does not operate by means of an externally or violently imposed force, but by means of the inmost, natural strivings of the individual animals themselves. For Augustine and Aquinas, as well as for Benedict XVI, God's all-comprehensive governance of the created universe is the most perfect example of such law or lawfulness. By virtue of God's eternal law, God leads all things to act for the sake of the common good of the universe. God does this not by any transitive or external action upon creatures, but rather by the intransitive, creative action that gives creatures their being in the first place, and thus also gives them their inmost natural desires and inclinations. In accordance with this view of law, Aquinas (1976, 3.122) argues that God, as creator, can never act upon creatures externally or violently; and furthermore that it is not possible for us humans to offend God except by acting contrary to our very own good.

Benedict XVI (2007a, #3) further develops some of the implications of this Augustinian-Thomistic account of law. He argues, for example, that the God-given "norms of the natural law should not be viewed as externally imposed decrees, as restraints upon human freedom"; instead, they should be welcomed as an invitation to satisfy the deepest desires that belong to us in accordance with a divine plan "inscribed" in our very nature. In a similar vein, he observes that the "natural moral norm" for human action does not have to be derived from any external or alien authority; it is discoverable by humans through the "inner logic of the deepest inclinations present in their being" (2008, #13). Furthermore, he argues that it is a mistake to regard our dependence on God and on the eternal law as a kind of "imposition from without" (2007b, 264-267); after all, we have our very being and desires only through God's providential governance of the created order within which we exist.

For Benedict XVI, these important implications of the Augustinian-Thomistic account of law are directly connected to a key lesson from Trinitarian theology. According to this theology, every created being, each in its own way, imitates the super-abundant, self-communicative activity of the triune God. Because God is love, "He does not live in splendid solitude," but is essentially self-giving, self-communicating, and relational. We can perceive this basic truth, says Benedict XVI (2009c),

> by observing both the macro-universe: our earth, the planets, the stars, the galaxies; and the micro-universe: cells, atoms, elementary particles. The "name" of the Blessed Trinity is, in a certain sense, imprinted upon all things because all that exists, down to the last particle, is in relation; in this way we catch a glimpse of God as relationship . . . .

In another context, Benedict XVI (2007b, 265) observes that human beings possess their very being and essence only by virtue of relationships with other beings and thus only within a larger, ordered whole: "Human beings are relational, and they possess their lives—themselves—only by way of relationship."

THE HUMAN BEING'S UNIQUE DIGNITY AND INTELLECTUAL NATURE

We saw earlier how Benedict XVI endorses the traditional Catholic view that being and goodness are convertible. Because of this, he is able to hold—as many deontological environmental thinkers hold—that every being can be said to have "intrinsic worth" or "inherent worth." But now we have also seen that Benedict holds that every being, including even the human being, can have its being and goodness only through relationships to other beings in a larger, ordered whole. Benedict's emphasis on the interconnectedness of all beings, and on the human being's necessary dependence on relationality, might seem to make him an ally of certain utilitarian environmental thinkers who tend to downplay the human being's unique status within the natural order. But it would be premature to draw any such sweeping conclusion. For Benedict XVI (2009a), it is true that humans can express their being and their goodness only within the context of a larger natural order; however, it is quite erroneous "to identify the person exclusively in terms of genetic information and interactions with the environment." Quoting approvingly from #347 of Pascal's *Pensées*, Benedict observes that the human being plays a distinctive role within the natural order because of the human being's unique intellectual nature:

> It must be stressed that man will always be greater than all the elements that form his body; indeed, he carries within him the power of thought which always aspires to the truth about himself and about the world. The words of Blaise Pascal, a great thinker who was also a gifted scientist charged with significance, spring to mind: "Man is only a reed, the most feeble thing in nature, but he is a thinking reed. The entire universe need not arm itself to crush him. A vapor, a drop of water suffices to kill him. But, if the universe were to crush him, man would still be more noble than that which killed him, because he knows that he dies and he knows the advantage that the universe has over him; the universe, instead, knows nothing."

While every created being is an instance of goodness and thus imitates God in some way, according to Benedict XVI, the human being is a unique instance of goodness, and thus imitates God in a unique way. This is because the human being is able to obtain intellectual knowledge (knowledge of the intelligible relatedness or togetherness) of things in the created order. Now the human being's act of intellectual knowing is an act that unifies things, or draws them together, in the mind of the knower. For Benedict XVI, the created order exists ultimately for the sake of being drawn together by God and to God. When the human being engages in acts of intellectual knowing, it apprehends things in their intelligible relatedness or togetherness, and thus cognitively *imitates* the act of divine drawing-together. One can say, then, that the realm of things that can be known by humans is ordered towards—and is in a way perfected through—acts of human knowing. While for Benedict the created order achieves its ultimate and complete perfection only in its being drawn together by God and to God, it achieves a partial perfection in its being drawn together (cognitively) in the mind of the human knower.

The basic point can be stated somewhat differently: the human being for Benedict XVI is a unique part within the whole of creation, since the human being's own perfection as a part within the created order (i.e., the human being's perfection in the act of knowing) is at the same time a partial perfection of the whole created order itself. While every created being possesses goodness or "intrinsic worth," the goodness or intrinsic worth of the human being is capable of including or comprehending the goodness or intrinsic worth of other created beings. Such inclusiveness or comprehensiveness is possible because human beings can engage in acts of intellectual knowing that cognitively draw together or unify the things that are known; such acts of human knowing imitate the all-inclusive, all-comprehensive act of God's knowing, which is the same as the act of God's being, which—in turn—is the final cause and final end of the entire created order. It is for this reason that the human being, while part of a larger created order, is also a being towards which other created beings might be ordered; and so the human being may make use of other created beings for the sake of satisfying legitimate human ends. Benedict XVI touches upon this important point frequently, when he reminds us (in a wide variety of contexts) that the human being is created in the image and likeness of God. The same point is developed by Aquinas, who argues that the perfection of the universe as a whole required the creation of finite intellectual beings such as human beings (Aquinas 1976, 2.45-46). Because the human creature—unlike lower creatures—is able to reflect the natural world's unity and goodness in a uniquely excellent way, we can say that "our duties toward the environment flow from our duties towards the person" (Benedict XVI 2010, #12).

It might appear that there is something problematically anthropocentric in the view that human duties to the environment are grounded in—or "flow from"—human duties to other human persons. The appearance of such anthropocentrism begins to dissolve, however, when one begins to recognize the fuller implications of Pope Benedict's "relational" account of human beings and other created beings. Human duties to the environment "flow from" human duties to other human persons, not because created beings in the environment have value only in their usefulness to humans, but rather because created beings in the environment can become truly themselves (can become truly perfected in their own being) "only by way of relationship."

Created beings in the environment can thus become truly perfected in their being, not insofar as they are *used*, but rather insofar as they are *understood and known*. When they are understood and known, created beings in the environment are cognitively drawn together and brought into relationship with other beings. Such drawing-together happens partially and imperfectly through acts of human knowing and loving; it happens fully and perfectly in the act of God's knowing and loving. Benedict XVI (2007b, 265) goes so far as to say that sin can be understood as the "rejection of relationality." Thus when human beings act as if created beings in the environment have value only in their usefulness to humans—i.e., when they act as if created beings in the environment have no inherent value or goodness of their own—they are acting as if human relationality can be subordinated or even denied in favor of human autonomy and self-assertion; and thus they are acting sinfully. Importantly, the view that sin consists in the "rejection of relationality" is fully compatible with the view that human beings nevertheless possess a "distinctiveness and superior role" within the created order (Benedict XVI 2010, #13).

The relational, "natural law" environmental vision offered by Benedict XVI also promises to clear up certain confusions in our contemporary thought and practices involving "rights," "duties," and "environmental justice." According to his relational, "natural law" account (which has its origins in the thought of Augustine and Aquinas), it makes sense to speak about "rights" and "duties" only where it makes sense to talk about "justice" or a "just ordering." Since justice consists in treating equals equally, and injustice consists in treating equals unequally (or unequals equally), it makes sense to talk about "justice" (and thus to talk about "rights") only where two or more individual beings can be said to be "equal" or "unequal" to one another in some relevant respect. In other words, it makes sense to talk about "justice" (and thus to talk about "rights") only where two or more individual beings stand in some kind of *relation* to one another.

Our contemporary discourses and practices involving "justice" and "rights" necessarily presuppose a kind of relationality, even though such relationality is often overlooked (or even denied outright). A key lesson to be drawn from Benedict's "relational" metaphysics is the lesson that "justice" and "rights" are necessarily relational, even though human rights are inviolable: while "rights" depend on "justice" and are thus relational, it remains the case that every act of injustice against a human being as such (i.e., every violation of a human right) is always wrong or intrinsically evil (thus human rights as such are inviolable). Contrary to many contemporary perspectives, then, Benedict (2007c, 345) teaches that it is possible to affirm the *inviolability* of human rights, yet without affirming that human autonomy is *absolute* or *non-relational*. There is no doubt that human freedom is a genuine good, but it can be the genuine good that it is, only within the context of an ordered "network of other goods." Accordingly, the "criterion of real right—right entitled to call itself true right, which accords with freedom—can, therefore, only be the good of the whole . . ." (Benedict 2007c, 349).

CONCLUSION

As we have seen, Benedict XVI's "natural law" environmental vision includes his endorsement of three metaphysical premises involving: a) the convertibility of being and goodness; b) the convertibility of being and order; and c) the uniquely intellectual nature of the human being. It is because of these three theoretical commitments that Benedict XVI can offer an environmental vision that is at once continuous with, and yet distinct from, certain contemporary accounts. In partial agreement with the contemporary deontological accounts, Benedict can assert that every individual being possesses "intrinsic worth" or "inherent worth." In partial agreement with the contemporary utilitarian account, Benedict can also assert that every individual being is also a part within some larger ordered whole. But going beyond both the deontological account and the utilitarian account, Benedict's "natural law" vision allows him to assert that the human being—by virtue of her/his unique intellectual nature—is able to apprehend the immanent orderliness and goodness of any aspect of the natural world, and thus is more capable than any other terrestrial being of reflecting God's wisdom and goodness. Since the perfection of the created universe requires the manifestation or reflection of God's wisdom and goodness, it follows for Benedict that the perfection of the created universe is made possible uniquely through the intelligent activity of human beings. It is for this reason that Benedict (unlike many contemporary thinkers) can assert that the human being is indeed part of a larger created order, and yet also unique and therefore uniquely justified in making use of other created beings for the sake of satisfying legitimate human ends.

Benedict XVI's ability to think beyond the limitations of contemporary utilitarianism and deontology also provides the key to appreciating his ability to think beyond naturalistic, reductionistic ecocentism (on the one hand) and arrogant, imperialistic anthropocentrism (on the other hand). He gives a clear, succinct summary of his position in *Caritas in Veritate* (2009b, #48), the encyclical which provided the starting point for our reflections in this chapter:

> Nature is at our disposal not as "a heap of scattered refuse," but as a gift of the Creator who has given it an inbuilt order, enabling man to draw from it the principles needed in order "to till it and keep it" (Gen 2:15). But it should also be stressed that it is contrary to authentic development to view nature as something more important than the human person. This position leads to attitudes of neo-paganism or a new pantheism—human salvation cannot come from nature alone, understood in a purely naturalistic sense. This having been said, it is also necessary to reject the opposite position, which aims at total technical dominion over nature, because the natural environment is more than raw material to be manipulated at our pleasure; it is a wondrous work of the Creator containing a "grammar" which sets forth ends and criteria for its wise use, not its reckless exploitation. Today much harm is done to development precisely as a result of these distorted notions.

REFERENCES

Aquinas, Thomas. 1976. *Summa Contra Gentiles.* Translated by Anton Pegis. South Bend, IN: University of Notre Dame Press.

———. 1981. *Summa Theologica.* Translated by Fathers of the English Dominican Province. Westminster, MD: Christian Classics.

Augustine. 1953. *The Nature of the Good against the Manichees.* In *Augustine: Earlier Writings,* translated and edited by J.H.S. Burleigh. Philadelphia: Westminster Press.

———. 1961. *The Enchiridion: On Faith, Hope, and Love.* Translated by J.F. Shaw and edited by Henry Paolucci. Chicago: Regnery Gateway.

———. 1964. *On Free Choice of the Will.* Translated by Anna S. Benjamin and L.H. Hackstaff. New York: Macmillan.

Austin, John. 1998. *The Province of Jurisprudence Determined and the Uses of the Study of Jurisprudence.* Indianapolis: Hackett Publishing Company.

Baur, Michael. 2012. "Law and Natural Law." In *The Oxford Handbook of Aquinas*, edited by Brian Davies and Eleonore Stump, 238-254. Oxford: Oxford University Press.

Benedict XVI. 2007a. "Message of His Holiness Pope Benedict XVI for the Celebration of the World Day of Peace." January 1. Accessed May 25, 2013. http://www.vatican.va/holy_father/benedict_xvi/messages/peace/documents/hf_ben-xvi_mes_20061208_xl-world-day-peace_en.html.

———. 2007b. "Sin and Salvation." In *The Essential Pope Benedict XVI: His Central Writings and Speeches*, edited by John F. Thornton and Susan B. Varenne. San Francisco: HarperCollins.

———. 2007c. "Truth and Freedom." In *The Essential Pope Benedict XVI: His Central Writings and Speeches*, edited by John F. Thornton and Susan B. Varenne. San Francisco: HarperCollins.

———. 2008. "Message of His Holiness Pope Benedict XVI for the Celebration of the World Day of Peace." January 1. Accessed May 25, 2013. http://www.vatican.va/holy_father/benedict_xvi/messages/peace/documents/hf_ben-xvi_mes_20071208_xli-world-day-peace_en.html.

———. 2009a. "Address to the Members of the Pontifical Academy for Life on the Occasion of the Fifteenth General Assembly." February 21. Accessed May 25, 2013. http://www.vatican.va/holy_father/benedict_xvi/speeches/2009/february/documents/hf_ben-xvi_spe_20090221_accademia-vita_en.html.

———. 2009b. *Caritas in Veritate*. June 29. Accessed May 25, 2013. http://www.vatican.va/holy_father/benedict_xvi/encyclicals/documents/hf_ben-xvi_enc_20090629_caritas-in-veritate_en.html.

———. 2009c. "The Name of the Holy Trinity is Engraved in the Universe." June 6. Accessed May 25, 2013. http://www.catholicculture.org/culture/library/view.cfm?recnum=9012.

———. 2010. "Message of His Holiness Pope Benedict XVI for the Celebration of the World Day of Peace." January 1. Accessed May 25, 2013. http://www.vatican.va/holy_father/benedict_xvi/messages/peace/documents/hf_ben-xvi_mes_20091208_xliii-world-day-peace_en.html.

Brennan, Andrew. 1988. *Thinking about Nature: An Investigation of Nature, Value, and Ecology*. Athens: Georgia Press.

Paul VI. 1967. *Populorum Progressio*. March 25. Accessed May 25, 2013. http://www.vatican.va/holy_father/paul_vi/encyclicals/documents/hf_p-vi_enc_26031967_populorum_en.html.

Ratzinger, Joseph. 1954. *Volk und Haus Gottes in Augustins Lehre von der Kirche*. München: Karl Zink Verlag.

Singer, Peter. 1979. *Practical Ethics*. Cambridge: Cambridge University Press.

II

Solidarity, Justice, Poverty, and the Common Good

FOUR

Human, Social, and Natural Ecology

Three Ecologies, One Cosmology, and the Common Good

Scott G. Hefelfinger

In *Centesimus Annus*, John Paul II (1991, #38) coined the phrase "human ecology" and brought it decisively into engagement with ecological concerns. Benedict XVI (2007, #8) continued this focus on both human and natural ecology, noting within the former a critical social dimension. This emphasis, and indeed priority, placed upon *human* ecology over *natural* ecology appears at first glance to fall victim to accusations of the same sort of alleged anthropocentrism that brought about the current crisis (White 1967, 1203-1207). Such an interpretation, however, would fail to take into account important elements of the operative philosophical and theological substructure—or so my argument will run. It is my hope to show that the notion of the common good, which often seems to operate in the background, provides the necessary foundation for these terms—human and natural ecology—and casts light on their meaning and relation to each other. Beginning from the concept of the common good in Aquinas, I would like to trace its contours, show its special relation to the human creature, and indicate how it is that human and natural ecology stand in intimate relations with each other. Finally, I would like to suggest that this cosmological vision is not only ontological in structure, but also of existential or narrative import, providing a workable and compelling story from a Catholic perspective upon which one can base one's life and into which one can insert one's own story.

COSMOLOGY AND THE COMMON GOOD

The notion of the common good is a venerable and rich one, and, as with all such notions, it admits of many interrelated uses. At the same time, this multiplicity of meanings can also give rise to misunderstandings, when one meaning is assumed where another is deployed. Clarifying our terms will help prevent just such a misunderstanding, and thus this inquiry into the common good will necessarily also involve identifying what is meant in speaking of the common good.

FROM A POLITICAL COSMOLOGICAL CONCEPTION OF THE COMMON GOOD IN *GAUDIUM ET SPES*

Nearly any discussion of the common good in Catholic theological circles will immediately conjure up the well-known passage of *Gaudium et Spes* (1965, #26) that describes the common good as "the sum of those conditions of social life which allow social groups and their individual members relatively thorough and ready access to their own fulfillment, [which] today takes on an increasingly universal complexion and consequently involves rights and duties with respect to the whole human race." This descriptive phrase is situated within the second chapter of the document's first part, a chapter dealing with the community of humankind.[1] Not surprisingly, therefore, it is preceded by a discussion of the human person's social nature (#25) and followed by a conclusion hitting on the point that "the social order and its development must invariably work to the benefit of the human person" (#26). This quick glance at the textual environment makes clear that what the council has in mind here is a certain political understanding of the common good as it relates to the human person in his or her social dimension. Of course it is not a matter of political policies but rather of politics in the classical sense of the word: the art of directing and safeguarding the goods and relationships shared among persons living in society (cf. Aristotle 1977, 1.1; Aquinas 1993, l.1.1-6). This understanding of the political common good is certainly of importance, and the conciliar text could easily be mined for insights and principles (cf. Porter 2005). Although this is without doubt a valid conception of the common good, it is neither the fundamental conception of the common good nor the one principally undergirding recent magisterial interventions related to ecological matters.

Towards a Pre-political Conception of the Common Good

There is another conception of the common good prior to the realm of the political. When Thomas Aquinas (cf. 1948, 2|2.26.3) speaks about common goods, he ascribes the pursuit of common goods to animals, even to rocks. Clearly, what is at stake is a pre-political common good—it is in fact a more egalitarian and in a certain sense less anthropocentric common good. More importantly, it is also more fundamental: the various political goods presuppose human nature and the inclinations it shares with non-human creatures. Speaking of a political common good may be the highest sort of ethical discussion (cf. Aquinas 2007, Prooem., #4-8), but because of this it is also last—other discussions are prior and more fundamental and among these is the foundational notion of the common good.

In trying to get at this more fundamental conception of the common good in Thomas' thought, I would like to take a tour of three important texts. The first is taken from the *Summa Theologica* [*ST*], the second from the *Summa Contra Gentiles* [*SCG*], and the third is again from the *ST*.[2] These are far from the only texts in Aquinas, but, in the interest of space, I will limit myself to considering principally these texts as taking three steps towards offering an understanding of the common good; along the way, other texts will be referenced but only to support the main points drawn from our principal texts. As we proceed step by step through these texts, we will see firstly how nature shows itself to be inclined to goods beyond the individual; secondly, that these goods are of various sorts and of differing degrees of breadth or commonality; and thirdly, how parts naturally desire the good of the whole in a way neither exclusively self-referential nor merely self-sacrificial.

Summa Theologica 1.60.5— Desire for Goods Beyond the Individual

Our first step places us in the *Prima pars* of the *Summa* where Thomas (1948) considers God firstly and then those things that proceed from God. Things proceed from God by way of creation, within which the angels by nature hold a position of preeminence. Thus, among created beings, Thomas begins by treating the topic of angels, devoting to them fifteen questions (qq. 50-64) containing seventy-two articles; of these I would like to consider only one: 60.5. Here the question in play is whether an angel by natural love loves God more than he loves himself.[3] In formulating his own position, Thomas presents us with three structural moments: first, he looks to nature; second, he notes how natural inclinations are related to the good; third, he relates his findings to God and the question at hand.

Nature as Model

The starting point for Thomas' (1948, 1.60.5) position is to "consider where natural movement tends in the natural order of things." The first move is already fraught with import: although the consideration at hand is somewhat lofty, pertaining as it does to angels, Thomas roots it nevertheless in the natural world as observed by humans. We will return to the significance of this point later, but for now we can note that Thomas takes nature as his starting point. He then gives us the reason he can take this starting point and then move to consider other related points: "the natural tendency of things devoid of reason shows the nature of the natural inclination residing in the will of an intellectual nature." Thus, although the human world is a world of reason and will, our understanding of this world begins in our experience of the pre-rational world of nature.[4]

Natural Inclinations and the Good

Finding our orientation within the natural world includes noting well the natural inclinations and tendencies unfolding dynamically before our eyes and prior to our interaction with or intervention in the natural world.[5] On this score, Thomas refers consistently to Aristotle's *Physics*, which serves as the *locus classicus* for a proper and profound philosophical investigation of nature. In the article at hand, Thomas (1948, 1.60.5, citing Aristotle 2004, 2.8 [199a8-10]) refers to Aristotle's observation that "according as a thing is moved naturally, it has an inborn aptitude to be thus moved"; this he does in order to show that because we observe *parts* in nature operating for the sake of *wholes*, sometimes even at their own expense, we can conclude that "in natural things, everything which as such naturally belongs to another, is principally and more strongly inclined to that other to which it belongs than towards itself." Thomas offers the example of the hand, which would naturally be raised up to defend the safety of the whole body before an incoming blow, even if this meant injuring itself in the process. In this way, natural inclinations are seen to have the good as their ends—they tend toward some good, some perfection, by nature. What is more, the goods they seek are proper but not exclusive to themselves; rather some such goods are truly goods of the whole, common goods. We begin here to catch a glimpse of the natural foundations of the notion of the common good.

Natural Inclinations, God, and Intellectual Creatures

Formulating the conclusion of his response, Thomas applies the analogy of the whole and part to the relationship between God and creatures: as the hand belongs to the body, so creatures belong to God. This is not to say that creatures are parts of God but rather that the ordering which creatures have to their Creator is similar to the way that parts are ordered to the whole—the being had by creatures comes only from God as a *part*-icipation in God's own being, and consequently every perfection had by creatures is a *part*-icipation in God's perfection (cf. Aquinas 1948, 1.6.4). It is for this reason that Thomas (1948, 1.60.5) insists here that, as the universal good, God comprises all creatures: "Consequently, since God is the universal good, and under this good both man and angels and all creatures are comprised—because every creature in regard to its entire being naturally belongs to God—it follows that from natural love, angel and man alike love God before themselves and with a greater love."

Although the last line speaks explicitly of "angel and man," it pertains in fact to all creatures[6] and, thus, the conclusion reached suggests the startling notion that all creatures love God more than themselves, which is to say they are more inclined to God's good than their own. This happens furthermore at the level of nature: when we consider the nature of created things, including humans and angels, it leads us to see that creatures are made to return to their Creator through the inclinations following from their natures. Such inclinations do not exclude, but rather provide the foundation for, the particular form of striving for God found in the realm of intellectual creatures: angels and humans can rise to a properly intellectual love, willing—in the proper sense of the term—God's good above their own.

Worth noting is the very last line of Thomas' response (1948, 1.60.5) in which he makes a strong claim about the importance of the foregoing discussion of nature and the good. Thomas writes that "if either [angels or humans] loved self more than God, it would follow that natural love would be perverse, and that it would not be perfected but destroyed by charity." In effect, what Thomas is asserting here is that a proper understanding of natural love—and thus necessarily also nature—stands as a preamble to the faith; it belongs to those topics accessible to human reason that are presupposed by the doctrine of the faith, as nature is presupposed by grace.[7] For, without this understanding of nature and its corresponding inclinations, the doctrine of charity would be nonsensical; it would posit an action of God the Redeemer—infusing a virtue ordering the individual beyond itself—that would be in conflict with an action of God the Creator, ordering natural inclinations exclusively to the individual. But this is not the case. Rather, there is a continuity found between the orders of nature and grace: because nature instigates a natural love for something beyond and above the singular creature, charity stands in continuity with this when it is understood as a supernatural love for something, or better Someone, beyond and above the singular person.

Putting It Together

Putting these three steps together, what we see in this lone article is an excellent primer or entryway to the Thomistic conception of the common good. This conception is firstly rooted in nature: looking to the natural world around us, we find that created things have a natural inclination to the good. In other words, what we have here is a classical and robust teleological conception of nature, one that puts nature in the position of being our tutor for "higher things." Secondly, the goods desired serve to perfect the one desiring; indeed it is precisely for this reason that they are both good and desirable. Thirdly, we see that the goods to which natural things are inclined are various and, more importantly, of various sorts: some are goods proper and exclusive to the singular creature (e.g., existence or food for living things); others go beyond the singular creature without thereby ceasing to be goods proper to the singular, as for example a child's existence is at once beyond its parent's singular good and yet a good truly perfective of or proper to the parent. Finally, undergirding this view is the notion that creatures participate in the goodness of the Creator, and it is for this reason that they desire both singular and common goods, but in such a way that even in the singular goods—the foundation of that good, and thus what is ultimately and most desired—is the divine good.

At this point, Thomas' cosmological view of the good begins to come into focus. The good is not a topic or concept pertinent only to rational creatures, to human beings; rather, to speak of it as we have seen Thomas doing indicates that somehow the whole of created reality is brought into play when we speak about the good. No element of reality is absented from striving for the good, whether its own or the greater good of the cosmos. Looking at how Thomas accounts for this and how the various parts are ordered to the whole thus presents itself as our next step along the way.

Summa Contra Gentiles 3.24—The Sorts of Goods Desired

If we set off in search of a more worked-out description of the sorts of goods that Thomas envisions, the third book of the *SCG* presents us with a sketch of precisely what we are looking for. As we have already established, each thing seeks its own good, its own perfection, but does so in a non-exclusive way. How can this be? What precisely constitutes a thing's *own* perfection? Thomas (1991, 3.24; see also De Koninck 2009, 75–77) clarifies firstly that "the good that is proper to a thing [*bonum suum*] can be taken in many ways," and he goes on to elaborate four sorts of goods that are truly proper to a thing, truly a thing's own.

Aquinas (1991, 3.24) writes, "One way is according to what is appropriate to the individual as such. It is thus that an animal seeks its good, when it desires the food by which it is kept in existence." This first sort of good is fairly straightforward, and it is the most limited sort of good.

A second way is "according to what is appropriate to the species as such. It is in this way that an animal desires its proper good inasmuch as it desires the procreation and nourishment of offspring, or the doing of any other work that is for the preservation or protection of individuals belonging to its species" (Aquinas 1991, 3.24). Here we find a wider conception of the good that remains nevertheless the proper good (*proprium bonum*) of the individual. In this category we find actions such as a mother seeking out food for her young or giving up her life for the sake of her young. Such apparently selfless activities do in fact constitute a good proper to the individual agent, even as it exceeds the lone individual in its scope.

"A third way is according to the genus as such. It is in this way that an equivocal agent seeks its proper good by an act of causation, as in the case of the heavens" (Aquinas 1991, 3.24). In this third case, the good extends beyond the bounds of a particular sort of creature to include others yet more remote from the individual agent. In Thomas' (cf. Aquinas 1948, 1.4.2) conception of the heavens, the sun is an equivocal agent causing all life on the planet for which reason humans, for example, owe a debt of gratitude to the light and warmth of the sun, without which we could not exist on the planet. Such a contribution made by an individual creature is quite diffuse in its effective range, and yet it nevertheless remains a good proper to that individual.

Finally, "a fourth way is according to the analogical likeness of things produced in relation to their principle. It is in this way that God, who is beyond genus, gives existence to all things on account of His own goodness" (Aquinas 1991, 3.24). Although this final category is exclusive to God, it nevertheless crowns the foregoing presentation of the expansiveness of the good: from the good of the individual as individual, Thomas widens his scope to encompass the good of the individual that diffuses itself not only to other members of the same species—not only to members of diverse genera, but to members taking their place in the whole of existing things.

This general schema begins from the individual and works its way to clearly cosmological proportions: created singulars desire goods ranging from very particular to the highest and most common goods, including the ordered whole of the cosmos itself and even God as the absolute highest good. In this way, the cosmos with the creatures it contains forms a complex and interwoven fabric of goods, particular and common. Jame Schaefer (2010, 81; see also Schaefer 2009, 21-24) has aptly summarized Thomas' cosmological vision of the common good:

> Aquinas established an understanding of the common good that was cosmic in scope. . . . What is the cosmic common good? Aquinas expounded systematically on the goodness of the universe that is demonstrated by the orderly functioning of its constituents in relation to one another. . . . The good of the whole—the common good—is the internal sustainability and integrity of the universe. From Aquinas' perspective, God instilled in each creature a natural inclination toward the common good. . . . At the root of this appetite for the common good is the natural inclination each creature has for God, who is the absolute common good of all creatures.

But this cosmological dimension never overrules or cancels out the properly singular dimension. This is to say, to the extent that it is built upon the natural inclinations of the individual, the broader or more common good never appears to the individual as an alien good (what Thomas calls a *bonum alienum*). Because it stems from a creature's very nature, it is proportioned to it as a perfection proper to that creature's being. It is for this reason that Aquinas insists in our text from the *SCG* that each of these four sorts of goods is in fact a *bonum proprium* or *bonum suum*. They are goods of the part precisely as a part of the whole.

Summa Theologica 2|2.26.3—Nature, Cosmos, and Common Good

From these two compact texts of Thomas, we have a seen a picture emerge that paints the contours of the common good on the broad canvas of nature in general, thus leading us to a fully cosmological conception of the common good. The common good is the good of the whole that is at once the good of the part; it is desired more than the good that belongs merely to the part, and thus, in the grand sweep of things created, individual beings have a natural desire for the good of the whole cosmos, both the intrinsic order and harmony of the parts, as well as the good that is God.

In a final step, Thomas (1948, 2|2.26.3.2) encapsulates the foregoing in an article on charity, where he asks whether charity compels us to love God more than ourselves. One objector puts forth an argument that gets right at the issue in question: "One loves a thing in so far as it is one's own good. Now the reason for loving a thing is more loved than the thing itself which is loved for that reason, even as the principles which are the reason for knowing a thing are more known. Therefore man loves himself more than any other good loved by him."

The objector reasons that humans love things precisely on account that the good involved is in some way one's own. What other reason could we have for finding something truly a good? If not a good *for us* in some way, how can we see it as a good? Thomas' (1948, 2|2.26.3.ad 2) reply to the objection is particularly terse: "The part does indeed love the good of the whole, as becomes a part, not however so as to refer the good of the whole to itself, but rather itself to the good of the whole." With this, Thomas (1948, 2|2.26.3) appears to be granting something to the objector: a part can only love the good of the whole in the way proper to a part, and this does in fact imply that the good of the whole is in some way the good of the part. But there is more to it than this, as the body of the article makes clear:

> The good we receive from God is twofold, the good of nature, and the good of grace. Now the fellowship of natural goods bestowed on us by God is the foundation of natural love, in virtue of which not only man, so long as his nature remains unimpaired, loves God above all things

> and more than himself, but also every single creature, each in its own way . . . because each part naturally loves the common good of the whole more than its own particular good.

As we saw in our first text from Thomas, the question here about human love for God is immediately broadened to encompass nature in general and the cosmos as a whole. Within this broader horizon, natural love—the love prior to intellectual or sensitive love[8] —springs from God's own gift of natural goods, i.e. bringing things into being with their proper natures, inclinations, etc. Seeking to return to its principle, natural love moves creatures to return to their Creator, the common good of the universe. Thomas (1948, 2|2.26.3) exemplifies this point with these words: "This is evidenced by its operation, since the principal inclination of each part is towards common action conducive to the good of the whole. It may also be seen in civic virtues whereby sometimes the citizens suffer damage even to their own property and persons for the sake of the common good."

Thomas points to civic virtues, which dispose citizens to the good of the whole body politic. Working to realize this good can involve sacrifice, but precisely because this broader good is also a good proper to the part, i.e. the citizen, it comprises the end and goal of the virtues of political life. The political example should not distract us from the original claim that *all* creatures desire the common good. Although we will turn momentarily to matters specifically human, at this point what we find is Thomas making the case that it is natural and proper to each creature to seek not only its own singular good but also broader goods as well: common goods or, we may even say, cosmological goods. Situated within the wide world of nature, Thomas' intellectual framework of the good is amenable to numerous strata of common goods, especially those of contemporary ecological concern. As Schaefer (2009, 24) summarizes,

> Aquinas' concept of the common good provides a "cosmological-ecological principle" for his ethical system. From the perspective of ecological degradation today, the good sought in common would be the good of ecosystems, of which humans are integral actors relying on other interacting *biota* (i.e., the fauna and flora of an area) and *abiota* (i.e., the air, land, and water with which and in which the biota interact) for their health and well-being.

The inclusive and flexible framework established by Thomas lends itself therefore to formulating a fully cosmological and ecological ethic. But to do so requires stepping into the properly human realm, where questions of morality and ethics come into play. Up to this point, our attention has been devoted to the created world precisely as natural, attending to the tendencies and inclinations found therein. Our next task is to link this natural realm with that realm proper to rationality, i.e., to the human realm, and to do this in a way that manifests continuity with all that we have established thus far. This is the task of linking natural ecology or cosmology with human ecology.

LINKING NATURAL AND HUMAN ECOLOGY

What has been sketched thus far might be called a philosophical description of natural ecology: the complex interrelations between beings in the natural world and beyond are relations established in the natures of things in the act of creation. These relations, making up this ecology, are based upon the notion of the good and in particular the notion of the common good. All creatures are profoundly interconnected and the ordering of one to the other both presupposes the good and comprises the intrinsic common good of the universe; taken altogether, these creatures are ordered to God, the extrinsic common good of the universe, and to manifesting divine goodness. There are certainly other crucial aspects to natural ecology, aspects taken up especially in the empirical sciences, but for the time being I am limiting myself to this particular philosophical substructure.

Up to this point, the human creature has received little particular attention, thus raising the question: what about human beings in this picture? What place and role to humans have in all of this? This is the juncture at which we begin to discern the links between natural and human ecology, a process that at the same time places us before the intersection between this philosophical analysis and recent magisterial teaching on the subject of natural and human ecology.

Natural and Human Ecology—A Problem?

Although the seeds for an understanding of "human ecology" are to be seen in John Paul II's 1990 *Message for the World Day of Peace*, the term itself was introduced a year later in his encyclical, *Centesimus Annus*: (1991, #38)

> Although people are rightly worried—though much less than they should be—about preserving the natural habitats of the various animal species threatened with extinction, because they realize that each of these species makes its particular contribution to the balance of nature in general, too little effort is made to *safeguard the moral conditions for an authentic "human ecology."* Not only has God given the earth to man, who must use it with respect for the original good purpose for which it was given to him, but man too is God's gift to man. He must therefore respect the natural and moral structure with which he has been endowed.

The Holy Father (1991, #39) went on to emphasize the family as the first and fundamental structure for "human ecology," since it is in the family that we learn "what it actually means to be a human person." Here we find implied two dimensions of human ecology: the first entails respecting human nature and human life in parallel to respecting nature and the natural world; the second, respecting human nature in the sense of understanding it and respecting the guidelines or moral intimations it reveals.

Making reference to this passage in *Centesimus Annus*, Benedict XVI picks up the term human ecology in his own 2007 *Message for the World Day of Peace* (#8), indicating that "all [of the foregoing] means that humanity, if it truly desires peace, must be increasingly conscious of the links between natural ecology, or respect for nature, and human ecology. This tracks the first dimension of the term indicated above: respect for nature ought to be matched, indeed outdone, by respect for human beings." Benedict (2007, #10) goes on to speak about being "guided by a vision of the person untainted by ideological and cultural prejudices or by political and economic interests." Not surprisingly, this correlates with the second dimension hinted at by John Paul II: respecting the moral intimations or guidelines discovered when attentive to human nature. Benedict (cf. 2008a, 2009, 2010, 2011a, 2011b, 2011c) will continue take up the theme of human ecology with great regularity, and each time he does these two ways of reading "human ecology" recur.

Still, the question lingers: why bother to speak of human ecology at all? Of what significance is it for our ecological concerns? To some, its significance appears in all likelihood to be of a negative rather than positive character. One finds this worry manifested in various ways. Lynn White, Jr. (1967, 1205) acknowledged that "human ecology is deeply conditioned by beliefs about our nature and destiny—that is, by religion." Considering the various belief systems, White (1967, 1206) proposes that "especially in its Western form, Christianity is the most anthropocentric religion the world has seen," where anthropocentric means that "no item in the physical creation had any purpose save to serve man's purposes." Precisely because of this unbalanced viewpoint, Christianity is likely to bear "a huge burden of guilt" for the current ecological challenges.

Sallie McFague (1993, 30) suggests a similar worry, taking as her point of departure an organic model of the universe. She observes, "The classic organic model is expressed in the phrase 'the Church as the body of Christ,'"[9] but she notes that this model suffers from, among other things, being both overly "spiritualized" as well as "anthropocentric and androcentric" (35-36). Such a model, therefore, comes to entail a highly curtailed view of the cosmos, leaving out such significant elements as matter, bodies, and women.

To consider a final example, John Haught (1990, 30) also worries that too much emphasis on humans leads to a devaluation of non-human creatures. It is especially the pairing of modern scientific materialism with anthropocentrism that "places a sharp discontinuity between the isolated human subject and the senseless universe investigated by science." Even religions fall into such dichotomies, Haught (31) argues, whether by anthropo- or theocentrism.

If true, such construals of human ecology, as setting up the human being as distant, above, and in competition with the rest of creation, would indeed be deeply problematic.[10] But there is a considerably better way of reading the relatively recent magisterial term "human ecology," one that enjoys much more continuity with the Church's intellectual heritage and presents a more nuanced and profound notion of "human ecology."

Towards a Solution: The Human Person and the Common Good

The notion of human ecology comes into its own when placed upon the robust cosmological account of the common good we have already explored. However, before tracing out those linkages, we find a helpful point of departure in Benedict's writings. We mentioned above the two aspects of human ecology as presented by John Paul II and Benedict XVI: respecting human nature in the sense of protecting it and being attentive to it. Now we can add that underlying these two interrelated dimensions is the viewpoint embodied in a short but profound passage penned by Benedict XVI (2009, #51) in his encyclical *Caritas in Veritate*:

> The book of nature is one and indivisible: it takes in not only the environment but also life, sexuality, marriage, the family, social relations: in a word, integral human development. Our duties towards the environment are linked to our duties towards the human person, considered in himself and in relation to others.

The key phrase here is the first: the book of nature is one and indivisible. Respecting nature necessarily involves respecting the human person who is also natural, and respecting the human person means respecting the moral "grammar" written into his nature (Benedict XVI 2010, #13). If, therefore, we cannot set an insurmountable chasm between the natural world and the human person, this suggests a certain continuity and interdependency between the two. Already we begin to see that a hasty charge of anthropocentrism is not quite adequate to the nuanced understanding put forward here. In order to explore this further it will be helpful to link this characterization of human ecology with the foregoing account of the common good and nature. In so doing, it will become apparent that both the similarities and dissimilarities between the human being and the natural world will defuse such an anthropocentric understanding and instead will open up into a cosmocentric perspective.

Nature, One and Indivisible

If the book of nature is one and indivisible, this immediately suggests a great degree of commonality found between the human person and the rest of creation. And the suggestion is not erroneous: indeed, as we have explored above, the picture painted by St. Thomas is one in which all natural beings—humans included—share certain properties, characteristics, and inclinations.[11] Among these, the most important for our consideration here is that, like every natural being, the human being also has natural inclinations or tendencies towards a variety of goods. In addition, as we saw above, these goods, while being all proper goods (*bona sua*), include both merely individual goods as well as common goods.[12] If there is something to be discovered about the human that sets it apart from other creatures, we do not come across it here: rather, the book of nature is one and indivisible.

Yet it is true that the human being is unique among created beings, and this in more ways than one.[13] But here we will limit ourselves to the topic we are already well into discussing: the good. In the unified book of nature, it is true that the human is wholly unexceptional when it comes to being naturally inclined to the good. Nevertheless, there are two decisive differences to note in our inquiry. The first pertains to the good in general, and the second to common goods.

Human Inclination to the Good

As with other natural things, human beings are also inclined to the good; yet, given our rational faculties, this inclination manifests itself in a new way and acquires a moral dimension. Thomas (1948, 1|2.91.2) outlines both this general similarity and the specific human difference in the context of his discussion on law. First, he adumbrates the common ground between human beings and other natural beings, writing, "[S]ince all things subject to Divine providence are ruled and measured by the eternal law . . . it is evident that all things partake somewhat of the eternal law, in so far as, namely, from its being imprinted on them, they derive their respective inclinations to their proper acts and ends." For our inquiry here, it is important to bear in mind the close affinity between ends and goods: Thomas is linking the natural inclination for the good to a participation in the eternal law, thus placing this discussion squarely in the context we have just discussed, i.e. the context of the good.

He (Aquinas 1948, 1|2.91.2) then identifies the distinguishing mark of the rational creature's participation in the eternal law:

> Among all others, the rational creature is subject to Divine providence in the most excellent way, in so far as it partakes of a share of providence, by being provident both for itself and for others. Wherefore it has a share of the Eternal Reason, whereby it has a natural inclination

> to its proper act and end: and this participation of the eternal law in the rational creature is called the natural law.

Thus, human participation in the eternal law is considered most excellent on account of the fact that humans share in Eternal Reason, humans are rational animals. This particular characteristic sets humans apart from other natural creatures inasmuch as humans can, through intellect, come to know the good *as* good. Whereas the other creatures of nature pursue the good in more limited ways, it is proper to the human person to see the good precisely as good and to desire it on this account. This desire for universal goodness is precisely the function of will, which is the rational appetite (cf. Aquinas 1948, 1.59.1). Thomas presents a short taxonomy of the various ways that inclination towards the good manifests itself in natural things. First, there is natural appetite which is an inclination to good apart from knowledge (e.g., rocks and plants); then there is sensitive appetite, an inclination towards good with some apprehension of the particular good, but not precisely *as* good (e.g. animals); and finally there is rational appetite, described by Thomas (1948, 1.59.1) in this way:

> Other things, again, have an inclination towards good, but with a knowledge whereby they perceive the aspect of goodness; this belongs to the intellect. This is most perfectly inclined towards what is good; not, indeed, as if it were merely guided by another towards some particular good only, like things devoid of knowledge, nor towards some particular good only, as things which have only sensitive knowledge, but as inclined towards good in general. Such inclination is termed "will."

This cognitive element, which is a participation in Eternal Reason, serves as the presupposition for the natural law. Human beings perceive the good in general and, under this general concept, the various goods attending upon their own human nature. Thomas (1948, 1|2.94.2) traces this line of reasoning beginning from speculative reason and its primordial apprehension of "being"; in parallel to this, practical reason has as its primordial apprehension "the good." Given the general notion of the good as that which all desire, practical reason formulates the first precept of law: "good is to be done and pursued, and evil avoided." This law, however, is specified into further precepts based on the variety of goods corresponding to human nature, thus providing the first intimations of the shape of human morality.[14]

In this way, natural law forms the basis for normativity and morality.[15] It is in the nature of the human person to share with non-human creatures an inclination towards the good, that first and most fundamental participation in the eternal law. At the same time, humans surpass these creatures in perceiving the good as good and thus entering into the field of moral responsibility. This field is built upon a conception of the human person as standing in continuity with the natural world and simultaneously transcending it: to be human is to be linked with and distinct from the natural world that surrounds us. Linked and distinct—these are both fundamental elements of our "natural location."[16]

If this anthropological conception grounds the realm of morality, then it is not surprising that an erroneous conception of the human will also lead to a denial of moral responsibility. These ties have been reaffirmed in recent magisterial statements of the Catholic Church, where it is often pointed out that, at root, what is at stake is an "anthropological error." In fact, two extremes characterizing two errors are suggested. The first involves the proposition that the human is in no way unique, a position ascribed to "deep ecology" (cf. International Theological Commission 2008, #81). This is to suggest that humans are linked but not distinct. The problem with this position is that it does away with moral responsibility, i.e. why should we be any more responsible than a rock? The second view is that the human is totally unique, is not linked but only distinct, which leads to an altogether different sort of denial of moral responsibility: the human can do whatever he wants with nature (cf. Benedict XVI 2011b). Both of these positions are problematic.

The *via media* is that the human is linked and distinct—unique in some respects, not all respects. When this continuity is acknowledged and then brought into the realm of the rational, where the universal good is perceived, only then does the project of attending to human nature move from being merely descriptive to being prescriptive as well. As we shall see momentarily, this applies as well to the term human ecology: affirming human ecology has the effect of exhorting to moral responsibility, including in this case the realm of natural ecology.

Human Inclination to the Common Good

The situation of common goods is parallel to that of goods taken generally: as with other natural beings, humans too are inclined to common goods but in their own unique way—precisely *as* common, that is, with the awareness of their commonality and desiring them on this account. The uniquely human posture towards common goods is to desire them in their abundance and in their communicability. Because this is a properly human awareness and willing, it follows that the human is by nature and vocation oriented to the common good of the whole of which it is a part. This is to say that the human is by nature and vocation *cosmocentric*.

With this, we are now in a position to summarize the foregoing in a few sentences. Firstly, it is the uniquely human task to care not only for singular goods, but also for common goods, including those of society, of the planet, and of the universe at large. Secondly, because this task is proper to humans, its fulfillment is an integral part of what it means to be human, which is to say, it belongs squarely in and presupposes the general realm of ethics, moral formation, and justice. Thus, human ecology, understood as the cultivation of *human* virtue particularly as guiding our *social* relations, is intimately bound up with *natural* ecology and the task of explicitly caring for the wide-ranging common goods with which we are surrounded. To fail in our task of caring for the natural world is to fail in being human—there is no other way around it. Finally, it is due to the breadth of this task that the human vocation sits comfortably with everything from self-preservation, to societal betterment, to liturgical worship in all of its cosmic richness.

CONCLUSION: LIVING OUT BENEDICT XVI'S ECOLOGICAL VISION IN THE UNITED STATES

The breadth of the human vocation resonates strongly with Benedict XVI's writings linking human with natural ecology. It also draws together in a way at once profound and catechetical the current *ecological* themes with earlier *liturgical* motifs close to the Holy Father's heart. This liturgical and catechetical emphasis counteracts the tendency to view the Catholic vision as static or ossified; instead, it provides avenues for noting and emphasizing the existential structure of this broad cosmological and anthropological vision. Through the lens of this narrative structure, this vision can be seen to correspond to life as lived.

At the conclusion of this largely speculative enterprise, I would like to draw four practical conclusions that are borne out by Benedict XVI's ecological vision centered around the priority of human ecology; these conclusions are by intention specific to the American landscape, and all pertain to education very broadly speaking. Because education is a social activity, I take up the three societies outlined by Pius XI (1929, #11-13) and then finally a particular place of education uniquely emphasized in the United States.

The Family, the Domestic Church

The first society is that of the family, and within the sheltered walls of the family there is much that can be done to live out and instill a rich ecological vision. Transforming Benedict XVI's vision into action in this realm can take some helpful cues from Wendell Berry. In particular, Berry's (1996, 17-25) eloquent writing on the self-sustaining independence of the family as an antidote to modern specialization brought about by hypercapitalism is highly pertinent. Conceiving of the family as a bastion from overly-capitalistic currents leads to a decidedly unspecialized, resourceful, and cultured domestic life. The ecological ramifications of this kind of implementation are enormous: attentiveness and respect for nature, seasons, harvest; food preparation and the joy of eating and dining; the beauty of creation and beauty of human labor; etc. The sacramental aspect of the Catholic faith takes up all of these elements into the vibrant para-liturgical life of the domestic Church.

Civil Society

The Church-State divide runs deep in the America, and perhaps rightly so when properly understood.[17] Nevertheless, given the importance of natural law, especially in the realm of ecological matters, this teaching must be rehabilitated and liberated from the common perception it garners as being sectarian. At the same time, and in accord with the unity of knowledge and the nature of law, this does not mean liberating the natural law from God. What is needed is a rediscovery of the tradition of natural law and a reappropriation in today's context, before today's questions. This theme is one that apparently has grown considerably in the thought of Benedict XVI—his own views appear to have changed significantly through the years, underscoring, it seems, the importance and urgency of rediscovering natural law thinking.

The Church's Liturgy

One of the best ways of appreciating and reverencing the cosmos anew is through the Church's liturgy—a theme that is especially close to Benedict XVI's heart. In *Spirit of the Liturgy*, Ratzinger (2000, 24-34) offers a profoundly cosmic view of the liturgical celebration, and this vision entails a shift from a "low" liturgy that is overly rationalistic. Actions, sights, smells, sounds, etc. are often downplayed in "low" liturgy, whereas the word—sign and instrument of human rationality—is overplayed. Although it is true that the word is of utmost importance, it is not of exclusive importance; indeed, to lose the trappings is to lose the human person and stray off into the realm of the angelic. Such liturgical celebrations might be "soulistic" but they are far from "holistic." What is needed in order to more adequately correspond to and portray a proper conception of human nature or human ecology is the cosmos, materiality, and—dare we say—animality. Rationality necessarily belongs here as well, but the anthropological and cosmological positions denoted above must be brought into play in the liturgical sphere; indeed, it is here that they find their crowning embodiment.

Catholic Schools

Turning finally to the unique situation of Catholic parochial schools in America, it is important to note that what is presupposed by human ecology is a unity of knowledge that welcomes all truth and relates it appropriately to God and the created order. This is a thematic concern of Pope Benedict (cf. 2006 and 2008b) that aligns nicely with his ecological thought. Perceiving this unity requires a contemplative spirit, one that is not overly eager to jump to action before each piece of knowledge is properly situated in its theoretical—and ecological—context. Thomas has a short treatment of just the virtue we need for cultivating this contemplative, ecologically situated sort of knowledge when he considers the virtue of studiousness (*studiositas*). Although much could be said about this surprising virtue, let it suffice to note that even a good such as knowledge and truth can be sought immoderately (cf. Aquinas 1948, 2|2.166.2, *corpus* and ad 3).

This immoderation can manifest itself in numerous ways, ways which Thomas mentions in his treatment of the corresponding vice, *curiositas*. Perhaps most importantly is when knowledge is sought immoderately by "[desiring] to know the truth about creatures, without referring [this] knowledge to its due end, namely, the knowledge of God" (Aquinas 1948, 2|2.167.1).[18] What Thomas has in mind is not a mere pious affirmation, but rather the unity of the knowledge in its ordination to God as beginning and end. This is, in effect, a properly contemplative disposition and approach to knowledge, seeing all things as they are, in their proper order, and thus also in relation to the highest common good, God himself. Studiousness can bring into its fold full-blooded theology of creation, which would include a nuanced and properly ecological account of natural law. Finally, and drawing on the material or bodily continuity humans share with the rest of creation, a rich aesthetic dimension would also be included, as suggested by John Paul II (1990, #14). Altogether, these would be the marks of a very distinctive and distinctively Catholic education that would in turn redound to society in general as well as the broader cosmos.

NOTES

1. Chapter two of *Gaudium et Spes* is entitled "The Community of Mankind" and is situated between a chapter focusing on the human person ("The Dignity of the Human Person") and one devoted to "Man's Activity throughout the World."

2. The three texts, all from Aquinas, are: ST 1.60.5; SCG 3.24; and ST 2|2.26.3. The basic line of thought I attempt to sketch here takes its inspiration from the work of Charles De Koninck, one of the least well-known but most profound members of the Thomistic revival of the last century, who dealt extensively with the question of the common good. See De Koninck 2009. Furthermore, I gratefully acknowledge that I was introduced to this topic by Susie Waldstein in a class on charity, where we took up the last text we will consider here, dealing with the order of charity.

3. If it seems slightly odd to consider a question on angels within a broader discourse on ecology, it is important to bear in mind the structural nature of the *Summa*: the question at stake here has to do with natural inclinations, one particular instantiation of which is will. Although Thomas has already considered the will earlier in his treatment of God (q. 19), he furthers his inquiry into the will here in the discussion of angels. The will returns once again in the consideration of human nature (qq. 82-3) and human action (*passim*, but especially qq. 8-10). As always with Aquinas, these

texts must be placed in relation to each other in order to catch a glimpse of the whole and its relation to a particular inquiry.

4. The ramifications of this point are numerous and weighty. For a constructive and convincing rehabilitation of this point, see Porter 2005.

5. After a subtle philosophical analysis, Aristotle arrives at this basic insight when he states that nature is an *intrinsic* principle of motion: its being intrinsic means, among other things, that it is there and operative prior to any human intervention; this contrasts of course with the realm of art, even though art imitates nature (see Aristotle 2004, II.1-2 [192b8-24 & 194a23]).

6. Note the emphasis in the *corpus* on "every creature." See also in this same article, ad 4: "God, in so far as He is the universal good, from Whom every natural good depends, is loved by everything with natural love."

7. In order to trace out this fruitful line of inquiry, see e.g. Aquinas, *De Trinitate* 2.3. See also, *De Potentia Dei* 4.2.15, where Thomas writes briefly and provocatively that natural knowledge is the foundation of supernatural knowledge: *cognitio naturalis praecedit cognitionem supernaturalem, tamquam fundamentum eius*.

8. Although I have omitted it for brevity, Thomas points out these gradations of love in the very same article: ". . . every single creature, each in its own way [loves God above all things], i.e. either by an intellectual, or by a rational, or by an animal, or at least by a natural love, as stones do, for instance, and other things bereft of knowledge . . . " (Aquinas 1948, 2|2.26.3).

9. The terminology employed by McFague at this point is clearly ecclesiological terminology. Precisely how this relates to or differs from cosmology is not made clear. This sort of disambiguation would greatly strengthen McFague's project, which for all its legitimate concerns is sometimes plagued by a lack of clarity.

10. Mark Wynn offers an insightful rejoinder to such problematic views of anthropocentrism as found in Aquinas; see Wynn 2010, 154-65. Wynn considers in particular the human's relation to non-human creatures as envisioned in Aquinas' account of the teleology of creation. At the same time, more attention could be given to the notion of the good, and particularly the common good, without which analyzing the teleology of creation does not yet answer the question of whether the good of a lower thing is compromised in its being ordered to the good of a higher thing. See, e.g., Wynn 2010, 157-8, where it is not entirely clear how non-human animals do not find their good compromised when "subordinated" to humans.

11. We have already seen this way of viewing things at work above, in considering how the inclinations we find in pre-rational nature help illuminate the inclination we find in properly rational nature, i.e. *voluntas* (see Aquinas 1948, 1.60.5). Aristotle goes so far as to point out that we ought even to identify those tendencies shared with lower creatures by the name of those lower natures rather than by our properly human nature, e.g. a human person does not fall *as* a human but rather *as a body*, likewise we move *as an animal*, though of course we think, love, procreate, etc. properly *as human beings* (see Aristotle 2004, II.1 [192b25]).

12. See Aquinas 1948, 1|2.94.2. Note that the goods listed include both individual (nourishment) and common (the truth concerning God) goods.

13. See Aquinas, 1948, 1.75, prol., on the human's unique ontological position; or 3.4.1.3, on the unique ability of the human person to repent; etc.

14. Note that the list of goods identified by Aquinas is a balance between goods held in common with non-human creatures and those specific to humans—if anything, the balance is tipped in the direction of goods shared with lower creation.

15. Natural law, however, does not exhaust the realm of normativity. As Thomas himself makes clear, the virtues play an indispensable role in "translating" precepts into concrete, practical action (see e.g. Aquinas 1948, 1|2.49.4 and 55.1). Thus both natural law and virtue theory are necessary components of any ecological ethic. From a Thomistic vantage point, Schaefer (2009, 269-80) has crafted a thoughtful and constructive account of the second of these components.

16. Noteworthy in this regard are the two creation accounts in Genesis, each of which, in their own differing ways, portrays the human person as "linked and distinct." The first account places the human squarely in the orderly progression of days, with the only (thought decisive) distinction being that *adam* is created in the image of God. The second account emphasizes the continuity between the human and the rest of creation in various ways: through having the first man be shaped from the earth, through having the animals brought before to be named (they are similar enough in *some* way to be considered as helpmates), which simultaneously marks his distinction as being able to name them.

17. Benedict XVI has even pointed to the separation of Church and State in America as something of an exemplar; see, his Apostolic Journey to the United States, Benedict XVI 2008c. See also Ratzinger 2008.

18. For a profound analysis of *studiositas* in this vein, see Hütter 2005.

REFERENCES

Aquinas, Thomas. 1993. *Commentary on Aristotle's* Nichomachean Ethics. Translated by C.I. Litzinger. Notre Dame: University of Notre Dame Press.

———. 2007. *Commentary on Aristotle's* Politics. Translated by Richard J. Regan. Indianapolis: Hackett Publishing, 2007.

———. 1991. *Summa Contra Gentiles*. Book III. Translated by Vernon J. Bourke. Notre Dame: University of Notre Dame Press.

———. 1948. *The Summa Theologica of St. Thomas Aquinas*. Translated by the Fathers of the English Dominican Province. New York: Benziger Bros.

Aristotle. 2004. *Physics, or Natural Hearing*. Translated by Glen Coughlin. South Bend: St. Augustine's Press.

———. 1977. *Politics*. Translated by H. Rackham. Cambridge, MA: Harvard University Press.

Benedict XVI. 2006. "Regensburg Address." Accessed May 27, 2013. http://www.vatican.va/holy_father/benedict_xvi/speeches/2006/september/documents/hf_ben-xvi_spe_20060912_university-regensburg_en.html.

———. 2007. "Message for the World Day of Peace." January 1. Accessed May 27, 2013. http://www.vatican.va/holy_father/benedict_xvi/messages/peace/documents/hf_ben-xvi_mes_20061208_xl-world-day-peace_en.html.

———. 2008a. "Christmas Greetings to the Members of the Roman Curia." Accessed May 27, 2013. http://www.vatican.va/holy_father/benedict_xvi/speeches/2008/december/documents/hf_ben-xvi_spe_20081222_curia-romana_en.html.

———. 2008b. "Lecture at La Sapienza University." Accessed May 27, 2013. http://www.vatican.va/holy_father/benedict_xvi/speeches/2008/january/documents/hf_ben-xvi_spe_20080117_la-sapienza_en.html.

———. 2008c. "Meeting with the Bishops of the United States of America: Responses to Questions." Accessed May 27, 2013. http://www.vatican.va/holy_father/benedict_xvi/speeches/2008/april/documents/hf_ben-xvi_spe_20080416_response-bishops_en.html.

———. 2009. Encyclical Letter *Caritas in Veritate*. Accessed May 27, 2013. http://www.vatican.va/holy_father/benedict_xvi/encyclicals/documents/hf_ben-xvi_enc_20090629_caritas-in-veritate_en.html.

———. 2010. "Message for the World Day of Peace." January 1. Accessed May 27, 2013. http://www.vatican.va/holy_father/benedict_xvi/messages/peace/documents/hf_ben-xvi_mes_20091208_xliii-world-day-peace_en.html.

———. 2011a. "Address to the Six New Ambassadors to the Holy See." Accessed May 27, 2013. http://www.vatican.va/holy_father/benedict_xvi/speeches/2011/june/documents/hf_ben-xvi_spe_20110609_ambassadors_en.html.

———. 2011b. "Bundestag Address." Accessed May 27, 2013. http://www.vatican.va/holy_father/benedict_xvi/speeches/2011/september/documents/hf_ben-xvi_spe_20110922_reichstag-berlin_en.html.

———. 2011c. "Message to the Brazilian Bishops for the 2011 Brotherhood Campaign." Accessed May 27, 2013. http://www.vatican.va/holy_father/benedict_xvi/messages/pont-messages/2011/documents/hf_ben-xvi_mes_20110216_fraternita-2011_en.html.

Berry, Wendell. 1996. *The Unsettling of America*. San Francisco: Sierra Club Books.

De Koninck, Charles. 2009. *The Writings of Charles De Koninck*. Vol. 2. Edited and translated by Ralph McInerny. Notre Dame: University of Notre Dame Press.

Haught, John. 1990. "Religious and Cosmic Homelessness: Some Environmental Implications." In *Liberating Life: Contemporary Approaches to Ecological Theology*, edited by Charles Birch, William Eakin, and Jay B. MacDaniel. Maryknoll NY: Orbis Books.

Hütter, Reinhard. 2005. "Intellect and Will in the Encyclical *Fides et ratio* and in Thomas Aquinas." *Nova et vetera*, English edition, 3: 591-96.

International Theological Commission. 2008. *The Search for Universal Ethics: A New Look at Natural Law*. Translated by Joseph Bolin. Accessed May 27, 2013. http://www.pathsoflove.com/universal-ethics-natural-law.html.

John Paul II. 1990. "Message for World Day of Peace." January 1. Accessed May 27, 2013. http://www.vatican.va/holy_father/john_paul_ii/messages/peace/documents/hf_jp-ii_mes_19891208_xxiii-world-day-for-peace_en.html.

———. 1991. *Centesimus Annus*. Accessed May 27, 2013. http://www.vatican.va/holy_father/john_paul_ii/encyclicals/documents/hf_jp-ii_enc_01051991_centesimus-annus_en.html.

McFague, Sallie. 1993. *The Body of God: An Ecological Theology*. Minneapolis MN: Fortress Press.

Pius XI. 1929. Encyclical Letter *Divini illius magistri*. Accessed May 27, 2013. http://www.vatican.va/holy_father/pius_xi/encyclicals/documents/hf_p-xi_enc_31121929_divini-illius-magistri_en.html.

Porter, Jean. 2005. "The Common Good in Thomas Aquinas." In *In Search of the Common Good*, edited by Patrick D. Miller and Dennis P. McCann, 94-120. New York: T & T Clark.

———. 2005. *Nature as Reason: A Thomistic Theory of the Natural Law*. Grand Rapids: W. B. Eerdmans Publishing Co.

Ratzinger, Joseph Cardinal. 2000. *The Spirit of the Liturgy*. San Francisco: Ignatius Press.

———. 2008. "Theology and the Church's Political Stance." In Ratzinger, *Church, Ecumenism, and Politics: New Essays in Ecclesiology*. San Francisco: Ignatius Press.

Schaefer, Jame. 2010. "Environmental Degradation, Social Sin, and the Common Good." In *God, Creation, and Climate Change: A Catholic Response to the Environmental Crisis*, edited by Richard W. Miller, 69-94. Maryknoll NY: Orbis Books.

———. 2009. *Theological Foundations for Environmental Ethics: Reconstructing Patristic and Medieval Concepts*. Washington DC: Georgetown University Press.

Vatican II. 1965. Pastoral Constitution on the Church in Modern World [*Gaudium et Spes*]. Accessed May 27, 2013. http://www.vatican.va/archive/hist_councils/ii_vatican_council/documents/vat-ii_const_19651207_gaudium-et-spes_en.html.

White, Lynn, Jr. 1967. "The Historical Roots of Our Ecologic Crisis." *Science* 155(3767): 1203-1207.

Wynn, Mark. 2010. "Thomas Aquinas: Reading the Idea of Dominion in the Light of the Doctrine of Creation." In *Ecological Hermeneutics: Biblical, Historical, and Theological Perspectives*, edited by David G. Horrell, et al. New York: T & T Clark.

FIVE

Commodifying Creation?

Pope Benedict XVI's Vision of the Goods of Creation Intended for All

Christiana Z. Peppard

Building upon the premise of anthropogenic changes to earth systems, especially climate, this essay focuses on the significance of Catholic social teaching (CST) with regard to the central principle of the universal destination of the goods of creation and the concept of fundamental human rights. Particularly significant at the current moment are air and water, two compounds that are necessary for life on earth. Gleaning key insights from papal encyclicals and magisterial documents from 1967 to the present, this essay argues that the papal stance is prophetic and challenging in its specification that some goods are so vital as to transcend market value and in its assertion that these are *right to life* issues. These insights require that industrialized nations such as the United States fulfill various duties but, too often, such duties are unpalatable in the American context. If taken seriously, however, those teachings reframe right to life issues and require changes to current economic systems.

FACTS AND VALUES IN THE ANTHROPOCENE

In the long sweep of historical memory, the late twentieth and early twenty-first centuries will be characterized by several major, interrelated realities. Among them are globalization (especially its economic, technological, and cultural manifestations) and environmental degradation. Never before has global, plural humanity's cumulative impact on the environment been so evident. Never before has humanity been so capable of recognizing our own complex forms of direct and indirect interconnection. Moral agency and moral responsibility take on new textures in such a milieu.

Perhaps the most prominent measure of climate change, at least in popular consciousness, is global warming. In a recent summation by climate advocate Bill McKibben (2012) —who wrote the first book on global warming, *The End of Nature,* in 1989—the heat is on. The temperature of the planet has increased by "just under 0.8 degrees Celsius, and that has caused far more damage than most scientists expected." Permanent snow pack and ice, especially in the polar regions, is rapidly depleting; permafrost is melting across northern latitudes, releasing massive amounts of the greenhouse gas, methane (CH_4).

The most prominent measure of climate change assesses atmospheric carbon dioxide (CO_2) saturation in relation to warming temperatures. Currently, global atmospheric CO_2 is at 390 parts per million (ppm). Recent projected levels by the National Center for Atmospheric Research suggest that if present trends and carbon emissions continue, the CO_2 concentration could reach up to at least 550 ppm and even beyond (possibly to 1100 ppm) by 2100. The catch is that "recent constructions of atmospheric CO_2 concentrations through history indicate that it has been ~30 to 100 million years since this concentration existed in the atmosphere," when temperatures were radically higher (Kiehl 2011, 158-59).

These numbers are not just data points on scientists' graphs. They mean something big. Science policy analyst Joe Romm (2012) puts the matter starkly: a temperature increase of seven degrees Farenheit "would devastate civilization." What might this look like, precisely? Romm summarizes available climate research and reports that, if present trends continue unabated, effects will include:

Staggeringly high temperature rise, especially over land—some 10°F over much of the United States;
Permanent Dust Bowl conditions over the U.S. Southwest and many other regions around the globe that are heavily populated and/or heavily farmed;
Sea level rise of some 1 foot by 2050, then 4 to 6 feet (or more) by 2100, rising some 6 to 12 inches (or more) each decade thereafter;
Massive species loss on land and sea—perhaps 50 percent or more of all biodiversity;
Unexpected impacts—the fearsome "unknown unknowns";
Much more extreme weather;
Food insecurity—the increasing difficulty of feeding 7 billion, then 8 billion, and then 9 billion people in a world with an ever-worsening climate;
Myriad direct health impacts.

He exhorts people to "[r]emember, these will all be happening simultaneously and getting worse decade after decade."

These changes are human-induced, tightly correlated to fossil fuel use, and will become irreversible. Thus, with a rather grave optimism Romm (2012) claims that for a very limited amount of time, humanity can make some changes in this trajectory of devastation unleashed by global climate change. It also means that "inaction is the gravest threat humanity faces."[1] The notion that climate change is a grave threat is not just a consensus of environmental scientists. A report to the Pentagon warned in November 2012 that "that the security establishment is going to have to start planning for natural disasters, sea-level rise, drought, epidemics and other consequences of climate change" (Goldberg 2012).

Industrial society profligately belches carbon, with developed nations such as the United States at the helm of emissions, while India and China close the gap through their ongoing industrializing processes. Meanwhile, climate effects develop in an ever-intensifying feedback cycle. Soon the possibility of even slowing this trend—much less stopping or reversing it—will be impossible. Effects are already being felt; these will diversify and intensify, leading to devastating human and ecosystemic consequences. This is not new information. As John Paul II (1990, #6) observed in his Message for the World Day of Peace:

> The gradual depletion of the ozone layer and the related "greenhouse effect" has now reached crisis proportions as a consequence of industrial growth, massive urban concentrations and vastly increased energy needs. Industrial waste, the burning of fossil fuels, unrestricted deforestation, the use of certain types of herbicides, coolants and propellants: all of these are known to harm the atmosphere and environment. The resulting meteorological and atmospheric changes range from damage to health to the possible future submersion of low-lying lands. While in some cases the damage already done may well be irreversible, in many other cases it can still be halted. It is necessary, however, that the entire human community–individuals, States and international bodies - take seriously the responsibility that is theirs.

Yet global climate treaties have been shunned by the United States. How do Americans, in particular, begin to incorporate the reality of climate change into common consciousness? One scholar has suggested that, "for adults, the task is one of relearning. For children, the task is one of building the foundation for future lifelong learning" (Bateson 2007, 282). For everyone, reality must be reframed.

What's in a Name?

At the turn of the twenty-first century, the Nobel-prize winning atmospheric chemist Paul Crutzen coined the term "Anthropocene," referring to—in his words—the "geology of mankind," a geological epoch shaped decisively by global human impacts, the origins of which can be traced to the early Industrial Revolution and are ongoing (Crutzen and Stoemer 2000). While many geological and tectonic dynamics persist independent of human influence, Crutzen's point is that for the first time in the history of the earth—indeed, the universe—human beings are now a prominent, decisive force that co-determines the earth's environmental state; furthermore, some of these changes may be permanent. Thus the term Anthropocene links the human prefix—"anthropos"—with the standard geological suffix as a way of denoting large-scale, human-caused alterations in earth systems.

Dimensions of anthropogenic change include shorter-term impacts on geomorphology (through urbanization, for example) and longer-term impacts on biology (biodiversity loss, for example) and earth chemistry (notably, climate change and its multiple impacts, such as sea level rise, ocean acidification, and more). Climate change is an umbrella term that is attested in a range of ways by experts across a wide range of environmental sciences; these experts also identify human activities—particularly in the past 300 years—as the primary cause of climate change.

In the official lingo of the Geological Time Scale (GTS), the current era is the late Holocene. It is within this framework that Crutzen and others have proposed adding the Anthropocene, an era of massive, anthropogenic changes to earth systems. In 2008, the Geological Society of London—a longstanding, august body whose historical membership list includes the likes of Charles Darwin—affirmed, "by a large majority, that there was merit in considering the possible formulation of this term [the Anthropocene]: that is, that it might eventually join the Cambrian, Jurassic, Pleistocene, and other such units on the [GTS]" (Zalasiewicz et al. 2010, 2228). Such a proposal is not without its challenges or its critics. Still, Crutzen and others (Zalasiewicz et al. 2010, 2231) claim that, "the Anthropocene represents a new phase in the history of both humankind and of the Earth, when natural forces and human forces became intertwined, so that the fate of one determines the fate of the other. Geologically, this is a remarkable episode in this history of this planet."

The term Anthropocene has been adopted by those who recognize descriptive and prescriptive significance of anthropogenic changes to the earth's systems. In 2011, the *Economist* summarized the impact of the neologism:

> It is one of those moments where a scientific realisation, like Copernicus grasping that the Earth goes round the sun, could fundamentally change people's view of things far beyond science. It means more than rewriting some textbooks. It means thinking afresh about the relationship between people and their world and acting accordingly ("Welcome to the Anthropocene").

A crucial impetus for adopting this language is not just descriptive but prescriptive. It signals an ethical concern about potentially negative effects of environmental disruptions. The trajectory of global environmental degradation, and climate change in particular, most significantly impacts the world's poorest people—precisely those who have, for the most part, neither caused these disruptions nor benefited from the industrial processes that led to them. They are also the least able to mobilize resources to respond effectively. The film *Sun Come Up* testifies to this reality in the Carteret Islands of Papua New Guinea.

Similarly, Bill McKibben (2012) invokes the plight of lowland villagers in Bangladesh, whose lands are flooded and freshwater sources contaminated by seawater intrusion and who are not only losing the basis of their economic vitality but also experiencing epidemics of dengue fever as a consequence of climate change.[2] It was when McKibben himself caught dengue fever in Bangladesh and found himself weak and thrashing with fever on a hospital cot that he realized the profound injustice entailed in the sea level's rise. If McKibben survived, he could leave Bangladesh and its deteriorating coastal economy. His companions in the sick ward could not.

Environmental historians such as J.R. McNeill (2001) have observed how, since the Industrial Revolution, people living in poverty bear the biggest burden of environmental changes. Crutzen and colleagues (Zalasiewicz et al. 2010, 2231) indicate that "the present and likely future course of environmental change seems set to create substantially more losers, globally, than winners." Thus environmental philosopher and winner of the 2003 Templeton Prize,[3] Holmes Rolston III (2006, xi), suggests that we *Homo sapiens* are "the one species in the history of the planet that, now in this new millennium, has more power than ever for good or evil, or justice and injustice; indeed, the one species that puts both its own well-being and that of life on Earth in jeopardy."

There is much at stake in climate change. The facts of environmental degradation are value-laden, for they result from actions premised upon certain kinds of values. They may also be fodder for articulating alternative values that necessitate revised courses of action. And it is here that the Anthropocene presents quandaries for consideration in moral and ethical theories. Among other things, for example, it remains to be seen how traditional act- and intention-based moralities may grapple with moral anthropology, and especially the concept of responsibility, in a world characterized by anthropogenic climate change. Indeed, it would be a mistake to assume that the questions prompted by anthropogenic climate change are not properly moral. Whether direct or indirect, proximate or remote, intended or not, human agency and responsibility in the Anthropocene have decidedly different textures than in previous epochs.

The Anthropocene is not a fringe idea conceived for dissemination by pantheistic tree-huggers. Nor is the fact of anthropogenic climate change in doubt for most scientists and thoughtful global citizens. To the great credit of one of humanity's longest-standing institutions, the Roman Catholic Church is itself well within the Anthropocene's conceptual and moral gambit: Paul Crutzen has, since 1996, been among the elite scientists who constitute the Pontifical Academy of Sciences.

Skepticism, Science and Theology

In the United States, conversations about climate change often dissolve into facile skepticism or agnosticism. This dissolution is motivated by corporate and political interests and abetted by American scientific illiteracy, to be sure, but it glimmers with the veneer of respectability because of the idea of uncertainty. If there is uncertainty in climate science—so goes the argument—then it is permissible to be a climate agnostic or even a climate denier. This is an error. To be sure, uncertainty is a part of climate science, but not in the way that climate deniers depict. In the real world, it is more difficult to assess the dynamics of open systems (with their infinity of variables) than it is to measure chemical values in a closed, laboratory system, where variables can be reduced or controlled entirely.

Here, uncertainty is by no means the same thing as error. Uncertainty is part of the quest for knowledge and as such is part of both scientific and theological inquiry. It is inquiry towards the unknown (or, more accurately, the partially known)—the refinement of our questions, methods, discoveries and new questions—that characterize human reason at its best (Firestein 2012). Hence our best inquiries in both science and theology are borne out of a humble uncertainty, an awareness that more needs to be observed or said better, a cognizance that our current perception and knowledge are limited but still reliable.

The scientific data and consensus on climate change are robust. Still, some uncertainties remain. Precisely how warm will it get in the coming century? What will happen if we get to 550 ppm or higher? How high will oceans rise, in what places? Which models offer the best predictions of future trends? These are live, unanswered questions. But in neither science nor theology does ongoing discernment justify inaction or inertia.

In what follows, I first discuss the climate/water nexus. I then summarize the documentary heritage and significance of the idea of the common destination of the goods of creation within CST, ultimately specifying how this principle manifests as a critique of economic practice and a defense of the fundamental human right to environmental goods. I conclude by highlighting four major challenges to the incorporation of CST on economy and environment in the American context.

THE CLIMATE-WATER NEXUS

How climate change looks depends, in part, on where one looks. Climate change is a global reality, but its manifestations are always contextual. They include: increased frequency of severe weather events; rising seawaters; sustained droughts in arid regions; shifts in bird species' migratory patterns; and much more. Still, some general truisms hold. One such maxim is that *there is a profound relationship between climate change and water*. In the words of the National Center for Atmospheric Research (2013):

> Changing climate will directly affect the global hydrologic cycle. Many of these effects will be felt regionally, with, for example, potential for flooding or drought increasing. In addition, changes to water quality, quantity, and supply reliability may have effects on human health, aquatic ecosystems, and agricultural and energy production, among other ecosystems and economic sectors.

It is essential to note water is one of the few things that human beings quite literally cannot live without. Water is both *sine qua non* and largely *sui generis* for human (indeed, biotic) life: that is, it is a baseline of human existence, and it is non-substitutable. Human beings will continue to find ways to purify water, but it is unlikely that we will find a workable replacement for it.

For this essay, several aspects of the relationship between climate and hydrology are particularly noteworthy. First, greenhouse gases amplify global warming. Atmospheric water vapor is a greenhouse gas. In 2008, NASA satellite data confirmed that "warming and water absorption increase in a spiraling cycle" (Hansen 2008). That is, water vapor amplifies temperatures because it holds more heat; moreover, "water vapor feedback can also amplify the warming effect of other greenhouse gases" such as CO_2. Climate models had long predicted this relationship; satellite data confirmed the models' accuracy.

Second, climate change alters the availability of surface water in predictable ways. Surface water constitutes less than one percent of all fresh water on earth, but it is still all of the water we have come to see and trust in lakes, rivers, seasonal gullies, and so forth. The catch with climate change is that, as temperatures rise, water evaporates into water vapor. The impacts—in general—mean that wet places will continue to get wetter, and dry places drier. This drives home the point of ecologist Travis Huxman, who suggests that, "water is the hammer [with which] climate change will hit the earth" (Hull 2009).

In both wet and arid regions, however, agriculture will be particularly hard hit. Established patterns of agriculture will cease to yield standard harvests. The Midwestern "breadbasket" of the United States experienced the latter with respect to corn crops during the summer of 2012. Of course the impacts are not limited to the United States. Catholic Relief Services, along with a constellation of scientists, announced in mid-October 2012 that "higher temperatures and erratic rainfall will likely threaten beans and corn in Central American nations" (Stipe 2012). The term bandied about by hydrologists is *megadrought:* drought conditions that persist for decades.[4] The catch is that decades-long drought is not so much an aberration, then, as a new kind of normal. Worldwide, increased temperatures may result in large areas becoming untenable for planting key subsistence crops.[5]

Finally, a third inflection of the climate-water nexus has to do with the negative effects of industrial processes and pollutants on air and water quality. This concern traces back to Rachel Carson's 1962 observation, in her trenchant book *Silent Spring,* of the deleterious effects of industrial chemicals. As Carson and many others since have observed, synthetic chemical compounds as well as an intensification of natural compounds in the air and water can have profound toxicological and public health consequences. Past minimal thresholds, many of those compounds are toxic to human and biotic life and are especially concentrated and pernicious in the development of fetuses and infants.

The remainder of this essay depicts a powerful way of considering air and water: the principle of the universal destination of created goods. It identifies papal critiques of, and recommended limitations on, economic practice with regard to these vital goods; and it demonstrates how the magisterium views the defense of environmental goods such as air and water as fundamental human rights and, indeed, even right-to-life issues. Papal teachings on air and water in an era of climate change pivot upon a value-laden fulcrum of faith, facts, and future generations. Such claims call for a radical re-assessment of American policy, economic practice, and rhetoric.

The Universal Destination of Created Goods from *Gaudium et Spes* to *Caritas in Veritate*

The concept of the universal destination of the goods of creation was first articulated in the Vatican II document *Gaudium et Spes.* It indicates that the materials of the earth are not meant for the gain of privileged individuals or groups at the expense of others.[6] Rather, created goods are for the benefit of all, including future generations.

> God intended the earth with everything contained in it for the use of all human beings and peoples. Thus, under the leadership of justice and in the company of charity, created goods should be in abundance for all in like manner. Whatever the forms of property may be, as adapted to the legitimate institutions of peoples, according to diverse and changeable circumstances, attention must always be paid to this universal destination of earthly goods. In using them, therefore, man should regard the

> external things that he legitimately possesses not only as his own but also as common in the sense that they should be able to benefit not only him but also others. On the other hand, the right of having a share of earthly goods sufficient for oneself and one's family belongs to everyone. The Fathers and Doctors of the Church held this opinion, teaching that men are obliged to come to the relief of the poor and to do so not merely out of their superfluous goods. If one is in extreme necessity, he has the right to procure for himself what he needs out of the riches of others. . . . Furthermore, it is the right of public authority to prevent anyone from abusing his private property to the detriment of the common good. . . . By its very nature private property has a social quality which is based on the law of the common destination of earthly goods (#69, 71).

Important, normative commitments reside here. Created (or "natural") goods are intended by God for the benefit of all, regardless of specific arrangements of political economy. Legitimate forms of private property must also be oriented for the benefit of others, which in situations of duress means that less privileged persons can take from the property of others (a position classically defined by Thomas Aquinas). Furthermore, public authority should prevent individuals from using private property in ways that injure the common good. Indeed, private property itself is derivative from "the law of the common destination of earthly goods."

Subsequent papal encyclicals have advanced these themes in various ways. Paul VI, in *Populorum progressio* (1967), observed that an ascendant emphasis on economic development should not imperil the livelihoods or integral development of peoples. His 1971 encyclical, *Octogesima adveniens,* articulated clearly the threat of environmental degradation as a result of incautious technological exploitation. One passage (#21) in particular resonates with the idea of the Anthropocene described previously:

> [A]nother transformation is making itself felt, one which is the dramatic and unexpected consequence of human activity. Man is suddenly becoming aware that by an ill-considered exploitation of nature he risks destroying it and becoming in his turn the victim of this degradation. Not only is the material environment becoming a permanent menace–pollution and refuse, new illness and absolute destructive capacity–but the human framework is no longer under man's control, thus creating an environment for tomorrow which may well be intolerable. This is a wide-ranging social problem which concerns the entire human family.

The Christian must turn to these new perceptions in order to take on responsibility, together with the rest of men, for a destiny which from now on is shared by all.

The 1970s and 1980s saw an intensification of concern in western, industrial societies about international development and interdependence in light of economic globalization, growing wealth disparities, and environmental degradation. Thus John Paul II (1987, #39), in his encyclical *Sollicitudo rei socialis,* wrote that "interdependence must be transformed into solidarity, based upon the principle that the goods of creation are meant for all." Strong words are reserved for countries like the United States (which, while unmentioned by name, fits squarely into his depiction):

> Surmounting every type of imperialism and determination to preserve their own hegemony, the stronger and richer nations must have a sense of moral responsibility for the other nations, so that a real international system may be established which will rest on the foundation of the equality of all peoples and on the necessary respect for their legitimate differences (#41). . . . It is necessary to state once more the characteristic principle of Christian social doctrine: the goods of this world are originally meant for all. The right to private property is valid and necessary, but it does not nullify the value of this principle. Private property, in fact, is under a "social mortgage," which means that it has an intrinsically social function, based upon and justified precisely by the principle of the universal destination of goods (#42). . . . God intended the earth with everything contained in it for the use of all human beings and peoples. Thus, under the leadership of justice and in the company of charity, created goods should be in abundance for all in like manner. Whatever the forms of property may be, as adapted to the legitimate institutions of peoples, according to diverse and changeable circumstances, attention must always be paid to this universal destination of earthly goods (#47).

Four years later, in *Centesimus Annus,* John Paul II (1991, #37) would again invoke ecological degradation, this time emphasizing the problem of consumptive economic patterns:

> Equally worrying is *the ecological question* which accompanies the problem of consumerism and which is closely connected to it. In his desire to have and to enjoy rather than to be and to grow, man consumes the resources of the earth and his own life in an excessive and disordered way. At the root of the senseless destruction of the natural environment lies an anthropological error, which unfortunately is widespread in our day. Man, who discovers his capacity to transform and in a certain sense create the world through his own work, forgets that this is always based on God's prior and original gift of the things that are. Man thinks that he can make arbitrary use of the earth, subjecting it without restraint to his will, as though it did not have its own requisites and a prior God-given purpose, which man can indeed develop but must not betray. Instead of carrying out his role as a co-operator with God in the work of creation, man sets himself up in place of God and thus ends up provoking a rebellion on the part of nature, which is more tyrannized than governed by him. . . . [H]umanity today must be conscious of its duties and obligations towards future generations.

Moreover, "It is the task of the State to provide for the defence and preservation of common goods such as the natural and human environments, which cannot be safeguarded simply by market forces" (John Paul II 1991, #40). And his next insight is particularly challenging to contemporary American ears:

> Just as in the time of primitive capitalism the State had the duty of defending the basic rights of workers, so now, with the new capitalism, the State and all of society have the duty of *defending those collective goods* which, among others, constitute the essential framework for the legitimate pursuit of personal goals on the part of each individual. . . .

Here we find a new limit on the market: there are collective and qualitative needs which cannot be satisfied by market mechanisms. There are important human needs which escape its logic. There are goods which by their very nature cannot and must not be bought or sold. Certainly the mechanisms of the market offer secure advantages: they help to utilize resources better; they promote the exchange of products; above all they give central place to the person's desires and preferences, which, in a contract, meet the desires and preferences of another person. Nevertheless, these mechanisms carry the risk of an "idolatry" of the market, an idolatry which ignores the existence of goods which by their nature are not and cannot be mere commodities.

These are strong insights and moral invectives. They call into question the lionization of private property and economic exchange, suggesting instead that some goods are so essential as to transcend both market value and private property regimes.

Reception, Redaction and the Bias of Privilege

What have Americans made of these teachings? On the whole, they have not been robustly received in the American context. To be sure, the USCCB's documents, *Economic Justice For All* (1986), *Renewing the Earth* (1991), and *Global Climate Change: A Plea for Prudence and the Common Good* (2001) have sought to remind American Catholics about the prophetic content of those economic and environmental insights. The work of John Carr, formerly at the USCCB, as well as the efforts of Dan Misleh and colleagues at the Catholic Coalition on Climate Change, have been crucial in retaining and spreading magisterial insights regarding climate change and justice. Still, it is far from clear that pontifical teachings on the inherent link between economy and environment—including the proper role of private property and the powers of the state in overseeing essential goods—have been popularly received. As the US Catholic Bishops wrote in 2001:

> Much of the debate on global climate change seems polarized and partisan. Science is too often used as a weapon, not as a source of wisdom.

Various interests use the airwaves and political process to minimize or exaggerate the challenges we face. The search for the common good and the voices of poor people and poor countries sometimes are neglected.

That this situation persists is a great tragedy. It may be postulated that it is also the result of selective attention on the part of many Americans.

Reception or Redaction?

The reception history of Catholic social encyclicals in the United States is varied, especially in the past few decades as the economy and environment have increasingly come into papal focus. Many Americans do not read encyclicals themselves but rely on prominent interpreters who have wide readerships but engage a kind of selective reading—some might call it "cafeteria Catholicism"—that sidelines key insights from papal teachings since 1965. This is most evident with regard to papal biographer George Weigel's (2009) now-famous redaction criticism of Benedict XVI's social encyclical, *Caritas in Veritate*.

Caritas in Veritate commemorated *Populorum progressio* and was promulgated after the start of the U.S. economic crisis as effects were rippling throughout the global economy. Its scope and overall content have been well described elsewhere. Notable for this essay is Chapter Four of *Caritas in Veritate,* where Benedict XVI (2009, #50) writes extensively about rights, duties, and the natural environment.

> Today the subject of development is also closely related to the duties arising from *our relationship to the natural environment*. The environment is God's gift to everyone, and in our use of it we have a responsibility towards the poor, towards future generations and towards humanity as a whole (#48). . . . This responsibility is a global one, for it is concerned not just with energy but with the whole of creation, which must not be bequeathed to future generations depleted of its resources. . . . It is likewise incumbent upon the competent authorities to make every effort to ensure that the economic and social costs of using up shared environmental resources are recognized with transparency and fully borne by those who incur them, not by other peoples or future generations: the protection of the environment, of resources and of the climate obliges all international leaders to act jointly and to show a readiness to work in good faith, respecting the law and promoting solidarity with the weakest regions of the planet. One of the greatest challenges facing the economy is to achieve the most efficient use—not abuse—of natural resources, based on a realization that the notion of "efficiency" is not value-free.

The wellbeing of the environment and its preservation for future generations is inseparable from patterns of economic globalization, as Paul VI saw nascently in 1967 and as many papal documents have since observed. At stake are several crucial, interlocking ethical issues: the value of the environment, the distribution of vital resources, and the operative economic framework.

What is the value of the environment? Benedict's invocation of "efficiency" is telling. Efficiency may be economically expedient, but it is not neutral. It is a value shaped by the priorities of the market in an era of late capitalism; and as such, it "is not value-free." It prizes profit over equity; it values shareholders over stakeholders. And there are environmental costs built into capitalist modes of production that are not sufficiently accounted for in economic terms. The value of the environment cannot be reduced to economic efficiency or, indeed, to economic mechanisms. Cost is a much broader term, for Benedict XVI, than price or market value. Thus in his 2010 Message for the World Day of Peace, Benedict XVI (2010, #8) explained that environmental degradation has a cost that is unjustly distributed:

> Future generations cannot be saddled with the cost of our use of common environmental resources. . . . Universal solidarity represents a benefit as well as a duty. *This is a responsibility that present generations have towards those of the future,* a responsibility that also concerns individual States and the international community. Natural resources should be used in such a way that immediate benefits do not have a negative impact on living creatures, human and not, present and future; that the protection of private property does not conflict with the universal destination of goods; that human activity does not compromise the fruitfulness of the earth, for the benefit of people now and in the future.

Indeed, the guiding principle for Benedict XVI is the universal destination of created goods, for present and future generations. The priorities and principles are clear—or so it would seem.

In 2009, Weigel published a redaction criticism of *Caritas in Veritate* in which he claimed to distinguish between what the Pope himself must have written and what a "*gauchiste*," benighted—and yet somehow pernicious—"peace and justice contingent" must have slipped into the document. He denoted the former, "Benedictine" contributions as written with a "gold pen," and the latter as "red pen" revisions by the *gauchistes* (Weigel 2009).[7]

Such a claim suggests that Weigel thinks that he knows better than the Pope what it is that the Pope actually meant to say. That is one issue. Most problematic is that, in its overt categorization of what is authentically papal and authoritative, versus what is communal and questionable, Weigel's interpretation is blotted by a series of misrepresentations, ambiguities, and errors. It does not accord with authentic Catholic teaching on essential points. In other words, Weigel's widely disseminated interpretations of *Caritas in Veritate* problematically distort the papal message.

Unfortunately, such interpretations successfully capitalize on well-honed American allergies to ideas of redistribution, government ownership and oversight of public goods. To be sure, the complexities of economic structures and environmental degradation are mind-boggling. It is easier—both pragmatically and psychologically—to focus on other things. However, such an evasion edits out the well-established Catholic vision of the nexus of environment, economics and justice. Papal teachings on the goods of creation may sound difficult or jarring to American ears. That is perhaps the strongest confirmation of their prophetic nature in the realm of political economy.

Catholic social encyclicals since 1967 contain incisive, sustained critiques of materialism and imbalanced patterns of global consumption. These encyclicals critique patterns of privilege and access that benefit the few at the expense of many others. Some, such as *Caritas in Veritate,* also contain pragmatic elements—for example, the insistence that developed nations must engage in technology transfer, the requirement that externalities be internalized, or the defense of air, water and food as fundamental human rights.

Much of the burden falls in the realm of economic practice. Both the 1990 and 2010 Messages for the World Day of Peace (by John Paul II and Benedict XVI, respectively), emphasized the importance of maintaining the goods of creation intended for all. Both identified industrialized nations as particularly responsible for curbing consumerism and enabling international solidarity. And in line with previous papal encyclicals, the *Compendium of the Social Doctrine of the Church* (2004, #470) insists that there are baseline requirements for economic practice and arrangements of political economy:

> *An economy respectful of the environment will not have the maximization of profits as its only objective, because environmental protection cannot be assured solely on the basis of financial calculations of costs and benefits.* The environment is one of those goods that cannot be adequately safeguarded or promoted by market forces. Every country, in particular developed countries, must be aware of the urgent obligation to reconsider the way that natural goods are being used. Seeking innovative ways to reduce the environmental impact of production and consumption of goods should be effectively encouraged.

Particular attention will have to be reserved for the complex issues surrounding *energy resources*. Non-renewable resources, which highly-industrialized and recently-industrialized countries draw from, must be put at the service of all humanity.

The *Compendium* (#466) adds that key goods must be valued "beyond the strict market mentality." This is a particular concern with regard to water resources.

Water as a Revelatory Locus in Catholic Social Teaching

That air and water epitomize the goods of creation meant for the benefit of all has been stated in a range of contexts within CST over the past three decades. For example, in 1971, the Synod of Bishops' statement, *Justice in the World,* suggested that "the precious treasures of air and water—without which there cannot be life—and the small delicate biosphere of the whole complex of all life on earth, are not infinite, but on the contrary must be saved and preserved as a unique patrimony belonging to all human beings" (#8). In an address to the diplomatic corps, John Paul II (2003, #4) described how "the problem of water resources" provides a prime example of the universal destination of created goods, for "all peoples are entitled to receive a fair share of the goods of this world and of the know-how of the more advanced countries."

Water, in particular, is *sui generis* and a *sine qua non* of human existence. It is a key resource that is only partially renewable.[8] *Caritas in Veritate* pulls no punches when it comes to the universal destination of the goods of creation and the resulting claim that air and water are fundamental human rights. This claim has come, in part, from the context of the rampant commodification of fresh water.

For the past two decades, a debate has been raging—often out of sight—about whether water should be viewed primarily as a commodity or as a human right. A commodity is something that can be owned, traded, and sold for profit. To describe water as a commodity is therefore to suggest that fresh water falls primarily under the aegis of the global market economy. (Bottled water is an obvious example; there are other examples of "privatization," in which water is treated as a means to profit.)

A human right, by contrast, is a fundamental entitlement due to all people, regardless of ability to pay. The language of rights suggests that equity is far more important than profit; that the commodification of fresh water leads to inequities in distribution; and, therefore, that fresh water should not be valued primarily as an economic good. Many people further insist that fresh water should be considered as a public good or as part of the "global commons."

The Catholic Church has pontificated on the value of fresh water every third year since 2003 in letters from the Delegation of the Holy See to the World Water Forum. Thus fresh water provides a particularly good example of the Church's critique of commodification of these universal, fundamental goods. In particular, the Pontifical Council for Justice and Peace (PCJP)—on behalf of the Holy See—reiterates time and time again that water is precisely a "common good of humankind" under the principle of the universal destination of the goods of creation: "the few, with the means to control, cannot destroy or exhaust this resource, which is destined for the use of all" (PCJP 2006). It maintains the centrality of justice and especially the preferential option for the poor: "The primary objective of all efforts must be the well-being of those people—men, women, children, families, communities—who live in the poorest parts of the world and suffer most from any scarcity or misuse of water resources." Furthermore, according to the *Compendium* (2004, #485), the right to water stands in contrast to "any merely quantitative assessment that considers water as a merely economic good."

These are morally rich insights. Notions such as exchange value or price are paltry approximations of the complex, multifaceted reality that is fresh water. Furthermore, there is a potentially insidious quality to the predominance of market logic. As Harvard philosopher Michael Sandel (2012a, 2012b) has recently observed, "[m]arkets leave their mark. Sometimes, market values crowd out nonmarket values worth caring about." The principle of the universal destination of the goods of creation sits uneasily with the logic of commodity, with some forms of private property, and with other common assumptions in the United States. Profit, growth, efficiency, and exchange value are insufficient and potentially dangerous frameworks when it comes to the fundamental good, the fundamental right, indeed the "right to life issue" of fresh water.

FOUR KEY CHALLENGES TO CATHOLIC CLIMATE DISCOURSE IN THE U.S. CONTEXT

The first section of this essay demonstrated how human responsibility for environmental degradation is particularly evident in the Anthropocene, and it depicted key aspects of the climate-water nexus. The second section depicted how air and water are considered by Catholic teaching to be fundamental goods of creation, that is, goods that are intended by God for the benefit of the entire human family, which therefore are fundamental human rights that resist the dominant logic of economic exchange. By way of conclusion, I specify four aspects of this nexus that seem particularly challenging for American ears to hear.

First and foremost, the theological and moral principle of the universal destination of the goods of creation is oriented toward the flourishing of all, both present and future generations. This principle is in direct tension with the growth-and-profit orientation of contemporary economic systems, within which governments value growth in GDP, and corporations value profit to shareholders but not to current or future stakeholders. In the current framework, growth and profit are lionized over and against the protection of fundamental public goods such as air and water.[9] This is a political, economic and philosophical problem that requires radical re-evaluation. The Church attempts to make this point through the assertion of universal human rights to air and water, as well as critiques of economic practice. But much more work is needed on the definition and implementation of the category of global public goods.

Second, Catholic teachings on climate change and justice challenge the U.S. context because a certain type of "cafeteria Catholicism" seems to redact economic critiques out of papal vocabulary and attempts to divide matters of dogma from matters of ethics. Yet the Catholic Church has long held to the unity of *fides et mores.* As John Paul II proclaimed to Catholics in 1990 (#15), the ecological crisis is a moral issue: "Christians in particular realize that their responsibility within Creation and their duty towards nature and the Creator are an essential part of their faith."

Third, Catholic teachings on climate change challenge comfortable American assumptions regarding "right to life" issues. Magisterial documents are clear that rights to air and water are part of the fundamental right to life—indeed, in the words of the Pontifical Council for Justice and Peace, fresh water is a "right to life issue." This is powerful language in an American context that consistently equates right-to-life rhetoric with the single issue of abortion.

Fourth, Catholic teachings on climate change in the Anthropocene are progressive and challenging. They suggest that we are all in this climate mess together, a basic insight that contradicts treasured, national self-understandings—such as American individualism, exceptionalism, free-markets, private property, and rights conferred by power and privilege. Indeed, according to Catholic Social Teaching, super-developed nations—most prominently, the United States—bear responsibility for mitigating some of the worst effects, particularly in the realms of air and water, in ways that may not conduce to maximizing profit or generating economic growth.

CONCLUSION

If every pope from John XXIII to Benedict XVI were to jointly articulate a maxim for the intersection of environment, ethics, and economics, it would be *the universal destination of created goods.* This principle is a pivot point of faith, facts, and future generations.

The amplification of global climate change—particularly its manifestations in air and water—demonstrates how we are consuming and altering the world in dramatic and even permanent ways. The stakes are very high and the shape of the future is uncertain. But these daunting levels of complexity and uncertainty are not excuses for skepticism or agnosticism. For, as Benedict XVI (2009, #51) remarked in *Caritas in Veritate*: "The Church has a responsibility towards creation, and she must assert this responsibility in the public sphere. In so doing, she must defend not only the earth, water, and air as gifts of creation that belong to everyone. She must above all protect mankind from self-destruction." Such is the ethical challenge presented by the Anthropocene.

NOTES

1. Romm's 2007 book on climate change was titled *Hell and High Water* (New York: William Morrow); he views terms like global warming and climate change as palliative euphemisms that distort the clear, present, and urgent danger.

2. "Climate refugees" is now a widely recognized term, as indicated by a recent documentary of that name and used in reference to persons displaced by the effects of climate change—including, most recently, people in coastal New York and New Jersey in the aftermath of Hurricane Sandy.

3. The Templeton Prize for Progress Toward Research or Discoveries about Spiritual Realities.

4. See the discussion of this concept and its significance for the United States in deBuys (2011).

5. For a discussion of this topic in a theological vein, see Miller (2010, 9-13).

6. The following treatment of the Goods of Creation is adapted from Peppard (2012, 325-52). That article lays out an eight-point framework for considering fresh water through CST; as I suggest therein, it may also be adapted towards other environmental goods.

7. For example: "But then there are those passages to be marked in red—the passages that reflect Justice and Peace ideas and approaches that Benedict evidently believed he had to try and accommodate."

8. The issue of renewability is complex and context-dependent. For the purposes of this essay, I consider fresh water and air to be non-renewable because, once used, they are not necessarily subject to recapture and reuse in the same form or in the desired location. Neither are they amenable to replacement by other substances.

9. See the important work of Speth (2008 and 2012).

REFERENCES

Bateson, Mary Catherine. 2007. "Education for Global Responsibility." In *Creating a Climate for Change: Communicating Climate Change and Facilitating Social Change*, edited by Susanne Moser and Lisa Dilling, 281-91. New York: Cambridge University Press.

Benedict XVI. 2009. *Caritas in Veritate*. Encyclical Letter on Integral Human Development in Charity and Truth, June 29. Accessed May 27, 2013. http://www.vatican.va/holy_father/benedict_xvi/encyclicals/documents/hf_ben-xvi_enc_20090629_caritas-in-veritate_en.html.

———. 2010. "If You Want to Cultivate Peace, Protect Creation." Message for the World Day of Peace, January 1. Accessed May 27, 2013. http://www.vatican.va/holy_father/benedict_xvi/messages/peace/documents/hf_ben-xvi_mes_20091208_xliii-world-day-peace_en.html#_edn6.

Carson, Rachel, Lois Darling, and Louis Darling. 1962. *Silent Spring*. Boston: Houghton Mifflin. Top of Form

Catholic Church. 1965. *Gaudium et Spes*. Pastoral Constitution on the Church in the Modern World, December 7. Accessed May 27, 2013. http://www.vatican.va/archive/hist_councils/ii_vatican_council/documents/vat-ii_const_19651207_gaudium-et-spes_en.html.

———. 2004. *Compendium of the Social Doctrine of the Church*. Cittá del Vaticano: Libreria Editrice Vaticana.

Crutzen, Paul J. and Eugene F. Stoemer. 2000. "The 'Anthropocene.'" *IGPD Newsletter* 41:17-18. Accessed June 29, 2013. http://www.igbp.net/download/18.316f18321323470177580001401/NL41.pdf.

deBuys, William. 2011. *A Great Aridness: Climate Change and the Future of the American Southwest*. New York: Oxford University Press.

Firestein, Stuart. 2012. *Ignorance*. New York: Oxford University Press.

Goldberg, Suzanne. 2012. "US Military Warned to Prepare for Consequences of Climate Change." *The Guardian*. November 9. Accessed May 27, 2013. http://www.guardian.co.uk/world/2012/nov/09/us-military-warned-climate-change.

Hansen, Kathryn. 2008. "Water Vapor Confirmed as Major Player in Climate Change." National Aeronautics and Space Administration. November 17. Accessed May 27, 2013. http://www.nasa.gov/topics/earth/features/vapor_warming.html.

Hull, Tim. 2009. "Science Under Glass." *High Country News* 41.9 (May 18): http://www.hcn.org/issues/41.9/science-under-glass. Accessed May 27, 2013.

John Paul II. 1987. *Sollicitudo rei socialis*. Encyclical Letter for the Twentieth Anniversary of *Populorum progressio*, December 30. Accessed May 27, 2013.http://www.vatican.va/holy_father/john_paul_ii/encyclicals/documents/hf_jp-ii_enc_30121987_sollicitudo-rei-socialis_en.html.

———. 1989. "Peace With God the Creator, Peace With All of Creation." Message for the World Day of Peace, January 1. Accessed May 27, 2013. http://www.vatican.va/holy_father/john_paul_ii/messages/peace/documents/hf_jp-ii_mes_19891208_xxiii-world-day-for-peace_en.html.

———. 1991. *Centesimus Annus*. Encyclical Letter on the Hundredth Anniversary of *Rerum Novarum*. May 1. Accessed May 27, 2013. http://www.vatican.va/holy_father/john_paul_ii/encyclicals/documents/hf_jp-ii_enc_01051991_centesimus-annus_en.html.

———. 2003. Address to the Diplomatic Corps. January 13. Accessed May 27, 2013. http://www.vatican.va/holy_father/john_paul_ii/speeches/2003/january/documents/hf_jp-ii_spe_20030113_diplomatic-corps_en.html

Kiehl, Jeffrey. 2011. "Climate Change: Lessons from Earth's Past." *Science* 221 (6014): 158-59.

McKibben, Bill. 2012. "Global Warming's Terrifying New Math." *Rolling Stone*, July 19. http://www.rollingstone.com/politics/news/global-warmings-terrifying-new-math-20120719.

McNeill, J.R. 2001. *Something New Under the Sun: An Environmental History of the 20th-Century World*. New York: W.W. Norton.

Miller, Richard W. 2010. "Global Climate Disruption and Social Justice: the State of the Problem." In *God, Creation, and Climate Change: A Catholic Response to the Environmental Crisis*, edited by Richard W. Miller, 9-13. Maryknoll NY: Orbis.

National Center for Atmospheric Research. 2013. "Taking a Systemic Look at Characteristics of the Global Hydrologic Cycle."Accessed May 27. http://ncar.ucar.edu/press/taking-a-systemic-look-at-characteristics-of-the-global-hydrologic-cycle.

Paul VI. 1967. *Populorum progressio*. Encyclical Letter on the Development of Peoples, March 26. Accessed May 27, 2013. http://www.vatican.va/holy_father/paul_vi/encyclicals/documents/hf_p-vi_enc_26031967_populorum_en.html.

———. 1971. *Octogesima adveniens*. Apostolic Letter, May 14. Accessed May 27, 2013. http://www.vatican.va/holy_father/paul_vi/apost_letters/documents/hf_p-vi_apl_19710514_octogesima-adveniens_en.html

Peppard, Christiana Z. 2012. "Fresh Water and Catholic Social Teaching: A Vital Nexus." *Journal of Catholic Social Thought* 9.2: 325-352.

Pontifical Council for Justice and Peace. 2006. "Water, an Essential Element For Life: An Update." A Contribution of the Holy See to the Fourth World Water Forum, March 16-22. Accessed May 27, 2013. http://www.vatican.va/roman_curia/pontifical_councils/justpeace/documents/rc_pc_justpeace_doc_20060322_mexico-water_en.html.

Rolston III, Holmes. 2006. *Science and Religion: A Critical Survey*, 20th Anniversary Edition. Philadelphia: Templeton Foundation Press.

Romm, Joseph J. 2007. *Hell and High Water: Global Warming–the Solution and the Politics — and What We Should Do*. New York, NY: William Morrow.

———. 2012. "An Illustrated Guide to the Science of Global Warming Impacts." *Climate Progress Blog*, October 14. Accessed May 27, 2013. http://thinkprogress.org/climate/2012/10/14/1009121/science-of-global-warming-impacts-guide/?mobile=nc.

Sandel, Michael. 2012a. "What Isn't for Sale?" *The Atlantic*, April 2. http://www.theatlantic.com/magazine/archive/2012/04/what-isn-8217-t-for-sale/8902/.

———. 2012b. *What Money Can't Buy: The Moral Limits of Markets*. New York: Farrar, Strauss and Giroux.

Speth, James Gustave. 2008. *The Bridge at the End of the World: Capitalism, the Environment, and Crossing from Crisis to Sustainability*. New Haven: Yale University Press.

———. 2012. *America the Possible: Manifesto for a New Economy*. New Haven: Yale University Press.

Stipe, Jim. 2012. "Report: Climate Change Threatens One Million Maize and Bean Farmers in Central America." *Catholic Relief Services Newswire*, October 9. Accessed May 27, 2013. http://newswire.crs.org/report-climate-change-in-central-american-threatens-one-million-maize-and-bean-farmers/.

Synod of Catholic Bishops. 1971. "Justice in the World." Accessed May 27, 2013. http://www.osjspm.org/document.doc?id=69.

United States Conference of Catholic Bishops. 1986. "Economic Justice For All." Pastoral Letter on Catholic Social Teaching and the U.S. Economy, November. Accessed May 27, 2013. http://www.usccb.org/upload/economic_justice_for_all.pdf.

———. 1991. "Renewing the Earth." An Invitation to Reflection and Action on Environment in Light of Catholic Social Teaching. November 14. Accessed May 27, 2013. http://www.usccb.org/issues-and-action/human-life-and-dignity/environment/renewing-the-earth.cfm.

———. 2001. "Global Climate Change: A Plea for Dialogue Prudence and the Common Good." Accessed May 27, 2013. http://www.usccb.org/issues-and-action/human-life-and-dignity/environment/global-climate-change-a-plea-for-dialogue-prudence-and-the-common-good.cfm.

Weigel, George. 2009. "*Caritas in Veritate* in Gold and Red." *National Review*, July 7. http://www.nationalreview.com/articles/227839/i-caritas-veritate-i-gold-and-red/george-weigel?pg=1.

"Welcome to the Anthropocene." 2011. *The Economist*, May 26. http://www.economist.com/node/18744401.

Zalasiewicz, Jan, Mark Williams, Will Steffen, and Paul Crutzen. 2010. "The New World of the Anthropocene." *Environmental Science and Technology* 44.7: 2228-31.

SIX

The Grammar of Creation

Agriculture in the Thought of Pope Benedict XVI

Matthew Philipp Whelan

In his 2009 encyclical, *Caritas in Veritate,* Pope Benedict XVI (2009a, #48) writes that "[nature] is a wondrous work of the Creator containing a 'grammar' which sets forth ends and criteria for its wise use."[1] This is a striking formulation, upon which many have helpfully commented.[2] This essay addresses the issue of creation's grammar, as well as the ends and criteria for its wise use, by concentrating on the practice of agriculture. It suggests that how we understand agriculture intimately relates to how we understand the created order and ourselves as creatures. What is the grammar of creation? How to read it rightly? What does it entail for the practice of agriculture? These are the central questions this essay explores.[3]

To speak of creation's grammar is to speak of creation as a kind of language—or as previous generations of Christian's have spoken of it, as a kind of book—which has a given structure.[4] In the *Philosophical Investigations,* when Ludwig Wittgenstein asks his readers to consider the grammar of an expression, he is asking them to attend to the logic of their language, the rules that govern the way they speak. Wittgenstein (2009, #496) writes: "Grammar does not tell us how language must be constructed in order to fulfill its purpose, in order to have such-and-such an effect on human beings. It only describes, and in no way explains, the use of signs." The therapy Wittgenstein offers is, among other things, an attempt to re-attune us to the givenness of our world, to remind us of our position in relation to the language we speak, and to disabuse us, in

Fergus Kerr's (1997, 132) words of "the inclination to isolate the effects that the world has on [us] prior to [our] having the public words, and prior to the world's being worded . . . as if [we] alone, untutored, self-reliant, lacking history and kin, could possess the world, untrodden, uninterpreted, yielding itself to [us] alone."

Benedict's approach to the grammar of creation in *Caritas in Veritate* (2009a) seeks to do something similar. He offers no constructive proposal for the refashioning of our language—the response of so many to Lynn White's indictment of Christianity in his 1967 article, "The Historical Roots of Our Ecologic Crisis"—which would be akin to an agricultural technician advising that the farmer abandon previous agricultural practice, raze the land of local vegetation, and modernize. Rather than refashion Christian teaching on creation, Benedict proposes the opposite: he seeks to remind us what the doctrine of creation entails for our relationship to the world, to one another, and to God, and to trace the transformations of these relationships, especially in modernity. According to Benedict (2009a, #34), to attend to the grammar of creation is therefore to attend to the language, and, following Benedict we could even say the *logos*—Christians are given about creation. This language is a gift, which places us before, as he writes in *Caritas in Veritate,* "the astonishing experience of gift" creation itself. This is his therapy for how we learn to see that God's love is both the redemptive love through which we are recreated in Christ and the creative love through all creatures have being in the first place.

HUNGER AND HUMAN NEED

In *Caritas in Veritate,* Benedict (2009a, #27) first addresses agriculture in the context of a discussion of hunger. Benedict begins that section by addressing the way "*hunger* still reaps enormous numbers of victims" and that feeding the hungry "is an ethical imperative for the universal Church, as she responds to the teaching of her Founder, the Lord Jesus, concerning solidarity and the sharing of goods." The scriptural text upon which Benedict meditates is Matthew 25:31-46, in which Christ not only calls his followers to feed the hungry and perform the other corporeal works of mercy, but in which Christ identifies with the hungry, binding his body and his voice to theirs. The hunger of the hungry is the hunger of Christ. Commenting on the same scriptural text, St. Augustine (1959, #239.6) says in one of his Easter sermons: "Though rich, He is in need until the end of the world."

In this same section of *Caritas in Veritate,* Benedict (2009a, #27) discusses in more detail the alleviation of hunger, why hunger persists, and the role of agriculture in addressing it. But here at the outset, it is important to comment on what is often presupposed in *Caritas in Veritate,* as well as in Benedict's other writings on the topic: an intimate association between agriculture and hunger. Discussion of one seamlessly leads into discussion of the other. This is because the need for food is among the most elemental of human needs, and it is deep within the tradition of Catholic Social Doctrine that, in Pope Pius XII's words, economy is ordered to the "lasting satisfaction of needs" (quoted in Calvez and Perrin, 1961, 175). Commenting on this tradition, Calvez and Perrin (174-75) write that economy "begins" with human need; it is the fundamental human reality on to which all economic and social activity are "grafted." Human beings are needy creatures—a need that bespeaks embodiment and dependence upon the wider created order for sustenance. Need therefore functions, in a fundamental sense, as "the standard and measuring rod of the economy."

Benedict follows Catholic thought more generally, then, not only in his understanding of the way "economy" is embedded within a wider created order, but also in his acknowledgment of the unique place of agriculture within economy. We see this, for instance, in Benedict's claim in his message commemorating World Food Day in 2006 that "not enough attention is given to the needs of agriculture, and this both upsets the natural order of creation and compromises respect for human dignity." Throughout these messages—which he delivered each year of his pontificate—Benedict underscores the centrality of agriculture to any consideration of economy, often lamenting, "the inadequate attention given to the agricultural sector and the effects this has on rural communities." He says in his 2011 message that agriculture "must not be seen as a secondary activity but as the focus of every strategy of growth and integral development." He even argues that "the rural family needs to regain its rightful place as the heart of the social order."

THE COMMON GIFT OF CREATION

Benedict's association of hunger and agriculture, as well his comments upon the imperative of the Church and her members to respond with solidarity and the sharing of goods to her founder's ongoing hunger in the hunger of others, can be helpfully illumined by Catholic teaching on creation as a common gift, given to meet the needs of all. Pope John Paul II (1991, #31) draws on that teaching when he writes in *Centesimus Annus*:

> God gave the earth to the whole human race for the sustenance of all its members, without excluding or favoring anyone. This is *the foundation of the universal destination of the earth's goods.* The earth, by reason of its fruitfulness and its capacity to satisfy human needs, is God's first gift for the sustenance of human life. But the earth does not yield its fruits without a particular human response to God's gift, that is to say, without work.

In *Laborem exercens*, John Paul (1981, #19) calls this understanding of created goods "the first principle of the whole ethical and social order." In *Sollicitudo rei socialis* (1987, #42), he calls it "the characteristic principle of Christian social doctrine." That the *Catechism of the Catholic Church* (1995, #2401-6) locates this teaching as a commentary on the Seventh Commandment, "You shall not steal," serves to remind us that the Church understands stealing not simply in terms of the property arrangements secured by positive law but primarily in terms of the failure to acknowledge the gift of the earth to the whole of humankind.

This understanding of creation as common gift led figures like St. Basil, St. Ambrose, and many others, to read land and its fertility as bearing the trace of the purpose for which it was created. For instance, in the homily, "I Will Tear Down My Barns," Basil (2009, #1, 3) not only states that God gives "fertile soil, temperate weather, plenty of seeds, cooperation of the animals, and whatever else is required for successful cultivation," but he also exhorts his congregants, "Imitate the earth … Bear fruit as it does; do not show yourself inferior to inanimate soil. After all, the earth does not nurture fruit for its own enjoyment but for your benefit." In *On Naboth*, Ambrose (1997, #7.37) likewise tells his congregants to imitate the way "the soil brings forth a richer yield than it received." At the heart of the Church's teaching on the common destination of created goods is the claim in *Gaudium et Spes* (Second Vatican Council 1965, #69) that we should regard the "external things" that we "legitimately possess" not only as our "own" but also as "common," in the sense that these things should benefit not only ourselves but "others" as well. Commonality is not an extrinsic imposition on an understanding of property that is most fundamentally private. The Church's teaching is that commonality is intrinsic to property as such, to what it means for property to be property at all. The daily bread for which we pray is always ours (Mt. 6:11). According to *Gaudium et Spes* (#71), property "by its very nature . . . has a social quality." Property is not, as we are so often encultured to think, the enclosure of what is common for our own exclusive use. Rather, it is a way of using what is common that, in a fundamental sense, seeks to preserve it as common.

The commonality intrinsic to property can be further discerned in the claim that, when some have goods in excess of their needs, the goods cannot properly be said to belong to them. Instead, the goods are most properly said to belong to those who need them. That is why St. Thomas Aquinas (1975) argues in the *Summa theologiae* 2|2.66.7 that "according to natural law goods that are held in superabundance by some people should be used for the maintenance of the poor."[5] Thomas proceeds to cite these words from Basil: "The bread you are holding back is for the hungry, the clothes you keep put away are for the naked, the shoes that are rotting away with disuse are for those who have none, the silver you keep buried in the earth is for the needy. You are thus guilty of injustice toward as many as you might have aided, and did not." Basil's words are not hyperbole; they are an accurate moral description given the account of we have been examining. This understanding of creation as common gift animates Pope John Paul's (1979) condemnation of those who "keep unproductive lands that hide the bread that so many families lack" in his address to indigenous communities in Cuilapan, Mexico.

John Paul's words also indicate that the Church's teaching on the common destination of created goods entails agrarian reform. In its discussion of the common destination of created goods, *Gaudium et Spes* (1965, #71) condemns situations in which "there are large or even extensive rural estates which are only slightly cultivated or lie completely idle for the sake of profit, while the majority of the people either are without land or have only very small fields" and calls for the "distribution" of these "insufficiently cultivated estates." Benedict (2009a, #27) reiterates this call in *Caritas in Veritate* when he writes, "the question of equitable agrarian reform . . . should not be ignored."[6]

In a beautiful little book entitled *Apostolic Farming*, Catherine Doherty (1991) attempts to articulate the Church's teaching on the common destination of created goods as it applies to the practice of agriculture. Doherty was foundress of the Madonna House Apostolate and its farm, St. Benedict's Acres, in Combermere, Ontario. Doherty (39-40) calls the farming she advocates "apostolic farming," which she defines in the following way: "I see the apostolic farmer as one who seeks to make a farm richly productive by using the simplest means . . . who feeds his brothers and sisters at the least possible expense, so that the money saved might go to other places who need it." She continues (46): "Holy poverty should be a constant meditation. We must continually ask ourselves, 'How can I do without this? How can I substitute something less expensive?"

Doherty's comments resonate with the way that Benedict, when addressing the question of the so-called ecological crisis, frequently speaks of the need for new ways of living that are more modest, that consume less, and that give more to others.[7] At a meeting with the clergy of the diocese of Bolzano-Bressanone, for instance, Benedict (2008a) says: "It is not just a question of finding techniques that can prevent environmental harms, even if it's important to find alternative sources of energy and so on. But all this won't be enough if we ourselves don't find a new style of life, a discipline which is made up in part of renunciations: a discipline of recognition of others, to whom creation belongs just as much as those of us who can make use of it more easily." The way of life about which Benedict speaks is one that must learn renunciation. But such renunciation finds its proper location within the affirmation of creation as common gift. Renunciation is a crucial part of the *askēsis* whereby we learn to open others' access to what belongs to them but what has been closed off to them by injustice.

SOLIDARITY

We have been examining Benedict's association of agriculture and hunger in relation to the common gift of creation. Because economy is ordered to the satisfaction of need, the persistence of hunger is a scandal, the antithesis of the Church's teaching on the common gift of creation. It does not follow, however, that any and all forms of economic production and organization for the satisfaction of need are valid. This is because, as Calvez and Perrin argue (1961, 179-80), "[w]e cannot satisfy man's needs without satisfying his need for liberty, the needs of his free personality taken its totality." Human dignity and freedom, in other words, are intrinsic to the satisfaction of needs. The need for food must be understood in light of the needs of the whole human person.[8]

This is why Benedict refuses to reduce the hungry to units of bare biological need—as if those who live in surfeit could simply find ways to slake their hunger and leave it at that, as if economy achieved its purpose once all elemental needs were met. As mentioned above, Benedict understands economy to be embedded within a wider created order. But he also understands the God who creates the world to be the God who redeems it in Christ. Underlying and informing even the root sense of economy as "household management" (*oikonomia*) is its meaning as the Lord's dealing with the world (cf. Eph. 1:8-11, 3:2-11). In the mysterious economy of God's salvation, the hunger of the hungry person is the hunger of Christ, which is why the Church must respond to her founder not just by sharing of goods but with solidarity.

In this regard, in *Caritas in Veritate*, Benedict (2009a, #27) makes the very important comment, "[H]unger still reaps enormous numbers of victims among those who, like Lazarus, are not permitted to take their place at the rich man's table." The reference here is of course to the Gospel of Luke, in which a beggar, Lazarus, lies at the gate of a rich man who fails to offer Lazarus even the scraps from his table. According to Benedict—and here he is explicitly commenting on Pope Paul VI's encyclical *Populorum progressio*—the scandal is not simply that Lazarus lies, hungry and covered with sores, at the gate of the rich man, or that the voice of the Lord is unheeded when the rich man fails to toss Lazarus even scraps from the table. According to Benedict (2009a, #27), as well as to Paul (cf. *Populorum progressio* 1967, #47), the scandal is especially that Lazarus is not permitted to take his place at the table, and that they do not eat together. Benedict (2007a, 33) thinks that conventional wisdom when it comes to these matters assumes the world of Bertold Brecht, who famously said, "Grub first, then ethics." Brecht's words, however, reflect a mindset profoundly inimical to the Christian faith—so much so that Benedict even sees them exemplified in Satan's first temptation Jesus: "If you are the Son of God, command these stones to become loaves of bread" (Mt 4:1-3, cf. Lk 4:1-3). According to this view, bread is what is most fundamental. The rest comes later. Satan's temptation resides precisely in the way he urges acquiescence to this ordering of goods—an acquiescence we see, according to Benedict, not only in Marxism but also in much Western aid to so-called developing nations. The continuity between Marxist and liberal-democratic states reside precisely in the way both regard God "as a secondary matter that can be set aside temporarily or permanently on account of more important things."[9]

To be clear, Benedict in no way means to deny the centrality of bread, much less to suggest corporeal works of mercy like feeding the hungry are not constitutive to the Church's very life. The Church, Benedict (2010a) insists, is "constantly at work" through her own members "to alleviate the poverty and deprivation afflicting large parts of the world's population." Indeed, she must be, for, in the words of Augustine, the head of her body is in need until the end of the world. Continuing to suffer from hunger, thirst, inhospitality, nakedness, sickness, and imprisonment in the bodies of others is how Christ feeds, slakes, welcomes, clothes, and visits the members of his ecclesial body. Benedict's point is simply that the need for food must be understood in light of the needs of the whole person. To neglect the whole person in the attempt to secure food for all does not result in, as he writes in *Jesus of Nazareth* (2007a, 33), "justice or concern for human suffering," but in "ruin and destruction" of people and of material goods themselves.

Benedict (2007c) thinks that so many in our world remain hungry because our efforts to alleviate hunger "tend to be solely and principally motivated by technical and economic considerations, forgetting the primary, ethical dimension of 'feeding the hungry'" and, as he indicates in his 2008 World Food Day Message "the meaning of the human person." What is most essential, what Benedict (2007c) is most concerned to cultivate, is the "compassion and solidarity proper to the human being, which includes sharing with others not only material goods, but also the love which all need."[10] "Solidarity," Benedict (2006) claims, "is the key to identifying and eliminating the causes of poverty and underdevelopment." Benedict (2012) emphasizes agricultural cooperatives precisely because of their potential to foster such solidarity. They offer us "a different way of economic and social organization," "an alternative vision to that determined by . . . measures which seem to have as their sole objective profit, the defense of markets, the use of agricultural products for ends other than food, and the introduction of new techniques without the necessary precautions." Cooperatives remind us that solidarity does not just come after economic activity but can take hold of it from within.

Turning now to a closer examination of Benedict's remarks on agriculture, this last is a particularly salient point: solidarity does not come into play only after food is produced. The way we produce food, as well as what we do with it, enacts or refuses solidarity in complex ways. A striking instance of this can be seen in the comments of two Kansas farmers in a recent cover story in *Harper's*, entitled, "Broken Heartland: The Looming Collapse of Agriculture on the Great Plains" (Hylton 2012). An exchange between the farmers, Jack Geiger and Donn Teske, suggests some of the ways the agriculture that presently dominates the Plains limits the possibilities of solidarity. Geiger says, "It used to be when one family was struggling, all the other families would help them out. But now, if somebody's in trouble, everybody else is looking to see if they can buy their land." Teske then adds: "These days, you've got to grow to survive. It changes how people relate" (27). These comments help us see, among other things, that we misunderstand the nature of the technologies we employ if we regard them merely as instruments we use to shape the world. This is because we do not just use our technologies; at the same time our technologies can "use" and shape us as well. As Michael Hanby (2011, 200) puts it, technology embodies a way of "understanding the world, of reflecting it back upon ourselves through our industry, and therefore of being in it." The crucial questions therefore become: What kind of understanding of the created world is enacted in agricultural technologies that, instead of helping to open access to what belongs to others, put inordinate pressure upon farmers, in the words of the prophet Isaiah, to "join house to house" and "add field to field" (5:8)? As Isaiah warns, the end of this process is solitude: those who so join and add end up "alone in the midst of the land" (5:8). What kind of agricultural technologies help us better perceive and live into the communion that is at the heart of all created reality—a communion we glimpse when the Lazaruses of the world are not just fed but permitted to take their place at the table?

AGRICULTURE IN *CARITAS IN VERITATE*

Thus far, this essay has offered some general reflections upon the way Benedict raises the issue of agriculture in *Caritas in Veritate* in order to address the kind of agriculture necessary to attend to hunger and need. He does not leave his comments at this level of generality, however, but proceeds to describe in more detail the kind of agriculture he envisions.

Benedict (2009a, #27) immediately clarifies that one important mark of this agriculture is an emphasis on the local. Agriculture must make good use "of the human, natural, and socioeconomic resources that are more readily available at the local level," and it must involve "local communities in choices and decisions that affect the use of agricultural land." As noted above, one of Benedict's major concerns in *Caritas in Veritate* and elsewhere is the way "development" tends to name exclusively technical and economic considerations, which reduces and instrumentalizes the human being, and erodes solidarity. Benedict does not reject such considerations, but he urges that they be adapted to "the particular conditions of each country and each community" (2005), that they take into account "the cycles and rhythms of nature known to the inhabitants of rural areas" (2007c), and that they "encourage cooperation with a view to protecting the methods of cultivating the land proper to each region" (2009b).

Related to the priority given to the local, another mark of the agriculture Benedict envisions is the priority given to the long view. Against any kind of presentism or emphasis on short-term economic gains, Benedict (2009a, #27) writes of the crucial need for "a long-term perspective," which can guarantee "sustainability over the long term" in the way land is used. An entailment of this is that an agricultural economy—or any economy for that matter—should not be like fire, consuming that which nourishes it. The suggestion seems to be that alleviating hunger today, as well as in the future, depends upon good agricultural work, which strives to live from the gifts of the land, not the land itself. While Benedict does not write of industrial agriculture or agribusiness, it is not a stretch to see that he is calling for alternatives to an agriculture that damages the sources upon which it depends and models itself on the business enterprises he critiques elsewhere in *Caritas in Veritate*.[11]

The question of sustainability—a notoriously difficult notion[12]—might helpfully be located within the understanding of creation as common gift that we examined at the outset of this essay. Creation, we saw, is a gift given for the sustenance of all. Just as this understanding of creation as common gift profoundly unsettles commonplace understandings of what belongs to me and what belongs to others in the present, it must also unsettle that distinction across time. What kind of work—agricultural and otherwise—and what kind of uses of things acknowledge that commonality applies not just to the present time but unfolds temporally, such that my work and my uses of things should seek to preserve created goods as common, open to the claims of those who will need them in times to come?

The question of temporality is also important for other reasons as well. The farming families I came to know when I lived and worked in the Bribri and Cabécar Indigenous Territories in Talamanca, near the Costa Rica-Panama border, practiced many different kinds of agriculture: from slash-and-burn basic grain production for subsistence, to organic banana and cacao production in agroforests for export, to chemical-intensive monoculture plantain production for export. I was struck by the descriptions farmers often used to describe chemically-intensive production—a relatively new form of agriculture, such that many of the farmers who practiced it had started to do so only within their lifetimes. Many had ample experience with and memories of an agriculture that was not chemically-intensive. The observations of Severino Morales, a farmer from Shiroles, were typical: "The land is tired but it is forced to produce through the use of chemicals. Without chemicals, the land works at its own rhythm, on its own strength. It is faster with chemicals (*La tierra está cansada pero le fuerza producir a punto de químico. Sin químico, la tierra trabaja a su propio ritmo, bajo su propia fuerza. Es más rápido con químico*)."

My point here is not to disparage the use of synthetic agrochemicals or to forswear an agriculture that would make any use of them. It is simply to attend to Severino's significant observation that land has its own rhythm and strength—that it can, for instance, be exhausted and so need rest, and that the use of chemicals can compel land despite its exhaustion.[13] Attending to observations like Severino's seems to me precisely what Benedict (2007b) has in mind when he said, at a meeting with the clergy of the diocese of Belluno-Feltre and Treviso: "[W]e must respect the inner laws of creation, of this earth, we must learn these laws and obey these laws if we wish to survive. Consequently, this obedience to the voice of the earth, of being, is more important for our future happiness than the voices of the moment, the desires of the moment. In short, this is a first criterion to learn: that being itself, our earth, speaks to us and we must listen if we want to survive and to decipher this message of the earth."

Such talk might sound ridiculous—the quaint relic of a by-gone age. According to Benedict, however, the pervasiveness of this sense only indicates the pervasiveness of the problems in which we find ourselves mired. Speaking from my own experience, it is certainly the case that what typically passes for agricultural development has no time for what Severino or Benedict has to say. The speed Severino mentions, which is part and parcel not only of chemically-intensive agriculture but the cultures that live from it, is a powerful force, which, in practice, effortlessly disregards the need to make the sorts of discriminations about land and its condition that came second-nature to Severino. The most crucial temporal determinant for agricultural practice was not the land's time but the time of the intermediaries that will buy and sell the plantains the farmer produces, and the calendarized chemical applications necessary in order to meet export requirements. In the Territories, in other words, the cost of growing and selling plantains is the submission of the farm, both in its management and in its design, to a time that, as far as I could discern, had little stake in the flourishing either of farms or the families that farm them.

GENEALOGY OF MODERNITY

At this juncture, it is helpful to step back and consider certain aspects of the genealogy of modernity Benedict offers us both before and during his pontificate.[14] As noted above, Benedict associates the emergence of modernity with transformations in the doctrine of creation. For instance, in *In the Beginning*, Ratzinger (1995, 82-83) writes: "The obscuring of faith in creation . . . is closely connected with 'the spirit of modernity' . . . [T]he foundations of modernity are the reason for the disappearance of 'creation' from the horizons of historically influential thought." Examining certain key moments in this genealogy enables us to situate our topic within a larger historical trajectory.

To turn briefly to one such moment, Ratzinger sees a particularly significant transformation in the figure of Galileo, in whom there is a return to Plato's reflections on mathematics, but in such a way that, in Ratzinger's (1995, 84) words, "the knowledge of God is turned into the knowledge of the mathematical structures of nature." The concept of nature as the object of science begins to take the place of the concept of creation. It "conceals" creation, which increasingly begins to be understood exclusively as the object of science (92). Associated with these developments is the prominence of the subject-object schema, such that only the object as defined by natural science is really objective and able to produce true knowledge. God becomes "little more than the formal mathematical structures perceived by science in nature" (85).

In his 2007 encyclical *Spe salvi*, Benedict turns from Galileo to his contemporary, Francis Bacon, in order to examine other features of this new understanding of science and the knowledge it produces. Benedict (2007d, #16) quotes from Bacon's *Novum Organum* (2000, I.CXVII) to describe Bacon's project more generally: "We . . . stake the whole race on the victory of art over nature," a reversal of Aristotle's claim that art imitates nature (cf. Aristotle 1984, 2.2 194a23). Benedict regards the novelty of Bacon's approach to reside in his unique correlation between science and *praxis*, and the development of an approach to science in which science knows nature principally through controlling it.

John Briggs (1989, 142) describes Bacon's experimental procedure as the experimenter acting as the "inquisitorial judge" who examines "the accused"—nature—"about what it must have hidden." According to Briggs, we can discern an important feature of the character and ends of this science in Bacon's identification of natural matter with the figure Proteus, a servant of Neptune, the god of water and the sea in Roman mythology and religion. Proteus is not just servant but also prophet, giving up his knowledge only to those who can successfully capture him. But to do so entails seizing and holding him while he writhes and takes on various shapes. In *De sapientia veterum*, Bacon employs the figure of Proteus to describe this science's relation to matter:

> [I]f any one wanted his [Proteus's] help in any matter, the only way was to secure his hands with handcuffs, and then to bind him with chains . . . if any skillful Servant of Nature shall bring force to bear on matter, and shall vex it and drive it to extremities as if with the purpose of reducing it to nothing, then will matter . . . finding itself in these straits, turn and transform itself into strange shapes . . . when, if the

force be continued, it returns at last to itself (quoted in Briggs 1989, 33-34).

In other words, Bacon models his experimental procedure precisely on this "binding" of matter with "handcuffs" and "chains." The "prophecy" Bacon's science seeks is what it discerns through matter's writhing under duress, until, at last, matter gives up its secrets.[15]

In *Spe salvi*, Benedict (2007d) also notes a crucial theological implication of Bacon's project: its animation by the hope that this science will reestablish dominion over creation that was lost as a consequence of the fall. Bacon's science therefore subtly seeks to usurp the person and work of Christ, for it is constitutive to Christian theology that the recovery of what was lost through the fall can only be recovered through faith in Christ.[16] According to Benedict, Bacon's science and its relocation of Christian hope has profoundly shaped the modern world, especially the way in which humankind is increasingly seen to be the real creator of its projects and its destiny. When the most determinative truth about human history is the march of progress, it becomes nearly impossible to distinguish between what is possible to do and what is moral to do. "The technically possible, the desire to do and the actual doing of what it is possible to do," Ratzinger (1995, 66-69) writes, is like a "magnet" to which we seem "involuntarily attracted." In such a world, humankind need not concern itself with "merely dominating or shaping nature; now [it] will transform nature itself."

Benedict thinks that this is the heart of the issue. What Hanby (2011, 204, 202) calls "modernity's conflation of nature and artifice," in which science increasingly understands nature "through the *act* of making and unmaking it," offers us a world in which nature has no voice in human deliberation. Not only is nature not allowed to speak; it is not thought to have the capacity to speak in the first place. We therefore need not listen and, over time, we have increasingly grown deaf. We no longer have ears to hear what Albert Borgmann (1993, 51) calls "the eloquence of things." In a world of such deafness, nature is not understood to be the presupposition of human activity but rather its object, and it is the prerogative of humankind to manipulate nature in whatever it wants, because it is precisely in such manipulations that hope resides.

Ratzinger reads these developments in terms of resistance to human creatureliness—resistance to being dependent upon standards other than those given by humankind itself. Such dependence is understood to be an imposition—a burdensome constraint upon human freedom. But from the vantage of Christian faith, such resistance also reflects the desire to assume the position of the Creator, the consequences of which can only be damage, not only to the relationship between human creatures and their Creator, but to the relationships between human creatures themselves, as well as the rest of the created order. It is therefore no coincidence, Ratzinger (1995, 81) observes, that it is precisely at this juncture in human history that our "creations," agricultural and otherwise, no longer appear as our hope but threaten human survival, and that we suddenly see ourselves as "imperiled as never before" as we proceed to saw "off the branch on which [we] sit."

THE GRAMMAR OF CREATION

As we saw above, one important aspect of the transformation of the doctrine of creation in modernity is its replacement by the concept of nature, which is understood principally as the object of science. This is what Benedict calls the "concealment" of creation by the scientific concept of nature. Given the degree to which contemporary theological engagement with the so-called ecological crisis presupposes precisely this understanding of creation, it is important to examine how Benedict understands the relation of the human creature to the rest of the created order.

In *Caritas in Veritate*, Benedict (2009a, #48) addresses the issue directly. The official English translation of the encyclical renders the Latin terms "*rerum natura*," "*natura*," and "*creatio*" as "natural environment," "nature," and "creation," respectively. The encyclical tends to use these terms interchangeably. In common English parlance, however, these terms resonate differently, and so it is necessary to highlight two ways of understanding the relationship between human creatures and the created order that the encyclical forecloses.

When speaking of creation, we easily fall into an understanding of our relation to it as something which merely environs or surrounds us—something that is therefore separate from us, with which we have nothing in common. This has everything to do with the way that the scientific concept of nature renders the world a complex, interconnected set of objects that we stand apart from and gaze out upon. It is very difficult to speak of creation without being captured by this picture because it is so deeply rooted within our language.

Such an understanding of creation occasions two predominant contemporary options, which tend to conflict with one another in debates related to these matters: nature is, in Benedict's (2009a, #48) words, either an "untouchable taboo" that must be protected at all cost from any human interference or it is a natural recourse that humankind has every right to exploit as it sees fit; nature is either "something more important than the human person" or it is something over which we can aim for "total technical domination."

What is important to see is that for both these options, the underlying conception of the human creature is identical: the human creature is fundamentally separate from the rest of the created order, surveying it as a subject over a world of objects. But, as Benedict (2009a, #48) writes, "Neither attitude is consonant with the Christian vision of nature as the fruit of God's creation." This is because we do not exist outside of or apart from creation. As human creatures, our location is inside the membership of creation. To use Robert Sokolowski's (1982, 23) language, the fundamental "Christian distinction" is not between human beings and their environment but between creatures and their Creator.

None of this is to deny, of course, that as creatures that image God, we play a unique role within the created order.[17] It is simply to say that we play that role as creatures within creation. Therefore, when Benedict (2009a, #50) exhorts us to exercise *"responsible stewardship over nature"* and to strengthen the *"covenant between human beings and the environment,"* we do so as creatures, and so we do so from inside, not outside, creation. It is for this reason, as Benedict (2009a, #51; cf. #47) points out *"the way humanity treats the environment influences the way it treats itself, and vice versa,"* and *"when 'human ecology' is respected within society, environmental ecology also benefits."*

It is this same section of *Caritas in Veritate* that Benedict (2009a, #48) introduces the notion of creation's grammar and how to attend to it:

> Nature expresses a design of love and truth. It is prior to us, and it has been given to us by God as the setting of our life. Nature speaks to us of the Creator (cf. Rom. 1:20) and his love for humanity . . . Nature is at our disposal, not as "a heap of scattered refuse," but as a gift of the Creator, who has given it an inbuilt order, enabling man to draw from it the principles needed in order "to till it and keep it" (Gen. 2:15) . . . It is a wondrous work of the Creator containing a "grammar" [*grammaticam*] which sets for ends and criteria for its wise use, not reckless exploitation.

What does it mean to speak of the grammar of creation—a grammar from which we can draw the principles to till and to keep it—with regard to agriculture? This essay has already indicated some lines of response, especially with regard to the ends and criteria for good use.

In this last section we consider some important affinities between Benedict's approach to agriculture and an approach associated with botanist Albert Howard, plant geneticist Wes Jackson, entomologist Miguel Altieri, soil scientist Ana Primavesi, among many, many others.[18] These figures offer us a concrete exemplification of how agriculture might learn to discern the principles with which to till and to keep land in a way that combines, in Benedict's (2009a, #27) words, "traditional as well as innovative farming techniques." In *An Agricultural Testament*, Howard (1972, 1-5; cf. 2006, 17-32) argues that agriculture should model itself upon the ecology of the locale in which it finds itself. The art of agriculture must learn to imitate the nature of the place in which it is practiced. Farmers must study "the methods of nature," who is "the supreme farmer." To this end, Howard attempts to derive certain "rules" for such imitation from close observation of the forest and the way it farms:

> Mixed farming is the rule: plants are always found with animals: many species of plants and of animals all live together. In the forest every form of animal life, from mammals to the simplest invertebrates, occurs. The vegetable kingdom exhibits a similar range: there is never any attempt at monoculture: mixed crops and mixed farming are the rule. The soil is always protected from the direct action of sun, rain, and wind. In this care of the soil strict economy is the watchword: nothing is lost. The whole of the energy of sunlight is made use of by the foliage of the forest canopy and of the undergrowth. The leaves also break up the rainfall into fine spray so that it can the more easily be dealt with by the litter of plant and animal remains which provide the last line of defense of the precious soil. These methods of protection, so effective in dealing with sun and rain, also reduce the power of the strongest winds to a gentle air current.

Howard proceeds to make careful, detailed observations about the conservation of moisture in the upper layers of the soil, as well as the ways in which the forest "fertilizes" itself by the breakdown of local animal and vegetable materials, and by drawing the minerals needed from the subsoil.

A similar approach to agriculture exists in the Bribri and Cabécar Indigenous Territories, which were mentioned above. The Bribri and Cabécar are especially known for their cacao and banana agroforests. For those unfamiliar with the term, agroforestry refers to practices that involve the integration of trees and other woody perennials into farms through the conservation of existing trees or by the planting of new ones.[19] To the untrained eye, agroforests are almost indistinguishable from the surrounding forest. To a forester or an agronomist of certain training, they look "unclean," "primitive," and "inefficient." Because of the prevalence of trees, agroforests are a fairly permanent feature of the landscape in the Territories, especially at higher elevations, such as in the foothills of the Talamanca mountain range. And because of the shade from the canopy, agroforests cannot accommodate basic grains like maize and rice that need more sunlight. So, in addition to agroforests, most households have parcels of land under less permanent uses, especially basic grains.

What is interesting about even these less permanent uses when examined temporally is that, while they often begin as forest or fallow before moving to grains, they, too, tend to become agroforests. In other words, the pattern of crop succession is toward similitude to the forest. It is a crop succession in which, as Clifford Geertz (1963, 25) writes, "A natural forest is transformed into a harvestable forest." In explaining this practice, people would tell me they farmed "like the forest (*como el bosque*)"; that they planted "a little of everything (*un poco de todo*)" or "all mixed up (*revuelto*)," rather than a single stand of crops; and that the soil is "like the skin of a human being (*como la misma piel de uno*)," which is why it must be protected with tree cover.

We have been speaking about the forest because the ecology of the locale in which Howard as well as the Bribri and the Cabécar worked was forest. Wes Jackson, however, is from Kansas, and so he studies the prairie, whose dominant feature is diversity of perennials: grasses, legumes, sunflowers, and so on. Against the model of annual grain crops grown in monoculture, which leaves the soil uncovered for periods throughout the year—a major cause of the erosion that afflicts the agriculture of this region—the work of Jackson and his colleagues at the Land Institute attempt to model agriculture upon the prairie by breeding perennial grains that can be grown in polyculture, and so keep the soil covered year-round.

It seems to me suggestive to read this approach to agriculture as a complex discipline of learning to receive the locale in which it is placed and practiced as a gift—as its standard—to which it attends, and from which it draws the principles to till and to keep land. Learning to farm in this manner is simply what it means to learn to practice agriculture all, which Berry (2004, 143-44) defines as "proper use and care of an immeasurable gift." In line with Benedict's urging, this approach to agriculture also necessarily seeks to make good use of what is most readily available locally, and it involves local communities in the use of agricultural land.

This agriculture likewise employs traditional as well as innovative farming techniques. It is not against science or technology, but it is against the ways they are often wielded: without consideration of what it means to adapt them to the place and to the task at hand. For the crucial issue at stake is not the use of science and technology, but, in Hanby's (2011, 203) words, their "integration into a more comprehensive order of human knowing and making, apart from which they remain endemically reductionistic and dehumanizing." Ellen Davis (2009, 126) suggests exactly this in her description of the work of Wes Jackson and the Land Institute: "This is in one sense high-tech farming . . . But it is science and technology disciplined by humility. That is to say, new technologies are applied in accordance with the recognition that our ignorance how nature works vastly exceeds our knowledge."[20]

This approach to agriculture and its use of a science and technology disciplined by humility is, needless to say, profoundly out of step with predominant forms of agriculture, which are animated by, in Benedict's (2009a, #68) words, the "Promethean presumption" that shapes so much of our use of science and technology at present: the hope that humankind, as Bacon taught, can re-create itself through wonders of its own making. However, if Benedict is right, the approach to agriculture he traces in *Caritas in Veritate* and elsewhere in his writings is an agriculture that attends to creation's grammar. It is an agriculture animated by the humility (*humilitas*) that befits a creature made from humus and destined to return to it—a creature that Christ alone re-creates.

CONCLUSION

This essay has just scratched the surface of many complex matters, and it likely raises many more questions than it offers answers. There is admittedly much more needed to be said in order to fill out the lines of thought suggested here. For instance, what about the manifold ways creation does not manifestly display any design, order, or grammar? What about all the ways in which creation groans, for instance, in droughts and earthquakes (Rom 8:22)? What about all the damage that bad work only deepens, and the ecological uncertainty and discontinuity that we are told will only increase in years to come? What does all this mean for the kind of agriculture that Benedict advocates?

This essay has simply tried to do justice to Benedict's approach to agriculture. As he insistently reminds us, the grammar of creation is the grammar of gift, and so the first principle of an agriculture that would draw from creation the principles needed to till and to keep land must be the good use of a gift. For we encounter God's love not only in the redemptive love through which we are recreated but in the creative love through which all creatures live, move, and have their being. It is this encounter that leads Doherty (1991, 40) to describe agriculture as a "witness" to a love that "spills itself into the earth"—a love that fashions creation itself.

If we are to take Benedict seriously, the task before us involves much more than lobbying our governments for saner agricultural policies. We especially need witnesses. We not only need the witnesses of Talamancan agroforestry, Albert Howard, Wes Jackson, and so forth. We also need the witnesses of Doherty, Vincent McNabb, the Catholic Land Movement, Virgil Michel, Peter Maurin, Dorothy Day, the National Catholic Rural Life Conference, and countless others, who have sought to illumine the relationship between land and altar in a world that darkens it. But most of all, as Benedict (2008a) reminds the clergy of the diocese of Bolzano-Bressanone, and us as well, we must ourselves learn to "witness." We must "demonstrate with our example, with our own style of life, that we are speaking of a message in which we ourselves believe, one which it is possible to live."

NOTES

1. Many thanks to Natalie Carnes, Paul Griffiths, Nathan Eubank, Pete Jordan, Aaron Riches, Miguel Romero, Tobias Winright, Joseph Wolyniak, and William Whelan for their comments on earlier versions of this essay.

2. Cf. Cloutier 2010.

3. Unless otherwise indicated all magisterial documents were found at the Vatican's official website: www.vatican.va. For the sake of clarity, I use the title "Benedict XVI" or "Benedict" for writings since Cardinal Joseph Ratzinger's election and "Cardinal Ratzinger" or "Ratzinger" prior to it. The writings are listed accordingly in the bibliography.

4. Cf. Harrison 2001.

5. Cf. Speltz 1945; Pontifical Council for Justice and Peace 1995, ##171, 72-84.

6. I must confess that I am not sure why Benedict seems to limit the need for agrarian reform to "developing countries." There is nothing about the Church's teaching on the common destination of created goods that would limit it to one part of the world rather than another. For an interesting discussion of land reform in the context of US agriculture, cf. Graham 2005, 142-146. Graham does recognize that "the political viability of reorganizing land ownership on a wide-spread basis . . . is exceedingly tenuous"—to say the least!—but he quite rightly goes on to say that "this is precisely where the American and Judeo-Christian traditions of land ownership diverge" (146).

7. Cf. Ratzinger 1997, 230; Benedict XVI 2009b; Benedict XVI 2011; Benedict XVI 2010b, 45-47.

8. Cf. Sen, 1987. For a fuller account of some of the points developed in this lecture, cf. Sen 1999. For an interesting discussion of the affinity between Sen's work and that of Catholic Social Teaching, cf. Clark 2007.

9. As Henri de Lubac reminds us, "The revolutionary socialists are the heirs of liberals who, in the Western school, embraced atheism. 'To annihilate God' is the first point in their program and the first watchword spread abroad by their tracts. They draw the inferences of that atheism. No longer contending themselves with a vague belief in progress, they undertake to build up humanity without God" (De Lubac 1995, 322-323). Nicholas Berdyaev also comments on the affinity between Marxism and liberal-democracy (cf. Berdyaev 1924, 258). James C. Scott explores, in his words, "an illuminating point of direct contact between Soviet and American high modernism: the belief in huge, mechanized, industrial farms" (Scott 1998, 193ff).

10. Cf. Benedict XVI 2007c. Here and elsewhere, Benedict's thought often strikes a profoundly personalist note, centering upon the freedom and dignity of the human person to live in accordance with the justice and charity of Jesus Christ, charging his Christian brothers and sisters to take personal responsibility for changing the conditions around them, rather than looking first to the state or other institutions to do so.

11. There he writes of the need for "*a profoundly new way of understanding business enterprise*," because of the tendency of so many businesses to restrict accountability to investors alone, which, in turn, restricts their vision to the short term (Benedict XVI 2009a, #40).

12. For an interesting discussion of sustainability, cf. Jenkins 2008, 231ff.

13. Compare this with Bacon's words in *Novum Organum*: "For even as in the business of life a man's disposition and the secret workings of his mind and affections are better discovered when he is in trouble than at other times, so likewise the secrets of nature reveal themselves more readily under the vexations of art than when they go their own way" (Bacon 2000, XCVIII).

14. For a good overview cf. Rowland 2009, 108ff. Rowland observes that in this genealogy Ratzinger typically focuses on certain intellectual moments in which the Hellenic aspects of culture were severed from the Christian, and moreover, in which the Christian aspects were damaged by distortions in the doctrine of creation.

15. A significant consequence of Bacon's approach, according to Briggs, is that "the new universal philosophy makes no distinction in principle between natural and violent motion . . . Violent motions such as those used in the new learning's experiments are in fact more natural than motions that do now show the nature of bodies," 144.

16. Though Bacon's project seeks to usurp the person and work of Christ, it should be said that does not seek to deny him—Bacon's work is, after all, replete with the

language of theology. Rather than deny Christ, Bacon simply relocates him to a private and other-worldly sphere.

17. Cf. International Theological Commission 2004.
18. Howard 1972 and 2006; Jackson 1980; Altieri 1995 and 1982.
19. Cf. Schroth 2004, 2-4.
20. Consider the case of Howard, who trained as a botanist in England before going to India as an agricultural advisor. Upon his arrival, however, he soon found many of the local farming practices he encountered to be superior to the agriculture he was sent from England to promote. His work changed accordingly. He did not jettison his scientific training; rather, he allowed what he encountered in India to question that training, reshape it, and redirect it. I would suggest that one way to understand the work of Howard, as well as the other figures I have mentioned, is an attempt, as scientists, to lift up, valorize, and collaborate with the kind of "traditional" agriculture one still finds, for instance, in places like the Bribri and Cabécar indigenous territories.

REFERENCES

Altieri, Miguel. 1995. *Agroecology: The Science of Sustainable Agriculture*. Boulder, CO: Westview Press.

Aristotle. 1984. *Complete Works of Aristotle*. Princeton: Princeton University Press.

Bacon, Francis. 2000. *Novum Organum*. Cambridge: Cambridge University Press.

Benedict XVI. 2006. Message of His Holiness Benedict XVI to the Director General of the Food and Agricultural Organization (FAO) for the Celebration of World Food Day 2006. Vatican City, October 16. Accessed October and November 2012. http://www.vatican.va/holy_father/benedict_xvi/messages/food/documents/hf_ben-xvi_mes_20061016_world-food-day-2006_en.html.

______. 2007a. *Jesus of Nazareth*. New York: Doubleday.

______. 2007b. Meeting of the Holy Father Benedict XVI with the Clergy of the Dioceses of Belluno-Feltre and Treviso. Church of St Justin Martyr, Auronzo di Cadore, July 24. Accessed October and November 2012. http://www.vatican.va/holy_father/benedict_xvi/speeches/2007/july/documents/hf_ben-xvi_spe_20070724_clero-cadore_en.html.

______. 2007c. Message of His Holiness Benedict XVI to the Director General of the Food and Agricultural Organization (FAO) for the Celebration of World Food Day. Vatican City, October 4. Accessed October and November 2012. http://www.vatican.va/holy_father/benedict_xvi/messages/food/documents/hf_ben-xvi_mes_20071004_world-food-day-2007_en.html.

______. 2007d. *Saved in Hope (Spe Salvi)*. San Francisco: Ignatius Press.

______. 2008a. Meeting of the Holy Father Benedict XVI XVI with the Clergy of the Diocese of Bolzano-Bressanone, Cathedral of Bressanone. Cathedral of Bressanone, August 6. Accessed October and November 2012. http://www.vatican.va/holy_father/benedict_xvi/speeches/2008/august/documents/hf_ben-xvi_spe_20080806_clero-bressanone_en.html.

______. 2008b. Message of His Holiness Benedict XVI to the Director General of the Food and Agricultural Organization (FAO) for the Celebration of World Food Day 2008. Vatican City, October 13. Accessed October and November 2012. http://www.vatican.va/holy_father/benedict_xvi/messages/food/documents/hf_ben-xvi_mes_20081013_world-food-day-2008_en.html.

______. 2009a. *Charity in Truth (Caritas in Veritate)*. San Francisco: Ignatius Press.

______. 2009b. Message of His Holiness Benedict XVI to the Director General of the Food and Agricultural Organization (FAO) for the Celebration of World Food Day 2009. Vatican City, October 16. Accessed October and November 2012. http://www.vatican.va/holy_father/benedict_xvi/messages/food/documents/hf_ben-xvi_mes_20091016_world-food-day-2009_en.html.

______. 2010a. Message of His Holiness Benedict XVI to the Director General of the Food and Agricultural Organization (FAO) for the Celebration of World Food Day 2010. Vatican City, October 15. Accessed October and November 2012. http://www.vatican.va/holy_father/benedict_xvi/messages/food/documents/hf_ben-xvi_mes_20101015_world-food-day-2010_en.html.

______. 2010b. *Light of the World: The Pope, the Church, and the Signs of the Times*. San Francisco: Ignatius Press.

______. 2011. Message of His Holiness Benedict XVI to the Director General of the Food and Agricultural Organization (FAO) for the Celebration of World Food Day 2011. Vatican City, October 17. Accessed October and November 2012. http://www.vatican.va/holy_father/benedict_xvi/messages/food/documents/hf_ben-xvi_mes_20111017_world-food-day-2011_en.html.

______. 2012. Message of His Holiness Benedict XVI to the Director General of the Food and Agricultural Organization (FAO) for the Celebration of World Food Day 2012. Vatican City, October 16. Accessed October and November 2012. http://www.vatican.va/holy_father/benedict_xvi/messages/food/documents/hf_ben-xvi_mes_20121016_world-food-day-2012_en.html.

Berdyaev, Nicholas. 1924. *The End of Our Time*. New York: Sheed & Ward.

Briggs, John. 1989. *Francis Bacon and the Rhetoric of Nature*. Harvard: Harvard University Press.

Calvez, Jean-Yves, and Perrin, Jacques. 1961. *The Church and Social Justice*. London: Burns & Oates.

Clark, Meghan J. 2007. "Integrating Human Rights: Participation in John Paul II, Catholic Social Thought, and Amartya Sen." *Political Theology* 8: 299-317.

Cloutier, David. 2010. "Working with the Grammar of Creation: Benedict XVI, Wendell Berry, and the Unity of the Catholic Moral Vision." *Communio* 37: 606-633.

De Lubac, Henri. 1995. *The Drama of Atheist Humanism*. San Francisco: Ignatius Press.

Dougherty, Catherine de Hueck. 1991. *Apostolic Farming: Healing the Earth*. Combermere ON: Madonna House.

Graham, Mark E. 2005. *Sustainable Agriculture: A Christian Ethic of Gratitude*. Cleveland: The Pilgrim Press.

Hanby, Michael. "*Homo Faber* and/or *Homo Adorans*: On the Place of Human Making in a Sacramental Cosmos." *Communio* 38 (Summer 2011): 198-236.

Harrison, Peter. 2001. *The Bible, Protestantism, and the Rise of Natural Science*. Cambridge: Cambridge University Press.

Howard, Albert. 1972. *An Agricultural Testament*. Emmaus PA: Rodale Press.

______. 2006. *The Soil and Health: A Study of Organic Agriculture*. Lexington: University Press of Kentucky.

Hylton, Wil S. 2012. "Broken Heartland: The Looming Collapse of Agriculture on the Great Plains." *Harper's* June-July: 25-35.

International Theological Commission. 2004. *Communion and Stewardship: Human Persons Created in the Image of God*. Accessed October and November 2012. http://www.vatican.va/roman_curia/congregations/cfaith/cti_documents/rc_con_cfaith_doc_20040723_communion-stewardship_en.html.

Jackson, Wes. 1980. *New Roots for Agriculture*. Omaha: University of Nebraska Press.

Jenkins, Willis. 2008. *Ecologies of Grace*. Oxford: Oxford University Press.

John Paul II. 1979. Apostolic Journey to the Dominican Republic, Mexico, and the Bahamas. Meeting with Mexican Indios. Cuilapan, Mexico, January 29. Accessed October and November 2012. http://www.vatican.va/holy_father/john_paul_ii/speeches/1979/january/documents/hf_jp-ii_spe_19790129_messico-cuilapan-indios_en.html.

Pontifical Council for Justice and Peace. 1995. *Compendium of the Social Doctrine of the Church*. Vatican City: Libreria Editrice Vaticana.

Primavesi, Ana. 1982. *Manejo Ecologico del Suelo*. Rio de Janeiro: Libreria 'El Ateneo' Editorial.

Ratzinger, Joseph Cardinal. 1997. *Salt of the Earth: The Church at the End of the Millennium*. San Francisco: Ignatius Press.

______. 2002. *God and the World*. San Francisco: Ignatius Press.

Rowland, Tracey. 2009. *Ratzinger's Faith: The Theology of Pope Benedict XVI*. Oxford: Oxford University Press.

Schroth, Gotz. 2004. *Agroforestry and Biodiversity Conservation in Tropical Landscapes*. Washington, DC, Island Press.

Scott, James C. 1998. *Seeing Like a State: How Certain Schemes to Improve the Human Condition Have Failed*. New Haven: Yale University Press.

Second Vatican Council. 1965. *Gaudium et Spes. Pastoral Constitution on the Church in the Modern World*. Accessed October and November 2012. http://www.vatican.va/archive/hist_councils/ii_vatican_council/documents/vat-ii_cons_19651207_gaudium-et-spes_en.html.

Sen, Amartya. 1987. "Food and Freedom." Sir John Crawford Memorial Lecture. Washington, D.C. October, 28, 1987.

______. 1999. *Development as Freedom*. New York: Anchor Books, 1999.

Sokolowski, Robert. 1982. *The God of Faith and Reason: Foundations of Christian Theology*. Notre Dame: University of Notre Dame Press.

Speltz, George H. 1945. "The Importance of Rural Life According to the Philosophy of St. Thomas Aquinas: A Study in Economic Philosophy." PhD diss., Catholic University of America.

White, Lynn. 1967. "The Historical Roots of Our Ecologic Crisis." *Science* 155.3767: 1203-1207.

III

The Sacramentality of Creation

SEVEN

The Way of Wisdom

"Keep hold of instruction; do not let go; guard her, for she is your life" (Prov 3:14)

Elizabeth Groppe

The accelerating crises of climate change and ecological degradation are an indication that we have not safeguarded wisdom well. Scientists first testified to the United States Congress about the potential of climate change to threaten human well-being in 1979; yet, thirty-four years later, we have yet to pass any significant climate change legislation. Meanwhile, scientific projections about the dangers of climate change have become a reality. A study conducted by the Global Humanitarian Forum found that climate change in 2009 was responsible for 300,000 deaths, the suffering of 325 million people, and economic losses of over $100 billion. Over 90 percent of the persons most severely affected were from developing countries that have contributed least to global carbon emissions, including the people of the Carteret Islands of whom Bishop Unabali (2012) spoke so eloquently. If climate change continues unabated, projections for the future include the disruption of agricultural production by drought, floods, and temperature extremes; continued elevations of sea level that will threaten highly populated coastal regions; an increase in the intensity and frequency of storms and severe weather; shortages of water in many regions including those that rely for their water supply on melt-waters from glaciers that may no longer exist; and the extinction of many species of life (Miller 2010; Thomas 2004).

Pope Benedict XVI (2009) identified one of the root causes of the climate crisis as our failure to respect the wisdom of the created order, which we have approached simply as a resource for our extraction rather than a gift that sacramentally bears witness to God. "Today as never before," he stated, "it is essential to help people grasp that Creation is something more than a simple source of wealth to be exploited by human hands. Indeed, when God, through creation, gave man the keys to the earth, he expected him to use this great gift properly, making it fruitful in a responsible and respectful way. The human being discovers the intrinsic value of nature if he learns to see it as it truly is, the expression of a project of love and truth which speaks to us of the Creator and of his love for humanity, which will find its fullness in Christ at the end of time." Scripture and Christian tradition testify that the cosmos is a realm created through the wisdom of God and that this wisdom is reflected in the splendor of oceans, forests, mountains, valleys, and prairies. In this essay, I will briefly survey the wisdom tradition, consider challenges posed to the tradition by modern science, and finally reflect on possibilities for its recovery today in the wake of our changing climate.

THE WISDOM TRADITION

Wisdom eludes definition. "There is in her a spirit that is intelligent, holy, unique, manifold, subtle, mobile, clear, unpolluted, distinct, invulnerable, loving the good, keen, irresistible, beneficent, humane, steadfast, sure, free from anxiety, all-powerful, overseeing all, and penetrating through all spirits that are intelligent, pure, and altogether subtle" (Wis 7:22-23). The biblical texts that comprise the corpus of wisdom literature portray *chokmah* with an evocative array of rich images and metaphors. Rooted in the experience of daily life, the books of Proverbs, Ecclesiastes, Job, Wisdom, and Ben Sira date from the sixth through the third century BCE and offer counsel for personal, family, and social relationships embedded in reflection on nature and the cosmos.

According to the book of Proverbs, Wisdom is present at the beginning of creation. "The Lord created me," she testifies, "at the beginning of his work, the first of his acts long ago" (Prov 8:22). Wisdom compares herself to a craftsman who skillfully carries out the plans of a master architect: "When he established the heavens, I was there, when he drew a circle on the face of the deep, when he made firm the skies above, when he established the foundations of the deep, when he assigned to the sea its limit, so that the waters might not transgress his command, when he marked out the foundations of the earth, then I was beside him like a master worker" (Prov 8: 27-30). Because creation is made through wisdom, contemplation of the natural order can be instructive to human beings: "Go to the ant, you lazybones; consider its ways, and be wise. Without having any chief or officer or ruler, it prepares its food in summer, and gathers its sustenance in harvest" (Prov 5:6-7). Wisdom is portrayed not only as creation's craftsman but also as a woman crying out at the entrance of the city gate in reproof of the foolish who scoff at knowledge (Prov 1:20-33). The student of wisdom lives in fear of the Lord and embraces instruction (Prov 1:1-7).

The author of the Wisdom of Solomon, an anonymous Hellenistic Jew living in diaspora sometime between 100 BCE and 1 CE, sought instruction through the contemplation of creation. He was attentive to "the structure of the world and the activity of the elements; the beginning and end and middle of times, the alternations of the solstices and the changes of the seasons, the cycles of the year and the constellations of the stars, the natures of animals and the tempers of wild animals, the powers of spirits and the thoughts of human beings, the varieties of plants and the virtues of roots; I learned both what is secret and what is manifest, for wisdom, the fashioner of all things, taught me" (Wis 7:17-22).

The book of Job, the wrenching lament of a righteous man who has lost his flocks and herds, his children, and his health, testifies to the dearth of justice in the world and the hiddenness of wisdom:

> But where shall wisdom be found?
> And where is the place of understanding?
> Mortals do not know the way to it,
> and it is not found in the land of the living.
> The deep says, "It is not in me,"
> and the sea says, "It is not with me."
> It cannot be gotten for gold,
> and silver cannot be weighed out as its price.
> It cannot be valued in the gold of Opir,
> in precious onyx or sapphire. . . .
> Where then does wisdom come from?
> And where is the place of understanding?
> It is hidden from the eyes of all living,
> and concealed from the birds of the air. . . .
> God understands the way to it,
> and he knows its place.
> For he looks to the ends of the earth,
> and sees everything under the heavens.
> When he gave to the wind its weight
> and apportioned out the waters by measure;
> when he made a decree for the rain,
> and a way for the thunderbolt;
> he established it, and searched it out (Job 28:12-16; 20-21; 23-27).

Wisdom is both established by God and something for which God searches within creation.

Jesus son of Eleazar son of Sirach, a scribe and scholar who lived in the early second century BCE, compiled instruction on wisdom for young men in Jerusalem preparing to take on adult roles in the community. He personified wisdom as a feminine figure, a gift of divine origin created before all things and lavished upon those who love God. Ben Sira found wisdom not only in the height and breadth of creation, the drops of rain and sands of the sea, but also in the Torah given through Moses. Wisdom counsels honor of parents, humility, care of the poor, honesty, and self-control. "All this is the book of the covenant of the Most High God, the law that Moses commanded us as an inheritance for the congregations of Jacob. It overflows, like the Pishon, with wisdom, and like the Tigris at the time of the first fruits. It runs over, like the Euphrates, with understanding, and like the Jordan at harvest time" (Sir 24:23-26).

Jesus of Nazareth was shaped deeply by these wisdom traditions. Although the scrolls of wisdom literature known in Judaism as the *Ketuvim* (Writings) were not collected and canonized until the second century, the Synoptic Gospels attribute to Jesus over 100 wisdom sayings, and in Luke he is identified as someone with wisdom greater than that of Solomon (Luke 11:31). Passages in Matthew, notes biblical scholar James Dunn (1999), suggest that Jesus is not simply a teacher of wisdom but Wisdom incarnate. Indeed, Dunn (1989) believes that it is this wisdom tradition that enabled the theology of incarnation to emerge. This is evident already in the Pauline literature and comes to full expression in the Gospel of John, where Jesus is identified as the incarnation of the Logos through whom creation was made (John 1:13), a theology that echoes Proverb's account of Wisdom as agent of creation (Prov 8:2-23). In John's Gospel as in Sirach and 1 Enoch, pre-existent Wisdom is manifest through the prophets and the law, rejected by some, accepted by others, and then returned to God.

God's wisdom, according to Paul, is a secret wisdom, hidden from the world and known only to the spiritually mature. The Greeks judge God's wisdom to be folly, for it is the wisdom of the cross (1 Cor 1:23). In a culture in which maintaining one's public honor was of supreme importance, Christ underwent a humiliating death for the sake of others in fidelity to God. In so doing, he "became for us wisdom from God, and righteousness and sanctification and redemption" (1 Cor 1:30). God raised Christ to new life (1 Cor 15) and through him all creatures are reconciled (Col 1:20).

Christian theological reflection carries forward Scripture's sapiential tradition. God's wisdom, according to Thomas Aquinas (1225–1274) is the exemplar through which the universe is created and the law that leads creation teleologically to its divinely intended end:

> Through his wisdom God is the founder of the universe of things, and we have said that in relation to them he is like an artist with regard to the things that he makes. We have also said that he is the governor of all acts and motions to be found in each and every creature. And so, as being the principle through which the universe is created, divine wis-

> dom means art, or exemplar, or idea, and likewise it also means law, as moving all things to their divine end (*Summa Theologiae*, 1|2.93.1).

This eternal law or wisdom of God is both acquired by study and a gift of the Holy Spirit; it is recognized by discernment in faith and expressed in human formulations of moral and natural law. Jesus Christ, Aquinas emphasized, is wisdom incarnate.

Saint Bonaventure (1217–1274) approached creation as a mirror (*speculum*) of the divine and the first step on a ladder of ascent to God. We cannot climb these rungs unaided, but those who humbly open their hearts in prayer can behold in the multitude, beauty, and order of creatures vestiges of the power, wisdom, and goodness of the triune God. The form of knowledge that Bonaventure cultivated was not knowledge for the sake of mastery and control but the intimate knowledge that is the fruit of love and a gift of the Holy Spirit (Edwards 2006, 110). Those who contemplate creation with these eyes can recognize its divine artist, aware that every creature is "nothing less than a kind of representation of the wisdom of God," Bonaventure (1970, 179) urged in *Hexameron*.

"Is it not true," Benedict XVI queries (2010, 124), "that what we call 'nature' in a cosmic sense has its origin in 'a plan of love and truth'? The world 'is not the product of any necessity whatsoever, nor of blind fate or chance . . . The world proceeds from the free will of God; he wanted to make his creatures share in his being, in his intelligence, and in his goodness' (*Catechism of the Catholic Church* #295). The *Book of Genesis*, in its very first pages, points to the wise design of the cosmos; it comes forth from God's mind and finds its culmination in man and woman, made in the image and likeness of the Creator."

WISDOM CHALLENGED

In 1672, John Ray's *The Wisdom of God Manifested in the Works of Creation* offered a sapiential reading of creation. However, the biblical vision of a created order permeated by the wisdom of God was challenged by ensuing scientific developments. Since Charles Darwin's voyage on *The Beagle* (1831–1836) where he recorded in his diaries evidence for a process of evolution by adaptation and natural selection over the span of geologic time, the theory of evolution has become foundational to the biological and ecological sciences, strengthened by the discovery of genes, mediators of mutation unknown to Darwin. There continues to be scientific debate about various aspects of the process of evolution including the question as to whether selection operates at the level of the gene, the individual, the species, or all simultaneously. Scientists are also exploring the intersection of evolutionary theory with chaos and complexity theories. Yet the broad principle of the development of diverse species through the selection of adaptive genetic mutation over the broad span of geologic time is scientifically uncontroverted. In a number of ways, this theory and its interpretations have been very challenging to traditional Christian convictions about a world ordered by graceful Lady Wisdom.

Whereas the book of Proverbs speaks of wisdom as a master-builder who skillfully actualizes the plans of a divine architect, evolutionary science highlights the role of random genetic mutation in the emergence of life's diversity of forms. This has led some scientists to the conclusion that neither an immanent divine wisdom nor even an extrinsic divine designer stands behind the beauty and variety of life. "Pure chance," writes Nobel laureate Jacque Monod (1972, 122), "absolutely free but blind, [is] at the very root of the stupendous edifice of evolution." According to the widely-read biologist Richard Dawkins (1987, 5), "the only watchmaker in nature is the blind force of physics, albeit deployed in a very special way." There are purposeful forms in nature, Dawkins explains, but no overarching purpose, vision, or plan. Life forms emerge in the interaction of random genetic mutations with the geophysical forces of earthquakes, volcanoes, falling asteroids, and changing climates. "Blind to good and evil," the philosopher Bertrand Russell (1918, 46) concluded, "reckless of destruction, omnipotent matter rolls on its relentless way."

Whereas classic Christian theologies of wisdom such as that of Thomas Aquinas depict a cosmos in which everything is created with an order and purpose (*telos*) that shape the moral fabric of the universe, evolutionary science challenges teleological assumptions. "The one thing about which modern authors are unanimous," writes Ernst Mayr (1983, 324-34), "is that adaption is not teleological." Early theories of evolution such as those of French naturalist Jean Baptiste Lamark (1744–1829) and the philosopher and biologist Herbert Spencer (1820–1903) described evolution as a progression from simplicity to complexity, an approach amenable to some form of teleological interpretation. Today, the Harvard biologist E.O. Wilson also finds in evolution a movement from simplicity to complex diversity, but he does not attribute this to divine providence. Wilson is, in his own terms, a "provisional deist" who believes there may be a God who set the universe in motion but that there is no evidence for divine guidance within the process of evolution nor the working out of human destiny (Roberts 2006). According to Nobel laureate and physicist Steven Weinberg (2008, 14), "[W]e do not find any point to life laid out for us in nature, no objective basis for our moral principles, no correspondence between what we think is the moral law and the laws of nature, of the sort imagined by philosophers from Anaximander and Plato to Emerson."

Not only the process of adaptation through random genetic mutation can lead to such conclusions. The evolutionary process leaves in its wake a trail of death and suffering. Scientists estimate that over 90 percent of the species that once populated the biosphere are now extinct. Many of those that continue to thrive do so in competitive predator-prey relationships. Annie Dillard watched a giant brown water bug in Virginia's Tinker Creek bite into a frog, inject the dumbstruck creature with an enzyme that reduced its inner organs to a liquid, and then suck out the frog juice, leaving nothing but a flap of skin at the bottom of the stream. "I gaped," Dillard wrote (1974, 8), "bewildered, appalled."

Holmes Rolston III (1987, 138; also McDaniel 1989, 19-20) wrestles with the fact that female white pelicans lay two eggs, one of which hatches several days before the second, gaining a slight advantage in development. Few pelican parents can provide for two chicks simultaneously, and the elder chick is able to beat its younger sibling to the food in its mother's pouch. Often, it will actually drive the second chick away from the nest. If the exiled sibling attempts a return, it will be rejected by its own parents. This pattern has proven to be a successful evolutionary strategy, enabling pelican survival for almost thirty million years; parental energy is concentrated in the first chick to hatch; should this chick die there is another, but if the first chick thrives, the "backup chick" is unnecessary. The philosopher David Hull (1991, 406) interprets the waste, suffering, and pain that accompanies the emergence and demise of the variety of life as "careless, indifferent, almost diabolical." According to Richard Dawkins (1995, 133), "Nature is neither kind nor unkind. She is neither against suffering nor for it." Rather, "[t]he universe has precisely the properties we should expect if there is, at bottom, no design, no purpose, no evil and no good, nothing but blind, pitiless indifference."

WISDOM REGAINED?

Today, important work is being done to bridge the disciplines of theology and science by many scholars including John Polkinghorne, Ilia Delio, Celia Deanne-Drummond, Terence Nichols, and others. John Haught (2008) responds to the challenge posed by the tremendous scope of death and suffering inherent in the evolutionary process with an eschatological theology. Jean Porter's (2005, 82-103) engagement with scientific literature in her work on natural law leads her to conclude that there is a directedness and consistency in nature that cannot be accounted for by chance alone and that does allow for an affirmation of teleology in the Aristotelian sense that each creature pursues ordered activity for its well-being. Anglican theologian Sarah Coakley has worked closely with Harvard's Program for Evolutionary Dynamics to gain an in-depth understanding of research on evolution that has led program director Martin Nowak to the conclusion that cooperation is as important as mutation and selection in evolutionary process. It is only after a period of stable cooperation, Nowak explains, that break-through moments in evolution such as the development of multi-cellular organisms from individual cells occur, and Nowak has developed mathematically-precise accounts of these mechanisms of cooperation and their stochastically-recurrent evolutionary patterns. This, Coakley (2012, 1) comments, "gives the lie to recent cultural mythologies of evolution which are wholly dominated by 'selfishness' and violence on the one hand and erratic 'randomness' on the other. Of course, competitiveness *is* at the heart of natural selection—there is no attempt to deny that here; but we now understand that there is also another principle at stake alongside mutation and selection in the full spectrum of evolution, one that it is not inappropriate to call *productively* sacrificial." These developments open up new possibilities for conversation between evolutionary science and the Christian wisdom traditions.

Today, one can also find echoes of the tradition's affirmation that there is wisdom to be found within creation among those working to address environmental degradation and the climate crisis. Practitioners in the fields of sustainable agriculture, ecological design, and biomimicry describe their efforts in a manner that resonates with Christianity's affirmation that the wisdom of the triune God is reflected in the created order. Although practitioners in these fields do not expressly give their work a theological interpretation, their witness makes an important contribution to an exploration of the contemporary meaning of the wisdom tradition in our time of climate crisis.

Farming in Nature's Image

We readily associate climate change with images of congested highways or the smokestacks of coal-burning power plants; for most people, farming is not the first thing that comes to mind when considering the causes of global warming. Yet the agricultural sector is responsible for an estimated 37 percent of U.S. carbon emissions. Tilling land to plant annual crops releases carbon from the soil, and carbon dioxide and methane gas are emitted by farm machinery and livestock. Cornell University ecologist David Pimentel estimates that industrial agriculture uses ten kilocalories of hydrocarbons to produce just one kilocalorie of food (Benyus 1997, 19). Moreover, the dominant forms of agriculture practiced in America erode soil, deplete fertility and biodiversity, and require large amounts of hazardous pesticides and herbicides.

On a small surviving patch of Kansas prairie, Wes Jackson glimpsed the possibility of a new agricultural paradigm. Westward moving American immigrants from Europe plowed up the prairie to plant the annual grains that are the staple of our diet, acting on the assumption that nature's bounty was endless. Today, in a region where prairie once covered 85 percent of the land, only one-tenth of 1 percent of the prairie remains intact. Jackson developed a deep affection for this remnant, cultivating what he terms a "conversation" with the land. Through this conversation, he learned that prairie systems are composed of perennial plants whose vast root systems build soil and sequester carbon dioxide. New growth springs from the roots each year such that plowing and planting is unnecessary and the land is never exposed to the eroding force of wind or heavy rain. When rain does come, the intricate horizontal spread of dense roots that have had many years to mature are able to make optimal use of the water, and in times of drought the vertical depth of the root systems can find water at depths of twenty-five feet. In the richly diverse mature prairie, as many as 250 varieties of plants may coexist in a synergy that provides defense against blights and pest infestations. Plants in the legume family fix nitrogen and the annual decomposition of organic plant matter further enriches the soil such that the prairie system is nutritionally self-sustaining. The soil, Jackson learned, is alive with over 5000 species of microorganisms. Soil is not simply a receptacle for seeds and petrochemicals, but a living, breathing organism.

The prairie in its wisdom builds topsoil, cultivates a great diversity of perennial species, and generates abundant fecundity using the energy of the sun. The dominant forms of American agriculture are the antithesis of the natural prairie system—they deplete and erode the soil, support a monoculture of annual plants, and require not only solar energy but vast petrochemical inputs. Jackson began to imagine what agriculture might look like if humans took instruction from natural systems. "Essentially," he explains, "we have to farm the way nature farms" (Benyus 1997, 21).

Jackson and his staff at the Land Institute in Kansas are in the process of perennializing wheat, sorghum, and sunflower. They are cultivating these plants in polycultures. Although they cannot recreate the dizzying diversity of a prairie's 250 species within plots designed for agricultural production, they are replicating a pattern he observed on the prairie. All of the prairie systems he studied had four functional groups of plants: cool season grasses, warm season grasses, legumes, and plants from the sunflower family. "I'm not sure why that works," Jackson comments (2002), "but I'm going to go with that." If this humble emulation of a prairie's wisdom can be carried forward in quick order on a large scale, it will make a major contribution to the reduction of atmospheric CO_2 and the prevention of famine.

Ecological Design

While Wes Jackson has been carefully observing prairie systems, architect William McDonough and chemist Michael Braungart (2002, 118) have been reenvisioning human industry and civilization in the conviction that we must "follow nature's design framework." They describe industrial civilization as a process in which humans have imposed through control and force our own designs on nature, and they are wary of responses to the ecological crisis that emphasize increased efficiency in our use of energy and materials. Although using resources efficiently is clearly preferable to wasteful and profligate consumption, an emphasis on efficiency overlooks the fact that the very structure of our buildings, cities, and industries is tragically flawed; using energy and materials more efficiently within this structure will only slow our demise. They call for a radically new paradigm for industry and civilization.

McDonough and Braungart (2002, 81) have put forward what they term an "eco-*effective*" design model as distinct from an approach that is simply more efficient. Turning to nature for inspiration to develop this approach, they describe themselves as "humbled by the complexity and intelligence of nature's activity." Like the book of Proverbs that counsels, "Go to the ant, you lazybones; consider its ways, and be wise" (Prov 5:6-7), they hold up the industrious social insect as a model for our emulation. The cumulative biomass of ants is greater than that of humanity, but with very different consequences for the biosphere. Industrial *Homo sapiens* puts billions of pounds of toxic material into the air, soil, and water each year; produces immense amounts of waste; buries or burns valuable materials; and erodes cultural and biodiversity. Leaf-cutter ants, in contrast, enhance the ecosystems of which they are a part. In the process of collecting matter from the Earth's surface and carrying it down into their colonies to feed the fungal gardens that provide their communities with food, the ants aerate the soil and transport minerals needed by plants and soil microbes, contributing to soil health and fecundity. All the materials made by every species of ant, including their most deadly chemical weapons, are biodegradable and return nutrients to the earth. The cherry tree, another source of inspiration to McDonough and Braungart, sequesters carbon, produces oxygen, cleans air and water, creates and stabilizes soil, and provides habitat for a diverse array of species and food for microorganisms, insects, and animals. The architect and chemist conclude that it is not precisely the fittest but the "*fiting-est*" who thrive in natural systems by living within "an energetic and energetic and material engagement with place, and an interdependent relationship to it" (120).

Can we design a building like a tree? A factory that operates with the eco-effectiveness of an ant colony? Yes, McDonough and Braungart maintain, if we transform our basic mode of relationship to the earth from one of control through brute force to one of engagement and participation in earth's natural energy flows and nutrient cycles. Human systems must be designed like living systems, embodying the same basic principles that foster growth and fecundity in nature: we must live on current solar income, cultivate diversity, and cycle all nutrients such that what appears to be the waste of one process becomes food for another. We must not only do no harm to the biosphere, but *contribute* something to the earth, fostering life and regeneration, just as the leaf-cutter ants improve the health of the soil and the cherry tree provides habitat for other species. At the invitation of David Orr, director of the Environmental Studies program at Oberlin College, McDonough designed an environmental educational building constructed with local materials that operates on solar power and cycles all the greywater from the restrooms through vats where it is purified by live plants and then recirculated through the system. In Germany, Braungart helped a textile company identify fibers and dyes that were nontoxic and compostable such that production scraps and worn carpets could be used as mulch in gardens, keeping the nutrients cycling in service of life. The Ford Motor Company, which opened the nation's largest vertically integrated automobile factory on 600 acres in Dearborn, Michigan, in 1923, employed McDonough to redesign this aging industrial icon now situated on contaminated soils. The plant now supports 454,000 square feet of green roof where sedum and other plants absorb carbon dioxide, produce oxygen, and provide habitat for killdeer and other species. Swales have been constructed around the factory and planted with indigenous species to filter rainwater and provide additional habitat, and experimental efforts to purify the soil with phytoremediation are in process.

Biomimicry

McDonough and Braungart's vision for a new form of human relationship with the earth is comparable to that of Janine Benyus who coined the term "biomimicry" in 1990 as she was searching for words to label a folder of articles on efforts to develop artificial photosynthesis. The term, she explains, comes from the Greek *bios* (life) and *mimesis* (imitation) and refers to the emulation of nature. Benyus (1997, 3) has a degree in forestry, but her education provided limited preparation for her current work. "In reductionistic fashion," she explains, "we studied each piece of the forest separately, rarely considering that a spruce-fir forest might add up to something more than the sum of its parts, or that wisdom might reside in the whole." Today her approach is holistic and she sees in the natural world cooperative relationships, self-regulating feedback cycles, and dense interconnectedness. She believes that by emulating nature humanity can learn to reinhabit the earth in a way that can lead to life's flourishing rather than its demise, and her 1997 book *Biomimicry* identifies people she discovered practicing this emulation independently of one another in a wide variety of fields including agriculture, energy, and materials science.

She summarizes biomimicry's approach to the natural world with three principles:

1. *Nature is our model*. Nature's designs offer inspiration that can help solve human problems.
2. *Nature is our measure*. Nature provides a standard as to what kinds of design may be right or appropriate that is based on 3.8 billion years of evolutionary experience.

3. *Nature is our mentor*. Biomimicry approaches nature not with the extractive relation of the industrial age but with the intention of valuing and learning from the natural world.

In 2006, Benyus founded the Biomimicry Institute to promote the study and imitation of nature's remarkable designs.

Of particular importance for climate change is biomimicry's potential to inspire new forms of energy production. Even our most sophisticated and ingenious photovoltaic systems cannot compare to the elegance and efficiency with which life forms that we have regarded as "primitive" harvest and store energy from the sun, where hydrogen fusion produces every hour as much energy as all of human civilization uses in an entire year (Krupp and Horn 2008, 15). In 1912, the Italian chemistry professor Giacomo Ciamician imagined that we might one day photosynthesize the fuel we need by mimicking the "guarded secrets" of plants (Benyus 1997, 62). Teams of scientists are now studying these secrets. At Arizona State University, for example, chemists J. Devens Gust, Jr. and Thomas and Anna Moore have been trying to replicate the molecular structures with which bacteria transfer electrons and prolong charge separation across a membrane, a key to the process of producing and storing energy from the sun. "We peered over nature's shoulder," Gust explains, "tried something, peered over nature's shoulder again" (Benyus 1997, 70). After twenty years of labor they have developed a molecular structure that can absorb light from a laser beam and then use this energy to turn ADP into ATP, the fuel of cellular life. "That was again a very exciting thing," Gust comments (2002), "that was in the sense the culmination of this program that had been going on for over twenty years of trying to mimic what goes on in those bacterial reaction centers." Capturing the light of a laser beam, however, is a very different matter than capturing the much more diffuse light of the sun. James Guillet at the University of Toronto has designed the means to collect actual sunlight through photozymes, enzymes constructed from molecular building blocks that mimic the antennae pigments of plants.

Biomimicry may assist us not only with the production of solar energy, but also with the sequestration of carbon. Nature, explains Ray Hobbs of Arizona Public Service, has managed the carbon cycle in part through algae. "You are looking at the origins of life," he says with reference to this single-celled organism, "an organism that has survived for three and a half billion years and created the conditions for other life to emerge. [Algae] are the root of the food chain. And so elegant" (Krupp and Horn 2008, 10). Algae's absorption of carbon in the Paleozoic era changed the climate in a manner conducive to the flourishing of myriad new forms of life, and Hobbs and others are endeavoring to cultivate algae on acreage at Arizona's Redhawk Power Plant that would absorb half of the plant's carbon emissions. Once saturated with carbon, the algae can be dried and converted into fuel.

Materials science has also turned to nature for instruction. The "heat, beat, and treat" processes used in industrial manufacturing to make ceramics, alloys, and plastics require high pressures, harsh chemicals, and extremely high temperatures generated by the combustion of fossil fuels. And yet, notes Benyus (1997, 97), "despite our colossal energy expenditures, we still can't make materials as finely crafted, as durable, or as environmentally sensible as those of nature." Spiders, for example, make silk at room temperature that ounce per ounce is five times stronger than steel. Mussels make a nontoxic biodegradable adhesive through which they can stick themselves firmly to a wet and sandy rock and resist the pressure of the rising and falling tides. The abalone makes a shell with a smooth, luminous mother-of-pearl inner lining that is not only beautiful but flexibly resistant to fracture and twice as strong as industrial high-tech ceramics, and it crafts this shell with no hazardous chemicals in a process friendly to life. Rich Humbert at the University of Washington and Paul Calvert at the University of Arizona Materials Laboratory are working to replicate the process in which an abalone builds nacre by secreting proteins that self-assemble into a framework of polymers that attract mineral crystals from sea water, a revolutionary approach to materials production inspired by nature's wisdom.

Interdisciplinary Cooperation

Learning from nature requires interdisciplinary cooperation. Benyus describes biomimicry projects that bring to the table people who do not often work together: engineers, materials scientists, microbiologists, protein chemists, geneticists, farmers, engineers, medical professionals, and business executives. Although none of the collaborations that she describes include theologians, one of the striking features of the literature on the emulation of nature is the occurrence of theological terms. We must act, writes Benyus (1997, 9), with "the humility and the spirituality that are needed to . . . take our seat at the front of nature's class." We must design *with* nature's wisdom, becoming students of nature rather than its conquerers. Marveling at the intricacies, effectiveness, and even poetry of nature, she writes that "biomimics develop a high degree of awe, bordering on reverence" (7). Fred Krupp and Marion Horn (2008, 112), in like vein, are struck by the awe that Ray Hobbs manifests in his work with algae.

Those deeply engaged in the practice of ecological design and biomimicry describe their work in a manner that transcends the disjunction of science and ethics that has characterized so much of post-Kantian western culture. Whereas Kant sharply distinguished the character of natural and moral knowledge, Benyus (1997, 22 and Preface) describes nature as a "standard" to which humans should conform and a means "to judge the 'rightness' of our innovations.'" Nature, in the words of Wes Jackson, is a "measure" by which we can assess the viability of our agricultural practices. Benyus emphasizes that for 3.5 billion years nature has been developing designs to support and sustain life, a position that is open to a teleological interpretation.

The insights of those emulating the generative intelligence of nature may help us to hear again "the wisdom of the ancients" to which Pope Benedict (2012, 125) appeals. This wisdom, he explained, "recognized that nature is not at our disposal as a 'heap of scattered refuse' (Heraclitis)" but rather that it is "a gift of the Creator, who gave it an inbuilt order and enables man to draw from it the principles needed to 'till it and keep it' (Gen 2:15)." Jackson, Benyus, and others who humbly take instruction from nature to develop new forms of agriculture, energy, and materials production exemplify one way in which we can embody the Christian wisdom tradition in a manner appropriate to the twenty-first century.

At the same time, a tension between the Christian tradition and the world as known to us through contemporary science remains. This is evident in the reflections of Claus Mattheck (1998, 27), a mechanical engineer at the Karlsruhe Research Centre in Germany who is studying trees to learn how to design systems that withstand stress. Noting that limbs distribute stress uniformly throughout a tree's structure, he is reminded of the biblical injunction to "bear one another's burdens" (Gal 6:2). Simultaneously, however, he is cognizant that the trees, antelopes, leopards, and other creatures that he studies to develop design principles have formed their imitable structures through "the hard struggle for existence, which occurs every day in nature with terrifying pitilessness" (25). The underside of the striking beauty and resilience of the pearly abalone shell is the death that is an inextricable part of the process of evolution.

A twenty-first century search for wisdom in the natural order is further complicated by the fact that "nature" no longer exists as an order untouched by humans (McKibben 1989). The epoch of earth's history in which we abide was officially known as the Holocene; according to the keepers of geologic time, we transitioned from the era of the Pleistocene into the Holocene in the year 10,000 BCE. In the 1980s, biologist Eugene F. Stoermer at the University of Michigan began using the term "Anthropocene" to articulate his insight that the activities of humanity have so fundamentally altered the planet that we have now moved from the Holocene into a new period of geological time shaped predominantly by human activity. Independently of Stoermer, atmospheric chemist and Nobel laureate Paul Crutzen coined the same term. "I was at a conference," he explained, "where someone said something about the Holocene. I suddenly thought this was wrong. The world has changed too much. So I said: 'No, we are in the Anthropocene.' I just made up the word on the spur of the moment. Everyone was shocked. But it seems to have stuck" (Pearce 2007, 21). Indeed, many scientists are now using this terminology, and the official theme of the 2011 meeting of the Geological Society of America was "Achean to Anthropocene: The Past is the Key to the Future." Crutzen dates the beginning of the Anthropocene to the Industrial Revolution in the eighteenth century, whereas others trace it back to the point in history at which human societies transitioned from hunting and gathering to agricultural production.

Climate change is one of the foremost signs that the era of the Anthropocene is upon us. The 2000 scientists gathered from around the globe at the 2007 assembly of the U.N.'s Intergovernmental Panel on Climate Change (IPCC 2007, 5) concluded unequivocally that the climate system is warming and that the observed increase in global average temperatures since the mid-twentieth century "is very likely due to the observed increase in anthropogenic greenhouse gas concentrations" (10). Climate change is already changing the shape of life on earth and scientists project that the warming of our atmosphere and oceans will continue to melt glaciers, raise sea levels, and change precipitation patterns in a manner that will have dramatic effects on the biosphere. Human activity has not only changed the earth's climate but has also spawned a hole in the stratospheric ozone that protects the earth from harmful radiation, altered the chemistry of the ocean in a manner that threatens marine life, leveled half of the earth's forests, released into the environment thousands of synthetic chemicals whose long-term consequences are unknown, and altered the nitrogen and phosphorus cycles in a way that is eroding the resilience of important Earth subsystems (Rockström 2009). Our degradation of ecosystems has already precipitated the extinction of species at a rate that is estimated to be anywhere from 100 to 1000 times the natural background rate of extinction; we are now, biologists explain, in the midst of the sixth mass extinction of life in earth's history. Climate change will greatly exacerbate this terrible loss. A research project by a team of scientists published in *Nature* in 2004 concluded that a climate warming in the mid-range of current projections will by the year 2050 lead to the extinction of fifteen to thirty-seven percent of the species examined in the study (Thomas et al. 2004). At the very same time that humans are beginning to recognize anew that there is a wisdom and intelligence to the created order worthy of our emulation, we are degrading and diminishing that order. A once exquisitely diverse and beautiful coral reef that is now a lifeless skeleton in the waters of an ocean acidified by the absorption of our carbon emissions reflects to us not the wisdom of the triune God but the folly of humanity in the era of the Anthropocene.

Theological Fragments

In a research laboratory at Harvard University, Martin Nowak is discovering the fundamental importance of cooperation in the process of evolution. In Kansas, Wes Jackson is cultivating an understanding rooted in affection for the prairie that is bearing fruit in a new form of agriculture. In the symbiotic relationships between leaf-cutter ants and their ecosystems, chemist Michael Braungart and architect Bill McDonough take inspiration for eco-effective buildings and factories that not only do no harm but also contribute something to the sustenance of other life forms. And in the cells of a species of bacteria that has thrived on the planet for three billion years, Devens Gust and others glean insight on molecular processes that may enable humans to power our own civilization with the abundant energy of the sun. Nature, Janine Benyus concludes, should be our model and standard. At the same time, we know that the evolution of life is inextricably shadowed by death and that in this era of the Anthropocene the biosphere bears not only the mark of its triune Creator but also the stark footprint of a form of human civilization that has ravaged the planet. In what sense, then, can we speak authentically of a spirit of wisdom that is "intelligent, holy, unpolluted, all-powerful, and penetrating through all" (Wis 7:22-23)?

In one sense, the dissonance between the holy wisdom to which the book of Proverbs testifies and our own earthly reality is to be expected. The Catholic tradition has emphasized that God's wisdom is not only imminent in the created order but also a transcendent reality. "Between the Creator and the creature," the fourth Lateran Council stated in 1215, "there can be noted no similarity so great that a greater dissimilarity cannot be seen between them" (Tanner 1990, 232). Aquinas gave classic expression to the analogical character of the relationship between created and divine wisdom. The divine perfections, he explained, are both truly manifest in the goodness and beauty we witness in creatures and also inconceivably greater.

Today, however, our theology of wisdom must account for not only the similarity-in-transcendent-difference between created wisdom and the wisdom of God but also the shadow side of evolution and humanity's devastation of the biosphere, realities with which medieval theology did not have to contend. The work of Jackson, Benyus, Braungart and others testifies that there is indeed wisdom to be found in "the varieties of plants and the virtues of roots" (Wis 7:17-22), but it is a wisdom shadowed by the death trail of evolution and threatened by anthropogenic habitat destruction, species extinction, pollution, and climate change. Given these realities, a theology of wisdom fitting to our time will be a theology of fragments, a paschal theology that is cosmic in scope, and a theology that impels us to sacrificial action on behalf of life.

FRAGMENTS OF WISDOM

Over the course of the past decade, David Tracy has moved from the method employed earlier in his career that emphasized the analogical and dialectical modes of theological knowledge to a method that emphasizes the Hidden-Revealed and Comprehensible-Incomprehensible God manifest to us not in the classical forms of order, harmony, and beauty but in fragments (Holland 2002). The form of the fragment, Tracy (2005) explains, emerged in literary theory and philosophy as a challenge to the totalizing system of modernity, a modern rationality of closure whose rules for the possible excluded the saturated religious phenomenon. The German Romantics Friedrich Schlegel and Novalis initiated this counter-modern genre in works such as *Philosophical Fragments from the Philosophical Apprenticeship* (Schlegel 1796) and *Fragments from the Notebooks* (Novalis 1798). Subsequently Søren Kierkegaard experimented with this genre in *Philosophical Fragments* (1844) and *Concluding Unscientific Postscript* (1846). The five-pointed star of Franz Rosenzweig's *Star of Redemption* (1921) broke through totalizing systems, influencing the work of Emmanuel Lévinas and Walter Benjamin. In North American theology, Tracy finds the form of the fragment in African American spirituals, blues, and slave narratives. He is particularly interested in apocalyptic and apophatic theological forms that fragment modern assumptions about the closure of history and language, breaking open the impossible reality of the Hidden-Revealed and Comprehensible-Incomprehensible God. Tracy's forthcoming *This Side of God* explores the form of the fragment and will be followed by the second volume *Gathering the Fragments*.

On a planet that has been literally fragmented by the human degradation of the biosphere, Tracy's reflections on the form of the fragment are very suggestive. The wisdom we can glean through contemplation of creation is limited not only by the ontological difference between created and divine realities but also by the difference between the biosphere as it was in the Holocene and the biosphere as it exists in the Anthropocene after 10,000 years of agriculture and 200 years of industrialization, hair-thin slivers of time on the scale of geological time in which humans have literally changed the face of the earth. The patch of Kansas prairie where Wes Jackson is seeking wisdom is only a fragment of the ecosystem that once carpeted the American plains. The abalone that Rich Humbert has been studying for twenty years to learn how humans might make more resilient materials with benign means is part of a population of sea creatures that has been depleted almost to extinction by overfishing. The trees that provide instruction to German mechanical engineer Claus Mattheck on the principle of uniform stress stand in forests threatened by climate change. Our degradation of the biosphere has compromised our very ability to know God's wisdom in creation; even in the joining of the best of our contemplative practices with the best tools that science can offer, we can perceive today only a fragment of the wisdom that once was there. Nicola Hoggard Creegan writes evocatively of a Trinitarian God who is not only hidden but also revealed in creation and redemption, a God who cannot be known objectively, a God whose face is veiled by the death that accompanies the process of evolution and by the terror of the cross, but a God who may nonetheless be glimpsed in edges and signs, a God reflected in the patterns and order and beauty of creation "that are suggestive of mind, perhaps even of love." Truly, writes Creegan (2007, 509), citing the words of the prophet Isaiah, "you are a God who is hiding" (Isa 45:15).

Today, God is not only hidden within the mysteries of creation and redemption but also banished by humanity's degradation of the biosphere. "The will of God is to be here, manifest and near," writes Abraham Joshua Heschel (1951, 153), "but when the doors of this world are slammed on Him, His truth betrayed, His will defied, He withdraws, leaving man to himself. God did not depart of His own volition; He was expelled. *God is in exile.*" The task of the theologian who would gather the fragments is not simply an intellectual or academic matter. It is the theologian's vocation to speak as truthfully as possible about God, which cannot be done if the mirror of creation in which Bonaventure countenanced a reflection of the divine artist has been shattered. "With the extinction of species and despoiling of places," writes Willis Jenkins (2003, 412), "we lose the ability to name and praise God. . . . The terrible paucity we are threatened by in ecological degradation is the loss of that by which to bless God, and so the increased likelihood of idolatry." Our losses are already inestimable. Theologians together with the church as a whole must actually gather the fragments, doing all that we can to protect and nurture creation. Kierkegaard, Tracy (2005, 104) explains, employed the form of the fragment, to attempt "to render present the one content modernity denied—the reality of the impossible—grace, Christ, God." The reality of ecological fragmentation, in contrast, does not evoke but exiles grace.

THE WISDOM OF THE CROSS

Paul found wisdom in the crucified Christ who appeared as foolishness to the Gentiles (1 Cor 23), the Christ who cried out in lamentation "let this cup pass" and then sacrificed his own life upon the cross. Hans urs von Balthasar (1990) describes Christ's descent into hell on Holy Saturday as a kenotic self-emptying into the abyss of personal, social, and cosmic sin, an entry into a second chaos of an unfinished creation. God's bodily resurrection of the crucified Christ broke the bonds of death and is our promise that "creation itself will be set free from its bondage to decay and will obtain the freedom of the glory of the children of God" (Rom 8: 21).

Following Peter and Paul, Pope Benedict (2010, 133) does not limit redemption to humankind but speaks of a "'new heavens and a new earth' (2 Pt 3:13)."

Christian discipleship requires, Jon Sobrino (1984) writes among the suffering people of El Salvador, those of us who are relatively privileged to take the crucified people down from their crosses. In a world in which all of creation is groaning (Rom 8:22), discipleship today also means acting on behalf of soils, abalone, prairie grasses, coral reefs, forests, and all of God's threatened creation. "[W]e are challenged," writes Celia Deane-Drummond (2006, 127), ". . . to begin to take account of the suffering of non-human creatures." This discipleship can take the form of making of our yards a sanctuary for a diversity of plants, insects, and birds rather than a monoculture of green grass; composting materials to build soil; supporting local organic agriculture; dramatically reducing our own consumption of materials and energy, and procuring the energy which we do use from non-carbon sources; and working together for the changes in local, state, national, and international policies and practices that are necessary to the transformation of our energy and agricultural infrastructures. Given the dysfunction our government has manifest thus far in response to urgent scientific warnings about climate change, this necessary transformation will not be easy.

Transformation may well require sacrifice. Sr. Dorothy Stang was steadfast in her work to protect the Amazon rainforest from illegal logging despite repeated death threats, and she ultimately gave her life for the forest and the people who dwell there. In Texas, Oklahoma, Tennessee, and Washington DC, people have been engaging in nonviolent civil disobedience to halt plans to construct the Keystone Pipeline that would enable production of 700,000 barrels of oil a day from Canadian tar sands. In the state of New York, people are blocking with their bodies equipment used in hydraulic fracturing for natural gas or taking out second mortgages on their homes to finance work to protect the integrity of our waters and underworld. At times, some forms of public witness to the moral urgency of a conversion from fossil-fuels to non-carbon sources of energy can be strident or polarizing in their tone.

Catholics can contribute greatly to the common good not only by offering sacrificial nonviolent witness to the urgency of changes in energy policy but also by making this public witness in a manner that is deeply prayerful and that testifies to God's beauty and love. Reflecting on the role of cooperation in the evolutionary process, Sarah Coakley (2009, 11) writes that phenomena such as the sacrificial practices of social insects may provide "a sort of evolutionary preparation for a higher and fully intentional human altruism that can arise only when the cultural and linguistic realm is reached." Evolution, she concludes, "delivers to us humans, made in '[God's] image,' the greatest possible inheritance of responsibility: to crown those regular intimations of evolutionary cooperation, long established and refined, with acts of intentional sacrificial altruism that now alone can save the planet."

CONCLUSION

The climate crisis is one of many indicators that our current form of human civilization is walking a path of folly. Pope Benedict XVI calls us to a radical conversion; he invites us to see creation not as raw material there for our appropriation but as an expression of God's plan of love and truth—an expression of wisdom. Developments in theology's dialogue with science and within the interdisciplinary fields of biomimicry and ecological design give new meaning to the Catholic affirmation that God's wisdom and goodness is manifest to us in the book of nature. We can take instruction from bacteria, algae, ants, and prairies that can help us address the climate crisis. At the same time, a theology of the sacramental presence of wisdom in nature is challenged by both the shadow side of evolution and humanity's desecration of the biosphere. A theology of wisdom fitting to our time of climate crisis will be a theology of fragments, a paschal theology that is cosmic in scope, and a theology that impels us to sacrificial action on behalf of life. "Keep hold of instruction: do not let go; guard wisdom, for she is your life" (Prov 3:14).

REFERENCES

Benedict XVI. 10 September 2009. "Address to a Group of Sponsors and Promoters of the Holy See Pavilion at the 2008 International Exposition in Zaragoza (Spain)." Accessed March 20, 2013. http://www.vatican.va/holy_father/benedict_xvi/speeches/2009/september/documents/hf_ben-xvi_spe_20090910_expo-zaragoza_en.html.

———. 2010. "To Cultivate Peace, Protect Creation." World Day of Peace Message. In *The Environment*, edited by Jacquelyn Lindsey, 121-33. Huntington, IN: Our Sunday Visitor.

Benyus, Janine. 1997. *Biomimicry: Innovation Inspired by Nature.* New York: Morrow.

Bonaventure. 1970. *Collations on the Six Days*. Translated by Jose de Vinck. Paterson NJ: St. Anthony Guild Press.

———. 1978. *The Soul's Journey into God (Itinerarium mentis in Deum).* Classics of Western Spirituality, translated by Ewert Cousins. New York: Paulist.

Creegan, Nicola Hoggard. 2007. "A Christian Theology of Evolution and Participation." *Zygon* 42:499-517.

Coakley, Sarah. 2009. "Evolution and Sacrifice." *Christian Century* 126:10-11, October 20.

———. 2012. *Ethics, Cooperation and Human Motivation: Assessing the Project of Evolutionary Ethics.* Lecture 3 of the 2012 Gifford Lectures on the topic "Sacrifice Regained: Evolution, Cooperation, and God."

Dawkins, Richard. 1987. *The Blind Watchmaker*. New York: W.W. Norton & Company.

———. 1995. *River Out of Eden*. New York: Basic Books.

Deane-Drummond, Celia. 2000. *Creation through Wisdom: Theology and the New Biology.* Edinburgh: T & T Clark.

———. 2006. *Wonder and Wisdom: Conversations in Science, Spirituality, and Theology*. Philadelphia: Templeton Foundation Press.

———. 2010. "The Breadth of Glory: A Trinitarian Eschatology for the Earth through Critical Engagement with Hans Urs von Balthasar." *International Journal of Systematic Theology*. 12:46-64.

Delio, Ilia. 2008. *Christ in Evolution*. Maryknoll NY: Orbis.

Dillard, Annie. 1974. *Pilgrim at Tinker Creek*. New York: HarperCollins.

Dunn, James D. G. 1989. *Christology in the Making*. 2nd ed. London: SCM Press.

———. 1999. "Jesus, Teacher of Wisdom or Wisdom Incarnate?" In *Where Shall Wisdom Be Found? Wisdom in the Bible, the Church, and the Contemporary World*, edited by Stephen C. Barton, 75-92. Edinburgh: T &T Clark.

Gregersen, Niels Henrik. 2001. "The Cross of Christ in an Evolutionary World." Dialog: *A Journal of Theology* 40:192-207.

Edwards, Denis. 2006. *Ecology at the Heart of Faith: The Change of Heart that Leads to a New Way of Living on Earth*. Maryknoll NY: Orbis.

Franck, Suzanne. 2011. "Sophia Wisdom & Climate Change." In *Confronting the Climate Crisis: Catholic Theological Perspectives*, edited by Jame Schaefer, 39-54. Milwaukee: Marquette University Press.

Gust, Devens. 2002. *Biomimicry*. Interview in DVD directed by Paul Lang. Bullfrog Films.

Haught, John. 2008. *God after Darwin: A Theology of Evolution*. Boulder CO: Westview.

Heschel, Abraham Joshua. 1951. *Man Is Not Alone: A Philosophy of Religion*. New York: Farrar, Strauss, & Giroux.

Holland, Scott. 2002. "This Side of God: A Conversation with David Tracy." *Cross Currents* 52:54-59.

Hull, David. 1991. "God of the Galapagos." *Nature* 352, August 8, 485.

Intergovernmental Panel on Climate Change (IPCC). 2007. Summary for Policymakers. In *Climate Change 2007: The Physical Science Basis. Contribution of Working Group I to the Fourth Assessment Report of the Intergovernmental Panel on Climate Change*, edited By S. D. Solomon, M. Qin, M. Manning, Z. Chen, M. Marquis, K.B. Averyt, M. Tignor and H. L. Miller. Cambridge: Cambridge University Press.

Jackson, Wes. 2002. *Biomimicry*. Interview in DVD directed by Paul Lang. Bullfrog Films.

Jenkins, Willis. 2003. "Biodiversity and Salvation: Thomistic Roots for Environmental Ethics." *The Journal of Religion* 83:401-420.

Krupp, Fred, and Marion Horn. 2008. *Earth the Sequel*. New York: Norton.

Mattheck, Claus. 1998. *Design in Nature: Learning from Trees*. Springer: Berlin.

Mayr, Ernst. 1983. "How to Carry Out the Adaptionist Program." *American Naturalist* 128:324-34.

McDonough, William, and Michael Braungart. 2002. *Cradle to Cradle: Remaking the Way We Make Things*. New York: North Point Press.

McKibbin, Bill. 1989. *The End of Nature*. New York: Random House.

———. 2010. *Eaarth: Making a Life on a Tough New Planet*. New York: Henry Holt.

Miller, Richard. 2010. "Global Climate Disruption and Social Justice: The State of the Problem." In *God, Creation, and Climate Change: A Catholic Response to the Environmental Crisis*, edited by Richard Miller, 1-34. Maryknoll: Orbis Books.

Moltmann, Jürgen. 2003. *Science and Wisdom*. Minneapolis: Fortress Press.

Nichols, Terence L. 2002. "Evolution: Journey or Random Walk?" *Zygon* 37:193-210.

Nowak, Martin A. 2011. *SuperCooperators*. New York: Free Press.

Pearce, Fred. 2007. *With Speed and Violence: Why Scientists Fear Tipping Points in Climate Change*. Boston: Beacon.

Porter, Jean. 2005. *Nature as Reason: A Thomistic Theory of the Natural Law*. Grand Rapids MI: Eerdmans.

Roberts, David. 18 October 2006. "E.O. Wilson Chats about His New Book on the Intersection of Science and Religion." *Grist*. Accessed March 20, 2013. http://grist.org/article/wilson2/.

Rockström, Johan et al. 2009. "A Safe Operating Space for Humanity." *Nature* 461:472-75.

Rolston, Holmes III. 1987. *Science and Religion: A Critical Survey*. New York: Random House.

Russell, Bertrand. 1918. *Mysticism of Logic and Other Essays*. New York: Longmans and Green.

Sobrino, Jon. 1984. *The Principle of Mercy: Taking the Crucified People from the Cross.* Maryknoll: Orbis.

Soule, Judith D. and Jon K. Piper. 2002. *Farming in Nature's Image: An Ecological Approach to Agriculture.* Washington DC: Island Press.

Tanner, Norman, ed. 1990. *Decrees of the Ecumenical Councils*, vol. 1. Washington DC: Sheed and Ward, Georgetown University Press.

Thomas, Chris et al. 2004. "Extinction Risk from Climate Change." *Nature* 427:145-48.

Tracy, David. 2005. "Form and Fragment: The Recovery of the Hidden and Incomprehensible God." In *The Concept of God in Global Dialogue*, edited by Werner G. Jeanrond and Aasulv Lande, 98-114. Maryknoll NY: Orbis.

Unabali, Bernard. 2012. Keynote Address by the Bishop of Bougainville, Papua New Guinea, at Catholic Consultation on Environmental Justice and Climate Change: Assessing Pope Benedict XVI's Ecological Vision for the Catholic Church in the United States. The Catholic University of America, Washington DC, November 8-10, November 8.

Von Balthasar, *Mysterium Paschale: The Mystery of Easter*. Scotland: T & T Clark, 1990.

Weinberg, Steven. 2008. "Without God," *New York Review of Books* 55:14, September 25. Accessed March 20, 2013. http://www.nybooks.com/articles/archives/2008/sep/25/without-god/.

EIGHT

The World as God's Icon

Creation, Sacramentality, Liturgy

Msgr. Kevin W. Irwin

A characteristic strength of Roman Catholicism throughout history has been its ability to reread, rethink, and reformulate its traditional beliefs in the light of contemporary issues and challenges—very often of a doctrinal and ethical character—as well as pastoral needs. This reflects the best of what it means that Catholicism is a "theological" as opposed to a fundamentalist religious tradition (Irwin 2002, 59-61).[1]

Today the environmental/ecological crisis in general and global climate change in particular offer such a challenge. Among the chief tenets of Catholic belief and practice (as well as what may be termed a Catholic worldview) that can be used to shape a substantial theological perspective on and ethical response to the environmental crisis are the following (as will be reflected in the balance of this paper). They are Catholicism's:

- theology of *creation* in general, the world as coming from God and sustained by God,
- the *incarnation,* that God so loved the world that he sent his Son into this world to save us by becoming human in order to die and rise,
- the principles of *sacramentality* and *mediation* as based on the theology of creation and the incarnation,
- *participation* in liturgy and sacraments as chief features of Catholic life in which we use the anthropologically appropriate means of human communication and the things from and of the earth to worship God,

- *common good* as a foundational principle of the way Catholics look at life, which prism includes both stewardship of and for creation, and
- *beauty* as reflective of God and the way things that are beautiful in themselves in nature or are the result of human artistry and productivity reflect the very being of God.

At the same time what is argued here is not denominationally particular to Catholics. Those churches which regularly celebrate the liturgy will find here resonances for what is hopefully a deepening unity and ecumenical approach to the environment.[2] More particularly, western Christian churches can rely here on the traditions and explanations of the liturgy as celebrated and understood in the Orthodox churches.[3]

The Liturgy Constitution from the Second Vatican Council asserts that the liturgy is the "summit toward which the activity of the Church is directed; at the same time it is the fountain from which all her power flows" (*Sacrosanctum Concilium* #10).[4] In a unique and privileged way the liturgy is where we are drawn again and again into God's saving actions of our salvation on the basis of the principles of mediation and sacramentality. We do not shun the world for worship, even as we deliberately enter into the sacred liturgy apart from the world for a time to be engaged in the very action of God among us through word, sign, symbol, song, gesture, etc. Through worship we are invited to embrace the world at a different level as we worship God through and with it. In articulating the meaning of worship from this perspective we rediscover and uncover our theological and ethical roots. In and through the liturgy we do not superimpose or force an idea to argue for the value of creation and our need to revere it. Rather the opportunity which liturgy and sacraments offer us—to articulate environmental concerns—comes from within the given strengths of the Roman Catholic theological and liturgical tradition. The adage *lex orandi, lex credendi* has seen a revival in recent theological and liturgical literature whose discussions often concern the relationship of texts and rites, the enactment of the work of our redemption through the use of human communication, things of this world and of human manufacture. Liturgy is an act of prayer and an act of theology, and as such it is an important theological source for what we believe.

RECENT PAPAL LEADERSHIP

As we enter into a brief discussion of the recent papal magisterium on ecology and environmental concerns, we must admit mention the kinds of documents which both Pope John Paul II and Pope Benedict XVI have written on a number of issues and the "relative theological weight" which can and should be assigned to them, especially when those documents are from the official church *magisterium*.

Hierarchy of Official Papal Teachings

In classical Latin, *magister* meant "master," not only in the sense of "school-master" or teacher but in the many ways one can be a "master" of an art, or a craft or a trade. Hence the term carries the connotations of the role or the authority of one who is a master. In modern Catholic usage the term *magisterium* has come to be associated almost exclusively with the teaching role and authority of the hierarchy. More commonly today *magisterium* refers not only to the teachings themselves but to those who exercise authority in teaching—namely the pope and the bishops. That there has been a proliferation of the number of documents which have come from modern popes and the Vatican is clear. What is not always clear is their relative weight and the influence they should have.

An important principle in interpreting documents from the Vatican is to be aware of a document's genre and the relative importance the church assigns to that document precisely because of its genre. To assert that "the Vatican says . . ." may be a convenient way to introduce "breaking news" on any of a range of possible topics in documents and other sources. But it can be sloppy theologically and ignore the fact that there is a clear and definitive "hierarchy" that the church itself assigns to its documents.[5]

With regard to the "hierarchy of documents" pride of place is always given to the statements from church councils, such as Trent and Vatican II, simply because they carry the highest theological weight because they are the result of a council. It is commonly held that among the "acts of the Holy See" as published in the Vatican's official organ *Acta Apostolicae Sedis* (in Latin starting in 1909) the highest rank is ascribed to the "acts of the Second Vatican Council." The same *Acta* publishes papal documents and pronouncements. After the "solemn profession of faith" it ranks "acts for the beginning and conclusion of the Second Vatican Council" as of highest authority followed by such documents as "decretals," "encyclical letters," "apostolic exhortations," addresses to consistories, apostolic constitutions, the *motu proprio*, and other papal pronouncements (Morrisey 1978; Sullivan 2003). What is less easy to define is the authoritative weight to be given to a post-synodal exhortation, given the comparatively recent appearance of this kind of document (recalling here that meetings of international synods only began after Vatican II, the first of which was held in Rome in 1967). While papal apostolic exhortations have traditionally been ranked after encyclicals and before apostolic letters "*motu proprio*," the fact that these post-synod exhortations are not only signed by the pope but are designed to explain and elaborate on the results of the Synod of Bishops, and are thus a distillation of the synodical process and an act of representatives of the whole episcopal college, would mean that they merit greater weight than an apostolic exhortation signed by the pope without the input of the synod process (Morrissey 1978, 21-22).[6]

Simply put, in order to interpret as adequately as possible what popes have said about ecology and the environment means that we take very seriously the genre of the document involved. A papal encyclical or post-synodal exhortation obviously ranks higher than a homily or weekly audience talk.

Other Papal Writing

What is less easy to define is the relative weight that should be assigned to papal writings that are not of the official magisterium, which issue is a relatively recent concern given the fact that both Pope John Paul II and Benedict XVI have written theological works while occupying the See of Peter. One clear distinction which can be borne in mind concerns what they wrote before becoming pope and after assuming the See of Peter. At the same time it is clear that what popes wrote prior to becoming pope clearly influenced at least some of what they taught and wrote as the Holy Father. This is to say that books like *The Acting Person* by Karol Wojtyla written in 1969 resonate with many of Pope John Paul II's official papal teachings, e.g. his first encyclical *Redemptor Hominis*. This is to say that with the publication of *Crossing the Threshold of Hope*[7] Pope John Paul II opened a category for recent papal teaching that goes beyond the usual categorization of weighing the genre of a papal document. In publishing this book during his papacy the pope responded to contemporary questions and responded as a pastoral theologian. But the lingering question is where does this book fall in trying to delineate its relative theological weight and import as coming from the pen of a pope?

If these questions and concerns can be raised about Pope John Paul II they can and need to be raised even more fully because of the volume of writing from the pen of (in succession) doctoral student, Professor, (Cardinal) Prefect Joseph Ratzinger and now Pope Benedict XVI. Keeping with the distinction I made above about writings prior to becoming pope and after becoming pope I want to raise some questions about Joseph Ratzinger's writings on the liturgy and Joseph Ratzinger/Pope Benedict XVI's writings on Jesus of Nazareth.

That Joseph Ratzinger has been concerned with the liturgy is reflected in his numerous writings on the subject, many of which are collected in his *Opera Omnia* the first part of which has become readily available in English translation as *The Spirit of the Liturgy*. In this first section of the *Opera Omnia* the then Joseph Ratzinger weighs in on a number of disputed and debated issues, among which are how to interpret the Gen 1:28 text about "subduing the earth," the dating of the feast of the Nativity, a certain caricature of "parody Masses" in a "secular" musical style, the moving of the sign of peace in the Mass to the rite for the presentation of the gifts, and raising the question of whether the eucharistic prayer ought to be proclaimed aloud (Ratzinger 2010, 99-100, 109-10, 141, 162, 202-203). While a number of questions could be raised about each of these issues (and countless others in the book) if one were to focus only on the issue of the location of the sign of peace it is clear that what is argued here has influenced Pope Benedict XVI's thought on the issue, and it influenced the debate among the Fathers at the Synod on the Eucharist. That this gesture should be characterized by a certain restraint is noted in the post-synodal exhortation *Sacramentum Caritatis* (Benedict XVI 2007, #49). The more fundamental issue concerns a possible erosion of confidence in the reformed liturgy if the then Cardinal Prefect has written about such a possibility for changing the rite.

With regard to theological writings composed while serving as the Holy Father the trilogy *Jesus of Nazareth* looms large. Among the debated and debatable issues on which the Holy Father has written concern his preference for the dating of the Last Supper (given the differences among the evangelists' accounts) and the date of the birth of Jesus on the basis of his weighing exegetical opinion based on the New Testament texts. What influence does this or should this have on other exegetical opinions? How easy is it to distinguish the "theological opinion" of Pope Benedict XVI from the magisterial teaching from the same pen?

Papal Teaching

The pressing theological and pastoral concern about ecology and the environment was addressed by Pope John Paul II in several places but none more poignantly or fully than his 1990 statement on "Peace with God the Creator, Peace with All of Creation."[8] Of particular import with regard to the liturgy are his assertions in #14:

> *Finally, the aesthetic value of creation cannot be overlooked.* Our very contact with nature has a deep restorative power; contemplation of its magnificence imparts peace and serenity. The Bible speaks again and again of the goodness and beauty of creation, which is called to glorify God (cf. *Gen* 1:4ff; *Ps* 8:2; 104:1ff; *Wis* 13:3-5; *Sir* 39:16, 33; 43:1, 9). More difficult perhaps, but no less profound, is the contemplation of the works of human ingenuity. Even cities can have a beauty all their own, one that ought to motivate people to care for their surroundings. Good urban planning is an important part of environmental protection, and respect for the natural contours of the land is an indispensable prerequisite for ecologically sound development. The relationship between a good aesthetic education and the maintenance of a healthy environment cannot be overlooked.

As will be seen this emphasis on respect for the beauty of creation and the works of human hands and care for them can and should be an important aspect of any argument about preserving the creation as a response to the ecological issues of our day.

More recently the ecological crisis has been addressed and in a number of statements by Pope Benedict XVI (who in one account has been called "the green pope").[9] One of the premises for these statements is that the world and all that dwells in it were made "good" (Gen 1:31) and are sustained in life by God (who in and through the liturgy can be described as the God of creation, the covenant, redemption and sanctification). That some of his statements occur at addresses on the occasion of meetings of the Food and Agricultural Organization (FAO) of the UN based in Rome is an indication of a papal overture to a wider audience than that represented in other of his writings. Among others, these include statements for the World Day for Peace (especially on January 1, 2010 on "If You Want Peace, Preserve Creation"), homilies (on Holy Thursday, Easter and Pentecost), general audience talks, the post-synod exhortations *Sacramentaum Caritatis, Verbum Domini* and parts of his encyclical *Deus Caritas Est.*

When it comes to assessing the *relative weight* of the documents in which the pope has addressed ecological concerns it is clear that the encyclical *Deus Caritas Est* is the most important, yet the post-synod exhortations would seem to be if an equal (if not higher) rank.[10]

With regard to the *contexts* for his statements certainly the World Day for Peace messages have in recent years become important days on which assertions about Catholic social teachings are reiterated. That some of his comments are addressed to the FAO on the occasion of days about food distribution goes to the heart of the relationship of celebrating the Eucharist and feeding the poor. In addition the content of homilies bears heavily in that follows because the setting is the very celebration of the sacred liturgy.[11]

With regard to the *content* of these papal teachings, certainly the repeated use of the phrase "human ecology" in the 2010 World Day of Peace Message (#12) and variations on it will be a lasting contribution of Pope Benedict XVI to this issue. Here he takes up the legacy of his predecessor in emphasizing the numerous important issues surrounding human life. One can only hold up this phrase as an important indication of a holistic and very "catholic" appreciation of the environmental crisis today.[12] When he takes up creation related themes in his preaching, for example specifically about water and water rights, he reflects the Holy See's concern for an equitable distribution of the world's resources.[13] When he writes about beauty and the cosmic rootedness of the liturgy he draws on a seminal essay in *The Spirit of the Liturgy.*[14]

Given the number of assertions the Holy Father has made about the environment and the ecological crisis could one at the same time offer the critique that a more focused document at the rank of an encyclical might serve us now and in the foreseeable future better than numerous assertions in a variety of papal texts? In addition one might raise the question about the possible limitation of his teachings on the environment, that it is highly anthropologically focused, emphasizing humans and their very legitimate responsibility for the environment. But might this approach mitigate a more holistic appreciation of the cosmos and all created things on their own merits? This is to say that while papal teachings presume that humans have "use" of things of this world and have the responsibility to care for, protect and preserve creation, do these approaches so emphasize humanity's role that they eclipse the rights and value of all of creation on its (and their) own merits? This is part of the basis on which some authors will critique the "stewardship" model and approach to environment (Cowdin 1994). More on this will be discussed below under the Liturgy of the Hours.

In effect this is to say that on the one hand the phrase "human ecology" from Pope Benedict XVI contributes to the contemporary debate as drawn from Catholic characteristics of integration and a holistic view of human beings living in the world. But on the other hand too much reliance on the "human" half of the phrase derogates from ensuring that the fullness of our belief in the goodness, beauty and value of all created things might be sustained. Pope Benedict's own appreciation for the liturgy does provide a way to remind ourselves of the wide angle lens which Catholicism puts on the world and *all* that dwell on it, humans and all of God's creatures.

LITURGY AND LITURGICAL THEOLOGY

In the "Introduction" to the bibliographical resource series entitled *Sacramenta*, the editor Maksimilijan Zitnik (1992, xvi) indicates that there is an inherent "interdisciplinarity" involved in studying the sacraments, as well as in categorizing studies about them for this rich and important four volume research tool. Indeed there are a number of ways of approaching the study of liturgy and sacraments and of appreciating their centrality in Catholicism (and, again, other "liturgical" churches).[15]

In addition I would argue that there are a number of ways to describe the reality that is the liturgy. Lest we venture too quickly into specific aspects of the liturgy, and principally what relation they have to daily live and ethical living, allow me to offer a working definition of the liturgy:

> Liturgy is the privileged means through which the baptized are immersed into the reality of the living God through Christ's paschal mystery in the power of the Spirit in the midst of the communion of the church through the use of our human faculties of mind, heart, will and imagination as experienced through human gesture, speech and silence through the use of the things of creation and of human manufacture in order to be shaped and formed by what we experience in order to

> enable us to live the lives more fully converted to the gospel in our world in a unique and yet provisional way until liturgy and sacraments cease and what we experience here through a glass darkly is made full, real and complete in the kingdom of heaven.

In what is presented here I want to avoid a didacticism which unfortunately plagues some celebrations of the reformed liturgy and at the same time to capitalize theologically on the liturgy's inherent multivalence.

Among the ways that one can approach this admittedly multifaceted reality that is "liturgy" is what is termed "liturgical theology" or "the theology of liturgy" which, as I will argue, are related but also distinct. The former is theology derived from the church's rites (e.g., ecclesiology, Christology, Trinity, creation, etc.) while the latter is a theological description of what happens uniquely in and through the liturgy (e.g., anamnesis, epiclesis, consecration, communion, etc.). That the phrase *lex orandi, lex credendi* is used to shape Pope Benedict's post-synodal exhortation *Sacramentum Caritas* (2007) is very significant. Its three parts are "A Mystery to be Believed," "A Mystery to be Celebrated," and "A Mystery to be Lived."

The premise *lex orandi, lex credeni* has patristic roots and can be said to have characterized a strength in the way Catholicism understands the liturgy from a *theological* perspective. For over four decades Catholics have implemented and become familiar with the most comprehensive adjustment ever undertaken to the liturgy in (western) Catholicism's history. Among the most obvious changes concern the lectionary of scripture readings for all the sacraments and the enhanced series of scripture reading at daily and Sunday Mass, the use of literally thousands of additional prayer texts (most from ancient yet recently discovered sources, with some modern compositions), the celebration of the liturgy largely in the vernacular and of the Mass celebrated "facing the people."[16] However, a major part of articulating what is proper and germane to Catholic worship is to explore the *theological meaning* of what it means to use the earth's primal elements—wind, air, fire, water—as well as the manufactured symbols of bread and wine, oil, chrism, etc. to worship God. Liturgical theology is also important to assist how we view the revised liturgical rites which sometimes made deliberate advances over previous rituals about the use of creation (Irwin 1994).

In effect emphasizing liturgical theology today reflects nothing less than a sea change from the pre-Vatican II liturgy's concern for accuracy in liturgical performance. This might be summarized as the move from "liturgy from the outside," i.e. rubrics and performance, to "liturgy from the inside," i.e. liturgical theology.[17] The liturgy presumes and articulates the inherent interdependence with all of creation in this world.

Among other possible premises (without wanting to "use" the liturgy in a preprogrammed or a utilitarian way), allow me to offer three on which liturgical theology is based with regard to its signs and symbols from creation.

Premise One: *Sacramentality* (or "The Sacramental Principle"[18]) Can Help Form a Catholic Contribution to Environmental Concerns.[19]

Sacramentality means that the celebration of liturgy articulates the Catholic belief that God is discoverable and discovered in human life, albeit partially this side of the kingdom, and that a theology of creation can be derived from the fact and way creation is used in the liturgy. The principle of sacramentality in general is that which undergirds Catholicism's uniqueness and which also undergirds the celebration of the liturgy. Sacramentality is a principle which is based on the goodness of creation, the value of human labor and productivity, and the engagement of humans in the act of worship. Sacramentality is a world view; it is a way of looking at life; it is a way of thinking and acting in the world. It is a world view that invites us to be immersed more and more fully in the here and now, on this good earth, and not to shun the things of this earth and on this good earth. It is a world view that asks us not to avoid the challenges which such earthiness will require of us. We do need to recover the paradigm of sacramentality, not only for the sake of liturgy and sacrament, but even more for the sake of sustaining one of Catholicism's chief tenets and ways of looking at and living in the world. Sacramentality deals with, values and reveres the things of this earth—earth, air, fire, water—these natural symbols, as reflective of God and revelatory of God. They are constantly used in worship as means of experiencing God, naming God, and worshiping God. But they are used in worship in relation to words and texts, lest their use be perceived to be pantheism of any sort. The God of creation, of the covenant, of revelation and of redemption is the very same God we worship through the liturgy. One needs all of these dimensions of sacramentality to try to be grasped by God and to attempt to "grasp" God. Liturgical prayers and texts help to keep that focus before us. But so do things of this world, things made by human hands, and the use of our bodies in worship—gestures, processions, seeing, listening, and responding by speaking and singing.

Sacramentality reflects Catholic liturgical practice that has always been connected to and rooted in the earth. There is a primalness to Catholic worship that stands alongside our use of prayers which contain concepts, images and metaphors about God and our very human condition. But to lose, or even to eclipse, the primalness of liturgy is to cut ourselves loose from what is a characteristic mooring for the way we have always worshipped God—through things from this earth, which earth was termed "good" in the book of Genesis.[20]

Put somewhat differently this is to say that there are inherent and underlying cosmic elements of the liturgy which need to be attended to as sources for understanding what liturgy is and what the cosmos is. Liturgy's use of the world helps us to continue to "name" created things (but not in the sense of dominating them). Given the vernacular translation of the post-Vatican II liturgy, prayers are readily understood as articulating appreciation for the cosmos. Yet it is at least notable that some prayer translations eliminate the theologically important word "creature" which is used to describe water as blessed at the Easter Vigil and salt that is poured into water at the Sunday blessing and sprinkling rite.[21] Do such usages help contextualize the danger of anthropomorphism in the liturgy?

Closely allied with "sacramentality" is a variation on it which is often termed "mediation." *Mediation* in liturgy means that we have direct access to the paschal mystery of Christ in and through the liturgy. But this direct access is through human means of communication, (words, gestures, signs), daily and domestic things in which humans engage (bathing, dining), in the community of the church (we are always intrinsically a covenanted people). These all form part of the way that the liturgy is the unique and privileged way of encountering God and being encountered by him.[22]

Premise Two: *Lex Orandi, Lex Credendi, Lex Vivendi.*

The traditional adage *lex orandi, lex credendi* "what we pray is what we believe" is expanded here to include *lex vivendi,*[23] meaning "how we live" as a result of and in relation to what we celebrate in the liturgy. Again the issue is not to superimpose but to draw from what is germane to Catholicism to argue for the important responsibility we bear for the environment. That there is always a life relation to the celebration of the liturgy is clear.[24] Recall above that in *Sacramentum Caritatis* Pope Benedict uses this triad—a mystery to be believed, celebrated and believed.

Liturgy as the law of prayer establishes the law of belief as well as the law of living, in the sense that it is the liturgy as the pivotal celebration of believers' participating in the paschal mystery of Christ that expresses belief and commits one to act on those beliefs. This includes word, signs, and actions.

Central to our argument is that the Catholic tradition has an important, imperative and integral contribution to make to the issue of the environment through the church's *lex orandi.* In general this means delineating how the liturgy combines systematic theology with ethics and how liturgical celebration always urges communities of faith to live the justice of God that it experiences in the liturgy. For the specific purpose of delineating an environmental/ecological theology and ethics this means appreciating how the liturgy articulates characteristic Catholic beliefs. This is to suggest that the insights from the way Pope Benedict XVI deals with the Eucharist—"to be believed," "to be celebrated," and "to be lived"—

should be applied to the things we hold dear in creation and use in worship and how we use and deal with them outside of the liturgy.

At the risk of over-generalization, there has been a chasm post-Vatican II with those concerned with social justice and those concerned with the liturgy to our collective detriment.[25] At the same time the term "social justice" should not be seen to be limiting but should be a way to underscore that engagement in liturgy should lead to growth in a number of things like charity, leading the virtuous life, etc. That has been made more obvious as a result of the discussions at the Synod on the Eucharist in the Life and Mission of the Church and of the initiatives of Pope Benedict XVI himself is clear in the additions he authored to the dismissal rite at Mass:

> Go and announce the Gospel of the Lord.
> Go in peace glorifying God by your life.

Experiencing God and God's Might and Power Through the Liturgy, in Which We Name and Address God:

- Through the proclamation of the scriptures and praying the Psalms.

Central to every act of liturgy is the hearing of scripture readings and praying of the Psalms. An important factor in understanding the scriptures as used in the liturgy is to appreciate their historical, ecclesial and cosmic meanings. This is to say that the scriptures reveals sacred events which shaped and continue to shape a covenanted holy people (historical meaning). That these events occur still through their proclamation in the liturgy underscores their anamnetic quality (ecclesial meaning). Some narratives also reflect creation and the cosmos directly such as the proclamation of the Johannine prologue at Christmas in the Western liturgical tradition and at Easter in the eastern liturgical tradition. The Word becoming flesh is set in the context of creation "in the beginning." The liturgical introduction to many scripture texts—*in illo tempore*—underscores the multivalence in the way we should appropriate the texts as proclaimed in the liturgy. As is often pointed out the readings at the Easter Vigil are salvation history in miniature, and they begin with the account of creation in Genesis. In the Holy Father's homily in 2011 he spoke explicitly about the responsibility of humans to care for creation.[26]

- Through liturgical prayers addressed to God (among other things) as both "creator" and "redeemer."

Among the prayers used at the liturgy those called "consecratory" (for example to consecrate chrism or to bless those ordained) and "blessing" prayers are ranked highest. These texts are rich in their recounting various times and places in salvation history when God acted in a special way to show his great mercy. The technical phrase is that these recount the *magnalia Dei* some of which are for creation and created things and persons. These texts are used in the Hebrew sense to bless God, to name God's attributes and nature and then in a derivative way to sanctify ("bless") the person or creature at issue, e.g. humans being anointed or ordained, water being blessed, etc. A review of each of these can help to underscore the way creation is revered as inherently coming from God and which need humans care.

Premise Three: The Integrity of Liturgy as It Is Celebrated in Enacted Rites and How This Integrity Can Help Delineate a Catholic Environmental Theology and Ethics.

Here the liturgy is understood to renew God's covenant with all creation, to incarnate the kingdom of God on earth and to express the wide expanse of God's grace by the use of the goods of creation as this world in its goodness and fullness leads to a new creation imaged as the completion and fulfillment—a "new heaven and a new earth" (Rev 21:1) of what already exists on this good earth as part of God's providential plan—the earth God made "good" (again Gen 1:31). Further, the liturgy, as this privileged experience of God through word, sign, symbol, and gesture is seen as explicitly challenging the community of faith to deepen its commitment to live what is celebrated ritually by working to preserve the creation. Because what now exists, is not always what should be, (e.g. polluted water, inequitable distribution of the land, water resources, etc). *liturgy linked intrinsically to life and mission,* in the sense that communities assembling for worship bring all of life with them in order to returning to human life with a different perspective on life and a challenge to live God's life in the human more deeply and fully.

Again the intent here is to avoid a politicization of the liturgy or to "use" the liturgy for predetermined ends. To keep the liturgy in balance and perspective. Not to overemphasize one aspect (e.g., the scripture readings) to the detriment of other aspects.[27] The value of the term *integrity* is that it does not impose or force anything but draws from what we experience through the liturgy.

THE CELEBRATION OF THE LITURGY

In addition to being far more than concerned with texts, the study of the liturgy and appreciation for the liturgy always leads to and is derived from its actual enactment in celebration. Allow me to turn now to *four* examples from the enactment of the liturgy's rites to illustrate and exemplify the argument that the liturgy's rites reflect a theology of and about creation.

Water, Water Rights, Pollution of this Natural Resource

One of the changes from the previous rites of baptism and the rite as revised after Vatican II is that clean water is to be used each time baptisms are celebrated, which water is blessed as part of the ceremony. There is therefore the presumption of clean water to be used at baptism.

What these prayers say about God and creation matters a great deal. The structure and contents of the blessing prayers now used to pray over the symbols to be blessed, e.g. water, chrism, oil of the sick of the catechumens, bread and wine, etc. focuses attention on how and when these elements were used in saving history (*anamnesis*) which paradigmatic events lead us to bless them now (*epiclesis*) to use them here and now for the same purposes of (re)creation and redemption (Tillard 1985). The use of these now-blessed signs ought to mean that we value and revere them outside the liturgy as well.

With regard to the use of water in baptism an advance in the post-Vatican II revised rites for baptism is the preference they give to immersion rather than infusion. The fuller appreciation of immersion in water cannot but help to articulate the value of water as God's gift even as we the use it in baptism to signify entrance into the community of faith. At the same time it should be recalled that the use of creation in liturgy involves respecting their inherent multivalence despite the words used to bless them. This is to say that the understanding of water as life giving, and the only element besides air without which we cannot live is juxtaposed with the reality that too much water through violent storms can cause death, or drowning. There is an inherent ambiguity involved in the use of such primal elements as water, light, etc.

However the reality of how water is allocated and used in our world may well diminish some of the kinds of liturgical theology that can be derived from these liturgical rites. Today the state of water is both tragic and paradoxical. Water covers nearly seventy-five percent of the earth's surface, but only three percent of it is fresh water and less than one percent of the earth's fresh water is accessible and usable by humans. This results in 3.5 million deaths each year from water related illnesses, of which eighty-four percent are children, i.e. approximately one child every twenty seconds (UNHDP 2006; Lui et al. 2007). The use and costs of water are also unequally distributed. For example, a person in the United States uses more water in a five minute shower than a person in the developing world uses all day, and the urban poor often pay five-to-ten times more per liter of water than their wealthy neighbors in the same city (UNHDP 2006, 10 and 35). The crisis is compounded by competing interests (agriculture, industry and domestic use); ineffective management by governments; increased efforts to privatize water supply and distribution; war and political instability; and a general lack of political will to address the crisis. Water is poised to be a major source (if not the major source) of international conflict for the next century resulting in failed states, intra- and inter-state conflict, famine, migration, and global economic pressures (Office of the Director of National Intelligence (USA) 2012).

One of the principal debates surrounding the global water crisis is whether or not water is a human right or a commodity. Water is a basic necessity for human existence as are food, air, shelter and health care. Without water, people die. Thus classifying water as a human right seems accurate. But what is often overlooked in the "human right or commodity debate" is that water requires infrastructure (purification and a delivery system) in order to be useable. Some of the basic necessities of human existence are readily available and are freely distributed (air), others are labor intensive and require infrastructure and labor (food, clothing, shelter, and health care). While water at first seems more like air in that it is often available free of charge (rivers, lakes, and rain), usable/potable water is not free. It requires expensive infrastructure and labor; as such water is more like the other basic necessities that are treated as both right and commodities. The Catholic Church's teachings on water reflect a more nuanced understanding of water as both a right and a commodity.

The *Compendium of the Social Doctrine of the Church* describes water as "a gift from God," to which "everyone has a right" and decries treating water as "just another commodity among many" or as "merely an economic good" (Pontifical Council for Justice and Peace 2004, #484-85). Pope Benedict XVI, on World Water Day 2007, clarified the Church's understanding of water by describing water as "a common good of the human family" and "an essential element for life." He went on to say, "Access to water is in fact one of the inalienable rights of every human" that ought to be managed according to the ethical principles of subsidiarity, participation, and the preferential option for the poor (Benedict XVI 2009, 11-12). Important to note here is that Catholic social teaching on water describes *access* to water as a right and decries treating water exclusively as a commodity, but it does not preclude the privatization of water (Pontifical Council for Justice and Peace 2006, #2-6). Instead it engages the Church's teaching on private property which distinguishes between "use" and "ownership."

The focus here is the intersection of water, ethics and sacrament (Allman 2011). In Catholic social thought, water is both a *common good* (given by God intended for all to use) and a *material good* that can be privately owned, so long as "access to safe water and sanitation for all" is guaranteed (Pontifical Council for Justice and Peace 2006, #6). If the water crisis were simply about questions of access and sanitation, then this would be principally a matter of social justice. But water ethics spills over into sacramental theology because water is used in liturgy and sacrament. The global water crisis is redefining water's meaning. What was once a symbol of purification and life has for many become a commodity they cannot afford and a symbol of filth and death. This makes the global water crisis not only a social justice issue, but one of sacramental stewardship as well (Allman 2011).

That care for the environment should be a presumed as a consequence of liturgical celebration should be obvious. This is concretized in how we use water and how water rights are determined. There is an important connection between using water in sacraments and how peoples on this earth are enabled to have sufficient and potable water for their very survival.[28]

Eucharist: Bread and Wine, the Work of Human Hands, Food Distribution

The central elements of the Eucharist—bread and wine—are not "natural" symbols like water and light but are manufactured symbols, the result of human ingenuity and hard work. This involves the cycle of planting, harvesting, milling wheat into flour and baking bread, and planting, harvesting, crushing grapes and fermenting them to become our spiritual food and drink. There is something very fitting about this cycle of dying and rising to produce the bread and wine, consecrated to become the means for our participating in the dying and rising of Christ, the paschal mystery.

The rite of presenting bread and wine is central to the celebration of the Eucharist. That there can also be a collection of gifts for the poor is part of the directives for the Evening Mass of the Lord's Supper. This reflects the classic Roman liturgy when gifts for the poor were also brought in procession with bread and wine. And the role of deacons was to help that ritual action and then to distribute the food after the liturgy ended. This is the meaning of the prayer at the very end of the Roman Canon (referring to a variety of foodstuffs):

> Through whom you continue to make all these good things, O Lord,
> You sanctify them, fill them with life,
> Bless them, and bestow them upon us . . . .

That food distribution is a hallmark of our charity reflects an inherent Catholic aspect of the Eucharist as always a celebration of the church and always concerned for those who do not have enough to eat. In light of Pope Benedict's addresses about food distribution and just wages one can easily argue that this is part and parcel of the meaning of the celebration of the Eucharist. Not unrelated to this issue is land distribution and how equitably the world's resources are shared. This may also call into question how we farm and the need for newer technologies to support the world's increasing population and need for food. In and through the celebration of the Eucharist we are drawn into and experience the paschal mystery of Christ made possible by "the work of human hands." Clearly part of a creation consciousness about the liturgy is that we experience "the works of our redemption" through "the work of human hands" then sanctified and given as the bread of life and cup of eternal salvation.

Liturgy of the Hours: Their Presumed Cosmic Context Is Morning, Evening, Light, Darkness

One of the reasons why the "Divine Office" is now termed the "Liturgy of the Hours" is the intent to restore their celebration to the hours of the day for which they were designed, as well as the emphasis placed on the two "hinges" of the Hours, Morning and Evening (*General Instruction of the Liturgy of the Hours* (GILOH), #37-54). The cosmic context for these hours is obvious. The reference in the *Benedictus* to "the dawn from on high shall break upon us" (Lk 1:78) is unmistakable and a poignant end to Morning Prayer. The fact that the days of creation are used in the hymns at Evening Prayer is normally (unfortunately) eclipsed in celebration because they are replaced by other hymns. But at least in theory these hymns articulate how creation is intrinsic to our celebration of the Hours.

Perhaps the clearest instance of praising God for creation is through the words of the Psalmist used so thoroughly and fully in the Psalms used in the Hours (e.g., by design the third psalm at Morning Prayer is always a psalm of praise (GILOH, #43). For example, Psalm 8 is at Saturday Morning Prayer week II).

While a rationale often given for using psalms in the liturgy of the hours (or divine office, as reflected in the original prayer book *The Taize* Office) is that through it humans "give voice to the praise of creation," cannot one raise the question that this relies too much on humans and not emphasize creation the way the Psalms do and the way the liturgy does? For example on Sunday at Morning Prayer we use the canticle of the Three Young Men (Daniel 3) in which the author says:

> Bless the Lord, all you works of the Lord…
> Sun and moon, bless the Lord.
> Stars of heaven, bless the Lord.
> Every shower and dew, bless the Lord.
> All you winds, bless the Lord.

This is followed immediately by an unexpurgated set of phrases:

> Fire and heat, bless the Lord.
> Cold and chill, bless the Lord.
> Frost and chill, bless the Lord.
> Ice and snow, bless the Lord.

The point to be made is that all aspects of creation themselves by their nature give praise to the Lord, their maker. The praise of creation is praise of creation. All creatures and all of creation have their own voice. That humans "name" them in singing psalms and canticles is for their sake, that humans do not forget what creation is and does all the time. Humans take time each day to offer thanks and praise. Created things and all of creation do this all the time. This means that humans do not to dominate other aspects of God's creation but that especially through the liturgy we remind ourselves that the world in all its fullness and facets give praise by their very being to their creator. We do not "use" them. We see them and are grateful for and with them and in awe offer our thanks and praise for and with them to God our maker.

Liturgical Year: Cosmic Context for Celebration

At the outset it is important to admit that the basis for this argument is the liturgy as celebrated in the northern hemisphere. That the southern hemisphere experiences a different cosmic context needs to be factored into this understanding of the church's feasts and seasons. Liturgical theologians have worked on this issue for decades with a variety of approaches and conclusions.[29] At the risk of being regarded as dismissive, my own sense is that the historical basis for Easter is Passover as celebrated at the time of the events of Jesus' betrayal, death and resurrection. There is also the important, and often neglected, eschatological pull of every liturgical celebration which can be enhanced by appreciating that when in the north we celebrate a long day's journey into night at Christmas time in December, they are celebrating a long night's journey into day in the southern hemisphere, in the south.

In pride of place is the Easter Vigil understood as recreation and restoration of the cosmos. As the Holy Father has remarked, the reading of the creation account from Genesis 1 has cosmic and re-creational overtones. The prayers are rich in the way they combine salvation history, creation and redemption here and now. Then is the Vigil of Pentecost, with enhanced scripture readings in the revised *Lectionary for Mass* among which are the texts of the tower of Babel (Gen 11) and Ezekiel's vision of dry bones which are factored into their accompanying collect prayers referring to "the first of the Spirit" and a people "washed clean at baptism." The Pentecost Day readings, especially Acts 2, draw on and draw out cosmic symbolism and universality. Among others, the structure and contents of the liturgical commemoration of the solemnity of the birth of St. John the Baptist in late June was determined by the beginning of the diminishment of light in the cosmos in the northern hemisphere to coincide with "He must increase, I must decrease" (Jn 3:30).

While these examples are rich in cosmic symbolism and earth-related meanings, in my opinion one of the weaknesses in the current, revised liturgy is the elimination of Rogation Days and the spring blessing of the fields for a fruitful harvest. This mitigates the earthiness and primalness of the liturgy. While I would argue against adding a feast to the calendar to honor creation and nature, I judge that losing Rogation Days was a decided loss in that it was a day on which creation was valued, blessed and revered.

The convergence of these traditional Catholic beliefs with the value which Catholic practice places on the celebration of liturgy and sacraments and with the way these tenets are articulated through sacramental liturgy specifically leads to the *crucial contribution* which Roman Catholic theology can and must make to the environmental crisis today. Taking the liturgy this seriously as a source for theology and spirituality leads to delineating a truly *integral,* Catholic vision of theology and reality inspired by the rites we celebrate. It is not a question of adding it on to ideas the themes, but of (re)discovering what is contained, presumed and undergirds the celebration of the liturgy.

THE WORLD AS GOD'S ICON

The image of the world as *God's icon* serves our purpose not in the sense of a stylized art form or a familiar component of some forms of liturgical celebration. Rather *icon* here is used to convey the characteristic Christian understanding (commonly accepted and presumed in Catholic theology and liturgy) that the divine is experienced in the human and that the arena for any divine revelation is the world—this good earth which God has made. The world is thus not merely a forum for God's revelation or a locus for it. The world itself is part and parcel of divine revelation and intervention. In Catholic theology both the medium and the message are significant. In this book the world and all that dwell therein are regarded as God's icon and this earth is a major part of the goodness and beauty of the divine revealed *in and through* it—not just *on* it. Essential to a theological appreciation of the world is the interdependence of *all that dwell in it.* Essential to our thesis is the interdependence of all in the world with all that is God, viewed specifically through the liturgy as the prism which gives us a proper perspective on (even) this good earth.

This image also prevents the world from becoming an *idol*—recalling how idolization is one of the classic temptations of organized religion from the golden calf in the Book of Exodus on. As *God's icon* the world is to be taken as seriously as possible and revered. But it is not to be (mis)taken for the divine in itself—understood as both the totally immanent and the totally other, as the fully incarnate and the fully eschatological.

The *icon* notion images a correct balance between these extremes. It also mediates a proper appreciation of how the totally other is in fact immanent here and now and how the one who will welcome us one day to the joys of God's kingdom is the God whom we encounter here and now and in whom even now we "live and move and have our being" (Acts 17:28).

Thus the liturgy is seen to be both *eschatological* and *prophetic* of what is "not yet" but what should be. As close as we can come, yet we need to intercede and pray for the veil to be lifted and when "sacraments shall cease." From through a glass darkly to seeing God "face to face." God revealed in and through creation yet is also a God who is totally transcendent and "other," "*tremendum et fascinans*" this side of the veil.

CONCLUSION

Paradoxically, the ecological crisis may make a significant contribution to delineating a proper method for the study of and reflection on the liturgy because this crisis places sacramentality and mediation at the forefront of our reflection. Though *ex opera operato /operantis*, intention, and other categories have been useful for describing how the sacraments work, they may not speak to an environmentally conscious generation. Reflecting on the world as God's icon provides a way for the church to rearticulate beliefs and tenets about the liturgy and the sacraments that are meaningful and helpful today.

NOTES

1. See the comments in the introduction to my essay where I cite Thomas O'Meara's *Fundamentalist: A Catholic Perspective* and "Fundamentalism and Catholicism: Some Cultural and Theological Reflections."

2. For three years I was privileged to serve on the USCCB and United Methodist Church-sponsored dialogue on ecology and the Eucharist whose final agreed statement is entitled "Heaven and Earth Are Full of Your Glory." I was honored to be one of the final editors of the statement and equally honored that citations from my 2005 work *Models of the Eucharist* were included in its footnotes. See http://www.usccb.org/news/2012/12-070.cfm.

3. The ecological initiatives of the Ecumenical Patriarchate began in 1986 and continue to the present. These include Patriarchal encyclicals, messages, homilies, interviews, results of Pre-Synodal Pan Orthodox conferences, assemblies of the Patriarchs and Primates of the Orthodox Church, seminars, regular and extraordinary meetings, publications of religious services on the environment, studies and books. Among others see the impressive publications from the Ecumenical Patriarchate of Constantinople: *So That God's Creation May Live,* The Orthodox Church Responds to the Ecological Crisis (1992); *The Environment and Religious Education* (1994); *The Environment and The Environment and Ethics* (1995); etc. through to the Sixth Symposium entitled "Religion, Science and the Environment" in 2007 to which Pope Benedict sent a "Message" and in 2012 entitled "L'Artico: Specchio della Vita" which appeared in *Per Una Ecologia dell'Uomo.* Antologia di Testi, ed. Maria Milvia Morciano, intro. Jean-Louis Bruges (Vatican: Libreria Editrice Vaticana), 46-48.

4. #10: "Nevertheless the liturgy is the summit toward which the activity of the Church is directed; at the same time it is the font from which all her power flows. For

the aim and object of apostolic works is that all who are made sons of God by faith and baptism should come together to praise God in the midst of His Church, to take part in the sacrifice, and to eat the Lord's supper."

5. For a useful contemporary overview of the complex topic of the various meanings of "magisterium" and the relative authority of magisterial documents, see: Avery Dulles (2007), *Magisterium: Teacher and Guardian of the Faith.* Naples: Sapientia Press; much of which expands on his (1988), "Lehramt und Unfehlbarkeit," in *Handbuch der Fundementaltheologie 4: Traktat Theologische Erkenntnislehre,* ed. Walter Kern, Josef Pottmeyer and Max Seckler, 153-178 (Friburg: Herder).

6. For an assessment of the importance of distinguishing among genres of magisterial documents and a delineation among them also see Francis A. Sullivan's *Creative Fidelity.*

7. Originally in Italian the book is a series of responses which the pope wrote in 1994 to questions posed by the Italian journalist Vitorrio Messori.

8. Also of note is the USCCB Statement "Renewing the Earth" (1991) which is available at http://www.usccb.org/issues-and-action/human-life-and-dignity/environment/renewing-the-earth.cfm.

9. See the helpful compendium of Benedict XVI's texts through November 2011: (2012), *Per Una Ecologia dell'Uomo: Antilogia di Testi,* ed. Maria Milvia Morciana (Roma: Libreria Editrice Vaticana). Among other recent books by Catholic theologians see Jame Schaefer (2009), *Theological Foundations for Environmental Ethics: Reconstructing Patristic and Medieval Concepts* (Washington: Georgetown University Press).

10. That they may well have a status above that of an encyclical could seem to be held given that they are the result of episcopal collegiality expressed in a synod for bishops.

11. See the helpful essay by Edward J. Kilmartin (1980), "The Sacrifice of Thanksgiving and Social Justice" in *Liturgy and Social Justice,* ed. Mark Searle, 53-71 (Collegeville: The Liturgical Press).

12. Among many others the website of the Natural Resources Defense Council is replete with information and advocacy about health hazards as a result of ecological degradation. Similarly the USCCB website indicates a number of bishops' conference statements about environmental health hazards.

13. See the statement from the Pontifical Council Justice and Peace (2012), *"L'Eau,Un Element Essentiel pour la Vie:* Instaurer des Solutions Efficaces," *Mise à jour,* Contribution du Saint-Siège au VI ème Forum Mondial de l'Eau (Marseille, France, mars). Available at http://www.vatican.va/roman_curia/pontifical_councils/ justpeace/documents/ rc_pc_justpeace_doc_20120312_france-water_fr.html.

14. Especially *Opera Omnia* (2000), pt. 1, chap. 2 on "Liturgy – Cosmos – History," 36-47; and Part Four, chap. two "The Body and The Liturgy," 162-75, trans. John Saward (San Francisco: Ignatius Press).

15. The literature on this is vast. Among others for a summary of the state of European and American authors see Kevin W. Irwin (1990), *Liturgical Theology: A Primer* (Collegeville: The Liturgical Press).

16. The General Instruction of the Roman Missal states (#299) that Mass should be celebrated "facing the people whenever possible" there are voices being raised that *ad Orientem* is the preferred posture. See Joseph Ratzinger, *The Spirit of the Liturgy,* "The Altar and Direction of Liturgical Prayer," 75-84. For a more complete argument favoring the *ad Orientem* posture see Uwe Michael Lang (2009), *Facing Towards the Lord,* Orientation in Liturgical Prayer (San Francisco: Ignatius Press).

17. As the Jesuit liturgical scholar Robert Taft says: "Liturgy . . . reminds us of the powerful deeds of God in Christ. And being reminded we remember, and remembering we celebrate, and celebrating we become what we do. The dancer dancing is the dance." Robert Taft (1986), *The Liturgy of the Hours in East and West: The Origins of the Divine Office and its Meaning for Today* (Collegeville: Liturgical Press), 345. And in another highly influential essay on method in liturgical studies, he says that "Liturgy it like a top. You cannot understand it unless you spin it." Robert Taft (1997), "The

Structural Analysis of Liturgical Units: An Essay in Methodology" in *Beyond East and West*, 2nd ed. (Rome: Pontifical Oriental Institute), 191-192).

18. More recently I have wondered whether "the sacramental principle" might be a better alternative phrasing in order to denote that it is really foundational and fundamental too all liturgy.

19. For indications of my own work in this area see: Kevin W. Irwin (1998), "Sacramentality and the Theology of Creation: A Recovered Paradigm for Sacramental Theology," *Louvain Studies* 23: 159-79; (2001), "Discovering the Sacramentality of Sacraments," *Questions Liturgiques* 81; Sacramentality, Eschatology and Ecology" in *Contemporary Sacramental Contours of a God Incarnate*, ed. Lieven Boeve and Lambert Leijssen, 111-123 (Leuven: Peeters); (1988), "Sacramental Theology," *New Catholic Encyclopedia*, Supplement 18, 447-52 (Washington: Catholic University of America); (2010), "Sacramentality: The Fundamental Language for Liturgy and Sacraments" in *Per Ritus et Praeces – Sacramentalita della Liturgia*, 131-60 (Rome: Studia Anselmiana).

20. Among the more accessible Orthodox voices is Alexander Schmemann (1997), *For the Life of the World: Sacraments and Orthodoxy* (New York: Herder and Herder).

21. The Latin phrases are "*creaturam aquae multis modis praeparasti*," "*hanc creaturam aquae benedicere dignetur*," and "*ut hanc creaturam benedicere*." See Kevin W. Irwin (1995), "The Theology of Creation in the *Missale Romanum* of Paul VI" in *Unum omnes in Christo. In unitatis servitio*, Miscellanea Gerardo J. Bekes OSB Octogenario Dedicate, 109-126 (Pannonhalam: Bences Foapatsag).

22. Mediated language is important so that we address God with modifiers "Lord," Savior," "Redeemer" etc. not "Jesus" alone and unmodified.

23. While others use the phrase *lex agendi* to indicate the requisite life relationship of the liturgy in my own work I have distinguished these term arguing that *lex agendi* means the actual enactment of the liturgy whereas *lex vivendi* is the life relation of the liturgy.

24. Among the more extensive treatments of this life relation see Raimondo Frattalone (2011), *Liturgia e Vita* in *Ogni Cosa Rendete Grazie; Questa e Infatti la Volunta di Dio in Cristo Gesu Verso di Voi* (Rome: Edizione Liturgiche).

25. See the important collection of essays published by Federation of Diocesan Liturgical Commissions (1980), *Liturgy and Social Justice*, ed. Mark Searle (Collegeville: Liturgical Press).

26. Pope Benedict XVI's homily at the Easter Vigil April 23, 2011, in *Per Una*, 170-75, at 172.

27. One of my suspicions is that in implementing the reformed liturgy that many American Catholic congregations adopted and adapted a Protestant model which emphasized the proclamation of the Word and preaching sometimes to the detriment of sign, symbol and the premise of sacramentality.

28. Among others, see the "Note" prepared by the Pontifical Council for Justice and Peace on water rights, accessible athttp://www.vatican.va/roman_curia/pontifical_councils/justpeace/ documents/rc_pc_justpeace_doc_ 20030322_kyoto-water_en.

29. Among others see Anscar Chupungco (1977), *The Cosmic Elements of Christian Passover* (Roma: Editrice Anselmiana).

REFERENCES

Allman, Mark J. 2011. "Theology H2O: The World Water Crisis and Sacramental Imagination." In *Green Discipleship: Catholic Theological Ethics and the Environment*, edited by T. Winright, 379-406. Winona MN: Anselm Academic.

Benedict XVI. 2007. *Jesus of Nazareth*. Vol. Two: *From the Baptism in the Jordan to the Transfiguration*. Rome: Libreria Editrice Vaticana.

———. 2007. *Sacramentum Caritatis: On the Eucharist as the Source and Summit of the Church's Life and Mission*. Accessed March 15, 2013. http://www.vatican.va/holy_

father/benedict_xvi/apost_exhortations/documents/hf_ben-xvi_exh_20070222_sacramentum-caritatis_en.html.
———. 2009. "Message of the Holy Father Benedict XVI on the Occasion of the Celebration of World Water Day 2007." In *Catholic Social Principles Towards Water and Sanitation*, ed. Dennis Warner. Baltimore: Catholic Relief Services.
———. 2011. *Jesus of Nazareth*. Vol. Three: *Holy Week: From the Entrance into Jerusalem to the Resurrection*. Rome: Libreria Editrice Vaticana.
———. 2012. *Jesus of Nazareth*. Vol. One: *The Infancy Narratives*. Rome: Libreria Editrice Vaticana.
Cowdin, Daniel. 1994. In *Preserving the Creation: Environmental Theology and Ethics*, edited by Kevin W. Irwin and Edmund D. Pellegrino, 112-147. Washington: Georgetown University Press.
Irwin, Kevin W. 1994. "The Sacramentality of Creation and the Role of Creation in Liturgy and Sacraments." In *Preserving the Creation. Environmental Theology and Ethics*, edited by Kevin W. Irwin and Edmund J. Pellegrino, 67-111. Washington: Georgetown University Press.
———. 2002. "The Development of Sacramental Doctrine in the Church." In *Recovering the Riches of Anointing: A Study of the Sacrament of the Sick*, edited by Jenifer Glen, 59-82. Collegeville: The Liturgical Press.
———. 2005. *Models of the Eucharist*. New York/Mahwah: Paulist Press.
John Paul II. 1990. "Peace with God the Creator, Peace with All of Creation." Message for the Celebration of the World Day for Peace. Accessed March 15, 2013. http://www.vatican.va/holy_father/john_paul_ii/messages/peace/documents/hf_jp-ii_mes 19891208 _xxiii-world-day-for-peace_en.html.
———. 1994. *Crossing the Threshold of Hope*. Edited by Vittorio Messori. Translated by Jenny McPhee and Martha McPhee. New York: Alfred A. Knopf.
Lui, Jie et al. 2011. *Water Ethics and Water Resource Management*. Ethics and Climate Change in Asia and Pacific Working Project, Working Group 14 Report. Bangkok, Thailand: United Nations Educational Scientific and Cultural Organization (UNESCO).
Morrisey, Francis G. 1978. *The Canonical Significance of Papal and Curial Pronouncements*. Washington DC: Canon Law Society of America.
Office of the Director of National Intelligence (USA). 2012. *Global Water Security*. Intelligence Community Assessment 2012-08, February 2. Accessed March 15, 2013. http://www.dni.gov/nic/ICA_Global%20Water%20Security.pdf.
O'Meara, Thomas. 1990. *Fundamentalist: A Catholic Perspective* (New York/Mahwah: Paulist Press.
———. 1996. "Fundamentalism and Catholicism: Some Cultural and Theological Reflections." *Chicago Studies* 35: 68-81.
Paul VI. 1963. *Sacrosanctum Concilium: Constitution on the Sacred Liturgy*. December 4. Accessed on March 15, 2013. http://www.vatican.va/archive/hist_councils/ii_vatican_council/documents/vat-ii_const_19631204 _sacrosanctum-concilium_en.html.
Pontifical Council for Justice and Peace. 2004. *Compendium of the Social Doctrine of the Church* Washington, DC: USCCB Communications.
———. 2006. *Water, An Essential Element, An Update*. Vatican Website. Accessed March 15, 2013. http://www.vatican.va/roman_curia/pontifical_councils/justpeace/documents/rc_pc_justpeace_doc_20060322_mexico-water_en.html.
Ratzinger, Josef. 2000. *The Spirit of the Liturgy*. Translated by John Saward. San Francisco: Ignatius Press.
———. 2010. *Opera Omnia. Theologia della Liturgia*, Vol. XI: *La Fondazione Sacramentale dell'Esistenza Cristiana*, edited by Gerhard Ludwig-Muller Citta del Vaticano: Libreria Editrice Vaticana. Translated by Edmondo Caruana and Pierluca Azzaro. From the German original, 2008.

Sullivan, Francis A. 2003. *Creative Fidelity: Weighing and Interpreting Documents of the Magisterium*. Eugene OR: Wipf and Stock. Originally published in 1996 (Mahwah, NJ: Paulist Press).

Tillard, Jean Marie Roger. 1985. "Blessing, Sacramentality and Epiclesis" In *Blessing and Power,* edited by Mary Collins and David Power, 96-110. Edinburgh: T and T Clark.

United Nations Human Development Programme (UNHDP). 2006. *Human Development Report 2006: Beyond Scarcity—Power, Poverty, and the Global Water Crisis.* New York: Palgrave MacMillan

"Water Facts." 2013. Water.org. Accessed March 15. http://water.org/water-crisis/water-facts/water/.

Wojtyla, Karol. 1979. *The Acting Person: A Contribution to Phenomenological Anthropology*. New York: Springer Verlag.

Zitnik, Maksimilijan, ed. 1992. *Sacramenta: Bibliographia Internationalis,* Vol. 1. Rome: Pontifical Gregorian University Press.

NINE

Pope Benedict XVI's Cosmic Soteriology and the Advancement of Catechesis on the Environment

Jeremiah Vallery

As a species, we have reached such a high degree of technological advancement that we have seemingly become masters of nature.[1] We have acquired promethean fire, but only at the cost of upsetting Earth's ecological equilibrium. It is evident that human beings exist within a whole characterized by a complex interacting between myriad species; it follows that we should cultivate a consciousness of our relatedness to and reliance upon the earth and the variegated life forms it supports. The theological corollary to this idea is that the salvation of human beings and the redemption of the earth (cf. Rom 8:19-25) are inextricably related. The notion that only human beings are capable of obtaining salvation has had deleterious repercussions for the earth. It is imperative that we consider anew the place of human beings in God's overarching plan for the cosmos.

While Catholic theology has evolved in its attentiveness to environmental concerns during the last half-century, Catholic catechesis on the environment has not kept up with the pace of this development. Many Catholics in America have an indifferent disposition toward the environmental crisis that is facing us, either believing that it does not exist, or that the issue is being blown out of proportion by propagandists. A catechesis that has a more theologically robust doctrine of creation that is based on Christ and that is capable of amplifying the ecological awareness within the People of God is needed in the Catholic Church today. American Catholics' lack of interest in the environment and the difficulty they have in relating the environment to their faith can be remedied by a renewal of catechesis on the environment inspired by Benedict XVI's cosmic soteriology. The pope emeritus recognizes the intrinsic interconnectedness between human beings and the universe, and affirms that the universe itself is a recipient of God's salvation. He teaches that the salvation of the cosmos flows from the Paschal Mystery, proceeds through liturgical worship, and culminates in the eschatological renewal of the universe. These insights build upon the foundations of contemporary catechesis on the environment, and prepare the way for a renewed Christocentric catechesis.

This chapter consists of three parts. First, I analyze the current state of catechesis on the environment in the United States, identify areas that can be strengthened, and present some of the attitudes Catholics have toward the environment. In the second part, I explain Benedict's cosmic soteriology and describe how Pierre Teilhard de Chardin influenced his thought. Third, I offer some suggestions on how Benedict's insights can be used by catechists to present to the faithful a doctrine of salvation that considers the role of the redemption of creation in our salvation, and the duties that flow from this relationship.

CATECHESIS ON THE ENVIRONMENT: *STATUS QUAESTIONE*

Since it would be beyond the scope of this chapter to provide an overview of all the catechetical literature that is used in the United States today, I will restrict my investigation to the *Catechism of the Catholic Church* (hereafter *Catechism* 2000) and to the *United States Catholic Catechism for Adults* (hereafter *US Catechism* 2006). The former has the distinction of being the first universal catechism to address environmental concerns, and the latter has faithfully followed in its footsteps.

The Catechism of the Catholic Church

There are three passages in the *Catechism of the Catholic Church* that explicitly mention humanity's duty to care for creation. The first states (2000, #339), "Man must . . . respect the particular goodness of every creature [and] avoid any disordered use of things which would be in contempt of the Creator and would bring disastrous consequences for human beings and their environment." According to the second passage (#2402), "the earth and its resources" are under "the common stewardship of mankind," and "are destined for the whole human race." Similarly, the third passage states (#2415):

> The seventh commandment enjoins respect for the integrity of creation. Animals, like plants and inanimate beings, are by nature destined for the common good of past, present, and future humanity. Use of the

> mineral, vegetable, and animal resources of the universe cannot be divorced from respect for moral imperatives. Man's dominion over inanimate and other living beings granted by the Creator is not absolute; it is limited by concern for the quality of life of his neighbor, including generations to come; it requires a religious respect for the integrity of creation.

The messages of these three passages can be reduced to three common themes: (1) the goodness of creation, (2) the "disastrous consequences" of the ecological crisis, if left unheeded, and (3) the social dimension of the ecological crisis. These are crucial components of the Catholic Church's teaching on the environment, and provide a sure foundation on which further catechetical reflection can be built.

In and of themselves, these passages lack a strong connection to Christology. While Christ is the heart of catechesis, since these passages are often taken out of context, the relation between Christ and the environment becomes obscured. Since many deists agree with the three themes of the above passages from the *Catechism,* i.e. that creation is good, that humanity is destined to commit a gradual mass suicide if the environmental crisis is neglected, and that the ecological crisis spreads suffering for present and future generations, it is apparent that the message of such passages is no longer distinctively Christian when said passages are extrapolated from their broader context, which is explicitly Christological. In a catechetical context, social justice issues must be explained in the light of Christ and his mission of salvation.

The United States Catholic Catechism for Adults

The *U.S. Catholic Catechism for Adults* (2006, 424) contains an important paragraph explaining the obligation of Catholics to conserve the environment:

> We show our respect for the Creator by our stewardship of creation. Care for the earth is a requirement of our faith. We are called to protect people and the planet, living our faith in relationship with all of God's creation. This environmental challenge has fundamental moral and ethical dimensions that cannot be ignored.

Another passage in the *U.S. Catechism* (2006, 422; citing *Catechism* 2000, #2432) mentions more explicitly the duties that businesses have toward the environment: "Those *responsible for business enterprises* are responsible to society for the economic and ecological effects of their operations. They have an obligation to consider the good of persons and not only the increase of *profits.*" The first passage (2006, 424), which speaks in broad strokes about the duty of Christians to care for the earth, is an improvement on the *Catechism of the Catholic Church* inasmuch as it makes explicit that "protect[ing] . . . the planet" is something that is inextricably related to the faith; however, it too lacks a Christocentric focus. The *U.S. Catechism* primarily explains *that* Catholics have a responsibility toward the earth and does not elaborate *why* they have such a responsibility, or how it is related to other elements of the Faith.

Some Catholic Attitudes on the Environment

Why are Catholics indifferent toward the environmental crisis? The reasons are multifaceted. While not exhaustive, the following list of five attitudes that some Catholics exhibit provides some insight into why they are apathetic: (1) some Catholics are very skeptical about the scientific evidence for climate change, (2) they have a tendency to think that all environmentalists are extremists, (3) they are convinced that environmentalism is a political issue that has nothing to do with religion, (4) in light of other moral issues (especially pro-life issues), they do not perceive the active care for our environment as something that is significant, and (5) they believe that the concern that we should have for the environment pales in comparison with their own salvation, and that therefore, they should be more concerned about the state of their souls than the state of the earth.

The first four of these attitudes are related to political perspectives. The current two-party-system in the United States forces Catholics to be politically heterogeneous if they are to advocate all of the social justice issues which the Catholic Church upholds. The Catholic Church seeks to protect the lives of unborn babies through the proscription of abortion and recognizes that protecting the most vulnerable among us and protecting God's creation are mutually related (Benedict XVI 2009, #51). The ways the two predominant political parties in the United States handle abortion and environmentalism demonstrates this division of values. On the one hand, Republicans tend to perceive the environmental movement as a monolithic entity which values animals more highly than people.[2] Many of them also are extremely suspicious of scientific evidence which indicates that global warming is a reality. On the other hand, while Democrats are much more concerned about the environment than Republicans, many of them support abortion laws which have led to the deaths of millions of innocents.[3] Our two-party system effectively pits the protection of the earth against the protection of unborn babies. Because citizens only have one vote, and no third-party candidate has been in the oval office since 1853, in a sense, American Catholics are forced to choose between defending the lives of the most innocent and preserving creation.

Catholics hold the last of the five attitudes mentioned above for doctrinal reasons rather than political reasons. In the grand scheme of the Christian understanding of redemption and the fullness of life in Christ, to some Catholics it seems as though the environment does not play a very significant role. After all, if salvation, which is only fully attained in the afterlife, is about the individual, why should one be that concerned about the earth? The infinite weight of the glory to be attained in the next life compared to this life often leads these people to overlook the significance of this earth for future generations. The view that the environment is not related to our ultimate end divorces human beings from the earth. Catholics need to be aware that eschatology and ecology are fundamentally linked (see Haught 1996).

The Catholic Church can begin to recalibrate these attitudes which reinforce indifference toward the environment through a renewal of catechesis on the environment that takes into account the cosmic extent of Christ's sacrificial death and resurrection, the relationship between liturgical worship and nature, and the common eschatological destiny of human beings and the earth. These elements, which constitute the basis of Benedict's cosmic soteriology, have the potential to change the hearts and minds of the faithful. By articulating these connections, the Catholic Church will be able to usher in an era of a heightened awareness of the religious significance of God's creation and of our obligation toward the environment.

BENEDICT XVI'S COSMIC SOTERIOLOGY

The roots of Benedict's cosmic soteriology predate Vatican II. On November 20, 1961, Cardinal Joseph Frings, the archbishop of Cologne, gave a speech in Genoa that had as its theme a comparison of the state of the world during the First Vatican Council (1869-70) with the state of the world at the dawn of the Second Vatican Council. Although Frings gave the speech, it was composed by Fr. Joseph Ratzinger, a young theology professor whom Frings had asked for assistance in preparing for the Second Vatican Council. The speech was published in German, French and Italian. After reading the Italian text, Pope John XXIII "summoned Card. Frings to a private audience to thank and commend him for setting forth ideas which agreed with ways in which he, Pope John, saw the situation and tasks of the coming Council" (Wicks 2008, 235).

The text is wide-ranging, covering social, philosophical, and theological issues. The following excerpt from the speech expresses Benedict's (Ratzinger 1961/1962, 172-3; my translation) early thoughts on cosmic soteriology:

> Perhaps the Christianity of the last century had actually restricted itself a little too much on [the issue of] the spiritual salvation of the individual found in the afterlife, and had not proclaimed loudly enough the salvation of the world, the universal hope of Christianity. Thus, it has acquired the task of thinking through these thoughts anew, and of simultaneously juxtaposing the fervor for the earth felt by modern peo-

> ple with a new, positive interpretation of the world as creation bearing witness to God's glory and, as a whole, destined for salvation in Christ, who is not only head of his Church, but is also the Lord of creation (Eph. 1:22; Col. 2:10; Phil. 2:9f.).[4]

This passage is notable since it explicitly mentions the salvation of the world (*Heil der Welt*). While this idea is certainly not new in Christianity, the call to reconsider this subject on the threshold of Vatican II seems to have anticipated much of the "fervor for the earth" (*der Inbrunst für die Erde*) which appeared after the council.

At the outset, it should be made clear that Benedict's doctrine of the salvation of the universe is different from the ones advocated by Origen of Alexandria and Hans Urs von Balthasar. Whereas Origen (*De Principiis* I.6.3) speculated that the demons, including Satan, will be saved and von Balthasar (1988) similarly proposed that every human being might be saved, Benedict (Ratzinger 2003, 34-38) affirms that God has given human beings the freedom to reject him; while God wants all saved, he respects each individual's freedom, and never forces salvation upon anyone. While the essence of his cosmic soteriology is that God's salvation applies not only to human beings, but also applies to the whole universe, in no way is it to be identified as *apokatastasis*, i.e. the salvation of all.

The Three Pillars of Benedict's Cosmic Soteriology

Pope Benedict XVI's cosmic soteriology consists of three main themes: the cosmic scope of Jesus' death and resurrection, the relationship between liturgical worship and the world, and the eschatological fullness of cosmic redemption. These themes are all interrelated in Benedict's thought. In his words (Ratzinger 2000, 60) prior to his pontificate:

> In the first stage the eternal is embodied in what is once-for-all [i.e., the Pasch of Jesus]. The second stage is the entry of the eternal into our present moment in the liturgical action. And the third stage is the desire of the eternal to take hold of the worshipper's life and ultimately of all historical reality. The immediate event—the liturgy—makes sense and has a meaning for our lives only because it contains the other two dimensions. Past, present, and future interpenetrate and touch upon eternity.

The passion of Christ and his resurrection inaugurate this movement of God's salvation, which is carried on through the ages by the liturgy, and which culminates in the eschatological unification of all creation. Benedict's soteriology is closely related to his liturgical theology since he sees the liturgy as the primary means through which human beings encounter the divine in this in-between stage of world history. Liturgy functions as a bridge between the event of the Paschal Mystery on the one hand, and the renewed universe on the other.

The Cosmic Ramifications of the Christ Event

The first theme in Benedict's cosmic soteriology is the cosmic extent of the sacrifice of Christ. His soteriology is different from most pre-conciliar Western soteriologies since he sees salvation as related to the cosmos. For Benedict (2011, 79), not only is Christ's sacrifice on the cross the means through which salvation is wrought for human beings, but it also has historical and cosmic ramifications. In his words, "His Cross and his exaltation is the Day of Atonement for the world, in which the whole of world history—in the face of all human sin and its destructive consequences—finds its meaning and is aligned with its true purpose and destiny."

Another basis for this all-embracing view of salvation is his insistence on the fundamental contextual nature of humanity. According to Benedict (Ratzinger 2004, 245), "man is himself only when he is fitted into the whole: into mankind, into history, into the cosmos, as is right and proper for a being who is 'spirit in body.'" A new way of seeing how human beings fit into the world calls for a new way of conceiving soteriology so that the context of those who are saved is also taken into account. "The universality of Jesus' mission . . . concerns not just a limited circle of chosen ones—its scope is the whole of creation, the world in its entirety," explains Benedict (2011, 100). "Through the disciples and their mission, the world as a whole is to be torn free from its alienation, it is to rediscover unity with God." Salvation is more than the personal individualistic salvation that was preached by the Catholic Church for centuries–it embraces all creation.

The Christological dimension of Benedict's cosmic soteriology resonates with Pauline Christology which, like Benedict's soteriology, also centers cosmic redemption on the cross. The letter to the Colossians states, "For in him all the fulness of God was pleased to dwell, and through him to reconcile to himself all things, whether on earth or in heaven, making peace by the blood of his cross" (1:19-20, RSV). Similarly, Benedict (Ratzinger 2004, 287) explains that "the hour of the Cross is the cosmic day of reconciliation . . . . [t]here is no other kind of worship and no other priest but he who accomplished it: Jesus Christ." The liturgy extends through time the cosmic form of worship inaugurated by Jesus Christ on the cross.

Liturgical Worship and the Cosmos

The blood of the covenant was sprinkled on the Israelites on the Day of Atonement (Ex 24:8). For Christians, Jesus' blood is the blood of the new covenant (Mt 26:28). The liturgical reenactment of the offering of the body and blood of Christ is an integral event in the life of the Church, precisely because it is a means not only of recalling Jesus' sacrifice, but also of participating in that sacrifice. For Benedict (Ratzinger 2005, 22), "The sharing in this 'pasch' of Christ . . . is consummated in the Liturgy." This participation of human beings in the self-offering of Jesus Christ has a transformative effect on Christians. There are two overarching themes in Benedict's cosmic soteriology in relation to the liturgy. The first is the connection between covenant and cosmos and the second is the relation between worship and the transformation of the world.

According to Benedict (Ratzinger 2000, 26), cosmos and covenant are very closely related. The nature of their relationship is that the covenant is the "goal of creation," an idea which he borrows from rabbinic theology. The covenant is the teleological antecedent for creation. Benedict (2011, 78) explains that in rabbinic thought, "[t]he cosmos was created, not that there might be manifold things in heaven and earth, but that there might be a space for the 'covenant,' for the loving 'yes' between God and the human respondent." For the Jews, the Day of Atonement is the apex of the liturgical year since it "restores this harmony, this inner meaning of the world that is constantly disrupted by sin." In fact, Benedict (Ratzinger 2000, 27) goes so far as to say that without the covenant, creation would be "an empty shell." The covenant is something greater than creation since creation is established for the unification of human beings with God, which is the end of the covenant. Without a possible relationship between God and humanity, the universe would serve no purpose.[5] In other words, the universe is the existential context of human beings, who have been endowed with the capacity to know and love God. For Benedict (Ratzinger 2000, 27), "Creation looks towards the covenant, but the covenant completes creation and does not simply exist along with it." He concludes that worship, which is a human activity meant to draw human beings toward the divine, "not only saves mankind but is also meant to draw the whole of reality into communion with God." This leads to the second liturgical theme in Benedict's cosmic soteriology: universal divinization.

For Benedict (Ratzinger 1996, 37), union with God, which is the purpose of the covenant and of worship, is not merely a vertical connection between God and the individual. All who partake of the body and blood of Christ "are . . . assimilated to this 'bread' and thus are made one among themselves—*one* body." In Benedict's view (Ratzinger 1996, 37), the Church, which is the body of Christ, is a unified entity that is saved collectively. He explains that "communion means the fusion of existences." Individual existences are fused together through the partaking of the body and blood of Christ since "[c]ommunion means that the seemingly uncrossable frontier of my 'I' is left wide open and can be so because Jesus has first allowed himself to be opened completely, has taken us all into himself and has put himself totally in our hands." Benedict (Ratzinger 2000, 28) expresses something similar when he describes the meaning of sacrifice:

> Belonging to God . . . means emerging from the state of separation, of apparent autonomy, of existing only for oneself and in oneself. . . . That is why St. Augustine could say that the true "sacrifice" is the *civitas Dei,* that is, love-transformed mankind, the divinization of creation and the surrender of all things to God: God all in all (cf. 1 Cor 15:28). That is the purpose of the world. That is the essence of sacrifice and worship.

Union with God consists of communion (*koinonia*) in which human beings are not only connected to God, but are also horizontally united with each other, as it were, in the Church through bonds of love. A human being's communion with God, therefore, is not merely a union between the individual and the Almighty; it is a union which, through sacrifice, extends beyond the individual to embrace other human beings as well as the entire creation. This unity ultimately leads to the apotheosis of the cosmos.

While unity depends upon God's action, individual Christians play a significant role in preparing creation for this cosmic transformation, a role that one could even describe as priestly. Since the universe would serve no purpose without humanity's capacity to be in union with God, the divinization of the universe cannot occur except through human beings. Just as the temple was made a holy place of worship for the ancient Israelites, so too is the universe to be made into a holy place where human beings, the "priests of creation," worship God (Zizioulas 2003, III; cf. Benedict XVI 2011, 92). Within this temple of creation, however, God is destined to dwell (Rv 21:22). The incarnation was "the beginning of a momentous movement in which all matter is to become a vessel for the Word" (Ratzinger 1997, 88). Benedict (92) elaborates on this point by explaining, "To spiritualize means to incarnate in a Christian way, but to incarnate means to spiritualize, to bring the things of the world to the coming Christ, to prepare them for their future form and thus to prepare God's future in the world." Through the liturgy, which impacts our actions outside of the liturgy, we are disposed to properly impart to the cosmos something of its final form.[6] We are empowered with freedom to make a lasting impact upon the state of the earth, thereby becoming cooperators with God in the eschatological renewal. Human beings are called to prepare creation for its destiny of salvation in Christ Jesus by caring for it, nurturing it, preserving it, holding it in dignity and respect, and using it in a sanctifying manner. Here, the liturgy provides a paradigm for environmentalist action: just as the bread and wine are prepared by human beings and is transformed into the body and blood of Christ, so too is the earth to be prepared for its transformation and divine inhabitation.

According to Benedict (Ratzinger 2000, 53), "Christian worship is surely a cosmic liturgy, which embraces heaven and earth." The liturgy is cosmic primarily because the universe itself is mysteriously "praying with us" and is "waiting for redemption" (70). Worship, in his view, is a "participation" in the Paschal mystery—as a result, through means of worship, human beings are drawn to the event of Christ's sacrifice on the cross (cf. Jn 12:32) (34). The cosmos also participates in this movement since God desires by means of the liturgy "to transform us and the world" (175). Benedict (2011, 238) explained that the liturgy "is always a matter of drawing every individual person, indeed, the whole of the world, into Christ's love in such a way that everyone together with him becomes an offering that is 'acceptable, sanctified by the Holy spirit' (Rom 15:16)." Far from being a merely anthropological phenomenon, worship involves the whole universe.

Eschatology and the Unification of the Cosmos in Christ

The liturgy has an eschatological orientation. The transformation of the world and of human beings is announced by the transubstantiated bread and wine. These common elements, by becoming divinized during the consecratory prayer of the priest, provide a glimpse of what is to happen at the end by prefiguring the transformation of humanity and of the entire universe. Benedict (Ratzinger 2000, 173) describes this in dramatic fashion:

> The elements of the earth are transubstantiated, pulled, so to speak, from their creaturely anchorage, grasped at the deepest ground of their being, and changed into the Body and Blood of the Lord. The New Heaven and the New Earth are anticipated. The real "action" in the liturgy in which we are all supposed to participate is the action of God himself. This is what is new and distinctive about the Christian liturgy: God himself acts and does what is essential. He inaugurates the new creation, makes himself accessible to us, so that, through the things of the earth, through our gifts, we can communicate with him in a personal way.

While, in one sense, the new creation is inaugurated in the "here-and-now," in another sense, the completion of the new creation is still to come. As Benedict (Ratzinger 2000, 50) explains, "The great gesture of embrace emanating from the Crucified has not yet reached its goal; it has only just begun. Christian liturgy is liturgy on the way, a liturgy of pilgrimage toward the transfiguration of the world, which will only take place when God is 'all in all.'" The liturgy will achieve its goal only at the end when all things are united in God.

According to Benedict (Ratzinger 2007, 238), heaven occurs in two phases. In the first, human beings are able to participate in the life of the Trinity by sharing in the life of Christ in the present life—in the second, the cosmos itself and saved humanity will be the locus of God's presence (cf. Rv 21:22). In this depiction of the two moments of heaven, the salvation of the individual and the redemption of the universe are related in such a way that human salvation is fulfilled only when the cosmos is restored: "The perfecting of the Lord's body in the *pleroma* of the 'whole Christ' brings heaven to its true cosmic completion . . . the individual's salvation is whole and entire only when *the salvation of the cosmos* and all the elect has come to full fruition" (emphasis added).

The three stages of cosmic soteriology in Benedict's thought (i.e. the Paschal Mystery, the liturgy, and the eschatological transformation of the universe) chronologically unfold into the final state of the salvation of all of creation. Each successive stage makes sense only in relation to the others. Christ's status as savior of the world receives a deeper significance than what has been attributed to it in the past—not only does he save humanity, but the universe itself is taken up by God, through Jesus Christ, and is redeemed in a process that is inaugurated by his self-sacrifice, is continued in the liturgy, and is completed in the divinization of creation at the end of time.

The Influence of Teilhard de Chardin on Benedict's Cosmic Soteriology

Pierre Teilhard de Chardin (1881-1955), a French paleontologist, geologist, theologian and philosopher, had a tremendous impact on Ressourcement theology (Williams 2012), in particular through his friendship and correspondence with Henri de Lubac. His thought was one of the sources of inspiration for the intellectual élan in Europe which heralded Vatican II (Ratzinger 1987, 334). While it is well known that Benedict (Ratzinger 2009, 226-29; Rausch 2009, 9-10) found elements of Teilhard's eschatology problematic, the positive influence Teilhard had on Benedict's thought is a subject that has not yet been thoroughly explored, at least in the English speaking world.[7] In this section, I will briefly explain how the theologies of Teilhard and of Benedict converge in certain respects and differ in others.

In the following passage, Benedict (Ratzinger 2004, 85) implicitly acknowledges his indebtedness to Teilhard for the cosmic emphasis in his theology:

> The East . . . has always sought to see the Christian faith in a cosmic and metaphysical perspective, which is mirrored in professions of faith above all by the fact that Christology and belief in creation are related to each other, and thus the uniqueness of the Christian story and the everlasting, all-embracing nature of creation come into close association . . . . today this enlarged perspective is at last beginning to gain currency in the Western consciousness as well, especially as a result of the stimuli from the work of Teilhard de Chardin.

Thanks to Teilhard, there is an influx of the Eastern emphasis on the relationship between Christ and cosmos into Western thought as well as Benedict's theology. While Teilhard's Christology is different from what the East traditionally held about Christ and the cosmos since he views creation from an evolutionary and modern cosmological vantage point, the East had a decisive influence on Teilhard's thought, which in turn acted as a stimulus for other thinkers in the West.

Benedict's original appropriation of Teilhard in *Introduction to Christianity* (1968) occurred at a time when there was a lot of enthusiasm for Teilhard's work in Germany. This might lead one to think that Teilhard's importance for him waned over the next several decades; however, this is not the case since his citations of Teilhard (2000) in *The Spirit of the Liturgy* indicate that Teilhard had a long-lasting impact on Benedict's thought.[8] The following passage from *The Spirit of the Liturgy* is Benedict's (Ratzinger 2000, 28-29) paraphrase of the main features of Teilhard's thought on the evolution of the universe and its relation to Christ:

> Teilhard de Chardin depicted the cosmos as a process of ascent, a series of unions. From very simple beginnings the path leads to ever greater and more complex unities, in which multiplicity is not abolished but merged into a growing synthesis, leading to the "Noosphere," in which spirit and its understanding embrace the whole and are blended into a kind of living organism. Invoking the epistles to the Ephesians and Colossians, Teilhard looks on Christ as the energy that strives toward the Noosphere and finally incorporates everything in its "fullness." From here Teilhard went on to give a new meaning to Christian worship: the transubstantiated Host is the anticipation of the transformation and divinization of matter in the Christological "fullness." In his view, the Eucharist provides the movement of the cosmos with its direction; it anticipates its goal and at the same time urges it on (cf. Teilhard de Chardin 1959, 254-72; 285-94).

The first thing to observe about this passage is that there is a remarkable resonance between Teilhard's thought on the eschatological convergence of all things into the Omega point (which he identifies with Christ) and Benedict's eschatology, in which everything is united in God. Secondly, it is clear from this passage that Benedict adopted the idea that the Eucharist is an anticipation of the *eschaton* from Teilhard.

Both Teilhard and Benedict hold that the universe is redeemed through Christ and believe that love is the essential unifying element in the *eschaton*. For Teilhard (1971, 39), "[T]he *whole* world has been corrupted by the Fall and the *whole* of everything has been redeemed. Christ's glory, beauty, and irresistible attraction radiate . . . from his *universal* kingship." Teilhard (1959, 265) defines love as the "direct trace marked on the heart of the element by the physical convergence of the universe upon itself." Love is essentially the movement of the universe toward a cosmic "Omega Point" (269):

> Expressed in terms of internal energy, the cosmic function of Omega consists in initiating and maintaining within its radius the unanimity of the world's "reflective" particles. But how could it exercise this action

> were it not in some sort loving and lovable *at this very moment?* . . . . To be supremely attractive, Omega must be supremely present.

Ultimately, Teilhard (272) identifies Christ as the Omega, who unites all things in "one . . . point of definitive emersion—that point at which, under the synthesising action of personalising union, the noosphere (furling its elements upon themselves as it too furls upon itself) will reach collectively its point of convergence—at the 'end of the world.'" These passages demonstrate how Teilhard views the love-driven eschatological unity of the universe.

Likewise, Benedict's view of universal redemption is founded on love. This especially becomes apparent when Benedict (2011, 238) speaks of the mystery of the Eucharist, which "is always a matter of drawing every individual person, indeed, the whole of the world, into Christ's love in such a way that everyone together with him becomes an offering that is "acceptable, sanctified by the Holy Spirit" (Rom 15:16)." For Benedict as well as for Teilhard, everything is drawn into the love of Christ in the end.

While there are quite a few similarities between these two thinkers, it is important to note two subtle differences between their thought. First, Teilhard seems to emphasize the continuity between this age and the age to come, associating technological advancement with human progress in the divine schema, whereas Benedict tends to separate more sharply the *hic et nunc* and the *eschaton*, and believes that technological progress is highly ambivalent (Wicks 2008, 256-59; Ratzinger 2009, 228). Second, for Teilhard, the end is very closely associated with the phenomenon of evolution of the human species as well as the physical universe; his view could be described as physicalist. In contrast, Benedict's eschatological view is more balanced and traditional insofar as he emphasizes the transcendent consummation of the universe over its immanent evolutionary consummation. While he is aware of the limitations of Teilhard's vision, Benedict has manifested a consistent appreciation for many of his insights. In short, Teilhard's impact upon the thought of Benedict was tremendous.

ADVANCING CATECHESIS ON THE ENVIRONMENT

So as not to mislead my readers, I want to make it clear that I have no formal training or experience as a catechist; my training is in systematic theology. As such, the proposals I offer in this last section are necessarily preliminary and incomplete. It remains for catechetical experts to formulate appropriate methods of renewing catechesis on the environment by highlighting Pope Benedict's cosmic soteriology. Having said this, I want to offer some initial suggestions.

One catechetical method of presenting Benedict's cosmic soteriology is to incorporate fundamental elements of his theology on the salvation of the universe into catechesis on the environment. This method can be advantageous especially when such catechetical applications of Benedict's work cite statements that he made while he was pope. It would be especially useful for catechists to sift through Pope Benedict's papal encyclicals and documents for important material on the environment.[9] Papal teachings are often much richer than catechisms since popes can go into much more detail on particular issues than catechisms. While encyclicals are not catechetical literature per se, they are addressed to all people of good will and carry a substantial magisterial weight, which legitimates their use in a catechetical context. It goes without saying that the presentation of such material must be age appropriate.

A second method is to present the Catholic Church's teaching on the environment as expressed in the *Catechism of the Catholic Church* with reference to the three pillars of Pope Benedict's cosmic soteriology, i.e. the Paschal Mystery, the liturgy, and the *eschaton*.[10] The theological justification for this method is that all aspects of the faith are interrelated. By utilizing the *analogia fidei* ("analogy of faith"), a hermeneutical principle that is mentioned in *Dei Verbum* #12c which presupposes the interconnectedness of all ecclesial doctrines and relates specific doctrines to the entire Faith, catechists will be able to show how a concern for the environment is related to the Catholic Church's teaching on Christ's passion and resurrection, the liturgy, and the *eschaton*.

Allow me to give some examples of how these methods might be implemented. The first method is to explicitly and directly adopt some of the important features of Benedict's theology. The *Youcat*, or *Youth Catechism of the Catholic Church*, includes the following quote from Pope Benedict (Schönborn 2011, 236): "Experience shows that disregard for the environment always harms human coexistence, and vice versa. It becomes more and more evident that there is an inseparable link between peace with creation and peace among men." This concise passage, which sheds light on the relationship between the environment and human beings, indicates that an assimilation of Benedict's writings in a catechetical setting can have the potential to be very fruitful and engaging.

As an example of how the *analogia fidei* might be implemented in the second method, a catechist might choose to read aloud the following passages from the *Catechism of the Catholic Church* (2000) to catechumens before they embark on a discussion of what the Catholic Church teaches about the environment:

> It is the mystery of Holy Saturday, when Christ, lying in the tomb, reveals God's great sabbath rest after the fulfillment of man's salvation, which brings peace to the whole universe (#624).

> *For the cosmos*, Revelation affirms the profound common destiny of the material world and man (#1046).

> "Christ died and lived again, that he might be Lord both of the dead and of the living." Christ's Ascension into heaven signifies his participation, in his humanity, in God's power and authority. Jesus Christ is Lord: he possesses all power in heaven and on earth. He is "far above all rule and authority and power and dominion," for the Father "has put all things under his feet." Christ is Lord of the cosmos and of history. In him human history and indeed all creation are "set forth" and transcendently fulfilled (#668).

> It is in the Church that Christ fulfills and reveals his own mystery as the purpose of God's plan: "to unite all things in him" (#772).

> The universe itself, which is so closely related to man and which attains its destiny through him, will be perfectly re-established in Christ (#1042, quoting *Lumen Gentium* #48).

These passages bring into relief the Catholic Church's teaching on the relationship between Christ and the cosmos and the eschatological relatedness of human beings and the renewal of the universe. After reading these passages, the catechist could ask the catechumens what they think these passages imply about the relationship between human beings and the earth and what they imply about how we should live our lives with respect to our environment. The catechist could then present the church's teaching on the environment within the context of this broader theological background, a background which constitutes the basis of the religious significance of environmentally responsible acts such as recycling and participating in local ecological initiatives.

I have adumbrated this preliminary sketch with the hope that these brief suggestions will spur further discussion on the ways Benedict's cosmic soteriology can be used to rejuvenate catechesis. In addition to Benedict's theological works, catechists have a vast amount of theological resources from which they can draw so as to increase the awareness among Catholics of their vocation to protect God's creation. By the power of the Holy Spirit, they will be able to cooperate in the Spirit's mission of renewing the face of the earth.

CONCLUSION

In the first section, I reviewed the current state of the question of contemporary catechesis on the environment in the United States. Catholic catechesis in the U.S. needs to develop a holistic panoramic view of human salvation within the context of the redemption of the universe. As demonstrated in the *Catechism of the Catholic Church* and the *U.S. Catechism for Adults,* the current state of catechesis on the environment emphasizes care for God's creation from a social justice perspective: Christians should care for the earth because it is good, neglecting it would be catastrophic, and suffering will be perpetuated if people do nothing. These reasons, however, could just as easily be held by deists, and perhaps even by atheists. What is distinctively Christian about such motivations to care for the earth? How is this motivation related to Jesus Christ? While this catechetical teaching on the environment is indispensable, it is begging to be more clearly related to its broader Christological context.

The two-party system existing in the United States today is partially responsible for the division that exists in many Catholics' minds between their preferred political party and the teachings of the Catholic Church. It is possible to rectify the misperceptions that some Catholics have on the importance of the environment by making more explicit the relation that exists between Christ and creation and by presenting the issue of the environment in light of the whole faith. The catechetical appropriation of Benedict's cosmic soteriology is able to revitalize catechesis on the environment by making such connections more explicit, which will enable Catholics to see just how closely a concern for the environment is related to their religious beliefs.

In the second part of this chapter, I focused on the three pillars of Benedict's cosmic soteriology, i.e. the Paschal Mystery and the cosmic scope of Christ's salvation, the liturgical representation of the Paschal Mystery through every epoch, and the renewal of the universe in the *eschaton*. I demonstrated how Benedict's vision of cosmic redemption gradually developed over a length of time that exceeded fifty years, and pointed out that one of the key theologians who impacted Benedict's thought on this issue was Teilhard de Chardin. While he did not accept all of the views of Teilhard, Benedict adopted many of his most important ideas, including the idea that the Eucharist foreshadows the eschatological divinization of the universe.

I offered in the third part of this chapter some catechetical suggestions on how to present Benedict XVI's cosmic soteriology to catechumens and those who are receiving religious education. By employing the hermeneutical principle of the *analogia fidei,* catechists will be able to present a symphonic rendering of interrelated themes within the *Catechism* instead of isolating the paragraphs in the *Catechism* that explicitly mention the environment. Such a rendition, precisely because of the multiplicity of other voices and instruments and the harmony that exists between them, is able to enhance catechumens' appreciation of the performance of the solo piece that is highlighted — in this case, the church's teaching on the environment. The way forward is to root catechesis on the environment in Christ and his salvific action, which is sustained in history through the liturgy and which culminates in God being all in all (cf. 1 Cor 15:28).

In this essay, I have only scratched the surface of Benedict XVI's cosmic soteriology; there is much territory that has yet to be explored. For example, in addition to strengthening catechesis among Catholics, Benedict's cosmic soteriology also has vast ecumenical and interreligious potential. In the area of ecumenism, the adoption of Benedict's thought on the redemption of the universe, especially since it is so closely connected to the Eastern concepts of *theosis* (deification) and priests of creation, has the potential to strengthen the ecumenical bonds between the Catholic Church and the Orthodox Church.[11] In the sphere of interreligious dialogue, Benedict's doctrine of cosmic redemption can help to demonstrate that Christianity is a religion that is concerned with the protection of this world since what we do now has implications for eternity. Christianity can reaffirm its solidarity with the other religions of the world in a united front against forces which perniciously degrade the dignity of human beings and of the earth. Pope Benedict's view on the redemption of the universe also gives the religions of the world a forum for discussing their experiences, values, responsibilities, and hopes for the future. Such dialogue is able to enrich cultures and provides the Catholic Church with unique opportunities to present the teaching of Jesus Christ to people of other faiths in ways that are not intrusive, intimidating or disrespectful.

Pope Benedict's writings are replete with hope not only for human beings, but for the entirety of God's creation, which is called to reflect God's glory in this life to human beings and to reflect to an even more eminent degree his majestic wonders in the next. The cosmos is charged with the latent potency of being renewed in the resurrected life of Christ. The common eschatological destiny between human beings and other creatures ought to inspire a profound respect towards creation. In addition to his teachings on the redemption of the cosmos, Pope Benedict XVI has left the Catholic Church a profound legacy of ecological teachings (see Lasher and Murphy 2010; also Benedict 2012). While his retirement spelled the end to a pontificate that worked hard to promote the protection of the environment, it prepared the way for Pope Francis (2013), who is continuing Benedict's legacy of promoting the protection of the environment.

Pope Francis chose as his namesake St. Francis of Assisi, the patron saint of ecology. By imitating Pope Francis's simplicity, which is modeled after Christ's, Catholics will be able to follow Christ more closely, stand in solidarity with the poor, and cut back on their consumption of resources. This model of ascesis and self-denial is something that the United States of America desperately needs. Only by picking up the cross and following Jesus can we hope to be filled with the Holy Spirit. Only by being filled with fire of the Holy Spirit will we be able to master ourselves and our technology, the promethean fire in our hands that threatens to destroy us. Only by mastering technology and ourselves will we be able to save the world from ecological destruction. The conclusion of this chain is something that the Catholic Church has professed all along: Salvation comes through the cross of Jesus Christ.[12]

NOTES

1. I am grateful to Daniel P. Scheid, Assistant Professor of Theology at Duquesne University, for reading the preliminary draft of this chapter and for offering me invaluable suggestions for improvement.

2. This view is propagated by several prominent Republicans such as Rush Limbaugh (see On The Issues: Every Political Leader on Every Issue 2013) and John Hawkins (see Rightwing News 2012). Since the environmentalists they cite are primarily misanthropes, Limbaugh and Hawkins give the impression that these are the only types of environmentalists there are. In contrast to the statements by these Republicans, who do not officially represent the Republican Party, the GOP has a more balanced view on the environment. It encourages local involvement in ecological issues as opposed to big-government legislation, and describes the Republican Party as the party in which the conservation movement originally developed in the United States. At the same time, the GOP downplays the ecological crisis stating, "The environment is getting cleaner and healthier. The nation's air and waterways, as a whole, are much healthier than they were just a few decades ago. Efforts to reduce pollution, encourage recycling, educate the public, and avoid ecological degradation have been a success" (2012-2013). Furthermore, while the GOP does not explicitly deny that human produced climate change exists, it does not mention climate change even once in its platform description. Compare this with the DNC's statement on the environment, "From efforts to restore the Great Lakes to safeguarding against oil spills, environmental protection is crucial not only to combat climate change but also to sustain the health of our ecosystems" (2012). The Republican Party portrays environmentalism as something that is essentially a nonissue since the game has been won, whereas the Democratic Party recognizes the urgent nature of climate change and of the environmental crisis.

3. "From 1973 through 2008, nearly 50 million legal abortions occurred [in the United States]" (Guttmacher Institute 2011).

4. The German text reads, "Vielleicht hatte sich das Christentum des vergangenen Jahrhunderts wirklich ein wenig zu sehr zurückgezogen auf das Seelenheil des einzelnen, das er im Jenseits finden soll, und nicht laut genug vom Heil der Welt, von der universalen Hoffnung des Christentums gesprochen. So würde ihm die Aufgabe zuwachsen, diese Gedanken wieder neu zu durchdenken und zugleich der Inbrunst für die Erde, die beim modernen Menschen spürbar ist, neu eine positive Auslegung der Welt als Schöpfung gegenüberzustellen, die von Gottes Herrlichkeit zeugt und als ganze zum Heil bestimmt ist in Christus, der nicht nur das Haupt seiner Kirche, sondern auch der Herr der Schöpfung ist (Eph. 1, 22; Kol. 2, 10; Phil. 2, 9f.)."

5. There is no question that this view is anthropocentric—the real question is whether this particular form of anthropocentrism is a hindrance to protecting the environment. Another related question is how deeply is this view entrenched in revelation?

6. Cf. Haught 1996, 54: "In transfigured status, then, the present cosmos will continue to remain deeply implicated in the world's eventually eschatological fulfillment. Without a hope that nature has such a future, our present ecological commitments might indeed have entirely too flimsy a footing."

7. The only account of Ratzinger's positive reception of Teilhard that I am aware of is Christian Modemann, "Vollendung in Zeit und Ewigkeit - Teilhard-Rezeption bei J. Ratzinger" in *Omegapunkt : christologische Eschatologie bei Teilhard de Chardin und ihre Rezeption durch F. Capra, J. Ratzinger und F. Tipler*, 75-84 (Münster: Lit., 2004).

8. Also see Benedict XVI 2009b in which Benedict mentions Teilhard favorably.

9. For an excellent compilation and analysis of papal statements on the environment made by John Paul II and Benedict XVI, see Lasher and Murphy 2011. This chapter is an assessment of the legacies of John Paul II and Benedict XVI, especially with respect to their anthropologies insofar as they address the issue of the relation that exists between human beings and the earth. Lasher and Murphy address various works by both pontiffs and trace their lines of thought in some of their encyclicals and other papal documents. They maintain that through advocating a relational anthropology, a concern for the rhythms of the cosmos, and the metaphysical insight that existence is a gift from God, John Paul II and Benedict XVI offer teachings on anthropology and the environment that have the potential to inspire social and ecological reform. This promise, however, is also marked by poignancy since globalization poses a great challenge to interior conversion. It is one thing to write eloquently about what can be done to ameliorate the ecological crisis - it is another thing for one to undergo a deep, interior conversion so as to change one's lifestyle for the sake of the common good.

10. Note that these two methods are not mutually exclusive–they can be used to complement each other.

11. The dialogue between Catholics and Protestants can also be strengthened through Benedict's cosmic soteriology. There has already been significant progress made in Methodist-Catholic dialogue on the subject of how the Eucharist is related to ecology (see Methodist-Catholic Dialogue 2008). Benedict's focus on the *eschaton* and its relation to ecology has the potential to be very appealing to the theological imaginations of our Protestant brothers and sisters.

12. One must distinguish carefully between an immanent salvation that can be experienced in this world and the transcendent salvation offered by God. We cannot fabricate the kingdom of God on earth. At the same time, while salvation *tout court* can never be reduced simply to one of the facets of salvation, the salvation offered by Christ necessarily includes every other type of salvation, whether it takes the form of liberation from oppression or of salvation from a specific threat such as the ecological crisis.

REFERENCES

Benedict XVI. 2009a. *Caritas in Veritate* (Charity in Truth). Vatican City: Libreria Editrice Vaticana. Accessed October 19, 2012. http://www.vatican.va/holy_father/benedict_xvi/encyclicals/documents/hf_ben-xvi_enc_20090629_caritas-in-veritate_en.html.

———. 2009b."Celebration of Vespers with the Faithful of Aosta (Italy): Homily of His Holiness Benedict XVI." Vatican City: Libreria Editrice Vaticana. Accessed March 15, 2013. http://www.vatican.va/holy_father/benedict_xvi/homilies/2009/documents/hf_ben-xvi_hom_20090724_vespri-aosta_en.html.

———. 2011. *Jesus of Nazareth: Part Two, Holy Week: From the Entrance into Jerusalem to the Resurrection*. San Francisco: Ignatius Press.

———. 2012. *The Environment*. Edited by Jacquelyn Lindsey. Huntington IN: Our Sunday Visitor.

Catholic Church. 2000. *Catechism of the Catholic Church*. 2nd ed. Vatican City: Libreria Editrice Vaticana.

———. 2006. *United States Catholic Catechism for Adults*. Washington DC: United States Conference of Catholic Bishops.

DNC. 2012. "Issues: Environment." Accessed March 28, 2013. http://www.democrats.org/issues/environment#more.

Francis. 2013. "Homily of Pope Francis [on the Occasion of his Inauguration Mass]" Vatican City, March 19. Libreria Editrice Vaticana. Accessed March 30, 2013. http://www.vatican.va/holy_father/francesco/homilies/2013/documents/papa-francesco_20130319_omelia-inizio-pontificato_en.html.

GOP. 2012-2013. "America's Natural Resources." Accessed March 28, 2013. http://www.gop.com/2012-republican-platform_america/.

Guttmacher Institute. 2011. "Facts on Induced Abortion in the United States." August. Accessed March 28, 2013. http://www.guttmacher.org/pubs/fb_induced_abortion.html.

Haught, John. 1996. "Ecology and Eschatology." In *"And God Saw That It Was Good": Catholic Theology and the Environment*, edited by Drew Christiansen and Walter Grazer, 47-64. Washington DC: United States Catholic Conference.

Lasher, Connie, and Charles Murphy. 2011. "'With Generous Courage' Promise & Poignance in the Legacies of Pope John Paul II & Pope Benedict XVI." In *Confronting the Climate Crisis: Catholic Theological Perspectives*, edited by Jame Schaefer, 365-88. Milwaukee: Marquette University Press.

Methodist-Catholic Dialogue. 2008. "Heaven and Earth are Full of Your Glory: A United Methodist and Roman Catholic Statement on the Eucharist and Ecology." Agreed Statement of the Seventh Round. Washington DC: United States Conference of Catholic Bishops. Accessed October 19, 2012. http://www.usccb.org/beliefs-and-teachings/dialogue-with-others/ecumenical/methodist/upload/Heaven-and-Earth-are-Full-of-Your-Glory-Methodist-Catholic-Dialogue-Agreed-Statement-Round-Seven.pdf.

On The Issues: Every Political Leader on Every Issue. 2013. "Rush Limbaugh on Environment." Last modified January 4. Accessed October 19, 2012. http://www.ontheissues.org/celeb/Rush_Limbaugh_Environment.htm.

Origen of Alexandria. 2009. *De Principiis*. Edited by Kevin Knight. Accessed March 15, 2013. http://newadvent.org/fathers/04121.htm. Originally published in *Ante-Nicene Fathers*, vol. 4, trans. Frederick Crombie, ed. Alexander Roberts *et al.* (Buffalo: Christian Literature Publishing, 1885).

Ratzinger, Joseph. 1961/1962. "Kardinal Frings über das Konzil und die Gedankenwelt." *Herder-Korrespondenz* 16: 168-174.

———. 1987. *Principles of Catholic Theology: Building Stones for a Fundamental Theology*. Translated by Mary Frances McCarthy. San Francisco: Ignatius Press. Originally published as *Theologische Prinzipienlehre* (Augsburg: Sankt Ulrich Verlag/Wewel, 1982).

———. 1996. *Called to Communion: Understanding the Church Today*. Translated by Adrian Walker. San Francisco: Ignatius Press. Originally published as *Zur Gemeinschaft gerufen: Kirche heute verstehen*, 2nd ed. (Freiberg im Breisgau: Herder, 1991).

———. 1997. *A New Song to the Lord: Faith in Christ and Liturgy Today*. Translated by Martha M. Matesich. New York: Crossroad Publishing. Originally published as *Ein Neues Lied Für Den Herrn: Christusglaube und Liturgie in der Gegenwart* (Freiburg im Breisgau: Verlag Herder, 1995).

———. 2000. *The Spirit of the Liturgy*. Translated by John Saward. San Francisco: Ignatius Press. Originally published as *Einführung in den Geist der Liturgie* (Freiburg: Herder, 2000).

———. 2003. *God is Near Us: The Eucharist, the Heart of Life*. Edited by Stephan Otto Horn and Vinzenz Pfnür. Translated by Henry Taylor. San Francisco: Ignatius Press. Originally published as *Gott ist uns nah. Eucharistie: Mitte des Lebens* (Augsburg: Sankt Ulrich Verlag, 2001).

———. 2004. *Introduction to Christianity*. Translated by J. R. Foster and Michael J. Miller. San Francisco: Ignatius Press. Originally published as *Einführung in das Christentum* (Munich: Kösel-Verlag GmbH, 1968).

———. 2005. "The End of Time." In *The End of Time? The Provocation of Talking About God*, ed. Tiemo Tainier Peters and Claus Urban, trans. J. Matthew Ashley, 4-25. New York: Paulist Press. Originally published as *Ende der Zeit?* (Mainz: Grunewald-Verlag, 1999).

———. 2007. *Eschatology: Death and Eternal Life*. Edited by Aidan Nichols. Translated by Michael Waldstein. 2nd ed. Washington, DC: Catholic University of America

Press. Originally published as *Eschatologie—Tod und ewiges Leben* (Regensburg: Friedrich Pustet Verlag, 1977).

———. 2008. "Biblical Interpretation in Conflict: On the Foundations and the Itinerary of Exegesis Today." In *Opening Up the Scriptures: Joseph Ratzinger and the Foundations of Biblical Interpretation*, edited by José Granados, Carlos Granados, and Luis Sánchez-Navarro. Translated by Adrian Walker, 1-29. Ressourcement. Grand Rapids, MI: Eerdmans Publishing Company. Originally published as *Escritura e interpretación. Los fundamentos de la interpretación bíblica*, ed. Luis Sánchez-Navarro and Carlos Granados (Madrid: Ediciones Palabra, 2003).

———. 2009. *Theological Highlights of Vatican II*. New York: Paulist Press. Originally published in 1966.

Rausch, Thomas P. 2009. "Introduction." In *Theological Highlights of Vatican II*, by Joseph Ratzinger, 1-16. New York: Paulist Press.

Rightwing News. 2012. John Hawkins. "Environmentalist Wacko Quotes." April 12. Accessed October 19. http://www.rightwingnews.com/quotes/environmentalist-wacko-quotes-2/.

Schönborn, Christoph Cardinal, ed. 2011. *Youcat: Youth Catechism of the Catholic Church*. San Francisco: Ignatius Press.

Teilhard de Chardin, Pierre. 1959. *The Phenomenon of Man*. Translated by Bernard Wall. New York: Harper & Row. Originally published as *Le Phénomene Humain* (Paris: Editions du Seuil, 1955).

———. 1971. *Christianity and Evolution*. Translated by René Hague. New York: Harcourt.

von Balthasar, Hans Urs. 1988. *Dare We Hope That All Men Be Saved? With a Short Discourse on Hell*. Translated by David Kipp and Lothar Krauth. San Francisco: Ignatius Press. Originally published as *Was dürfen wir hoffen?* (Einsiedeln: Johannes Verglag, 1986) and *Kleiner Diskurs über die Hölle* (Ostfildern: Schwabenverlag, AG, 1987).

Wicks, Jared. 2008. "Six Texts by Prof. Joseph Ratzinger as *Peritus* Before and During Vatican Council II." *Gregorianum* 89.2: 233-311.

Williams, A. N. 2012. "The Traditionalist *Malgré Lui*: Teilhard De Chardin and *Ressourcement*." In *Ressourcement: A Movement for Renewal in Twentieth-Century Catholic Theology*, edited by Gabriel Flynn and Paul D. Murray, 111-24. New York: Oxford University Press.

Zizioulas, John (John of Pergamon). 2003. "Proprietors or Priests of Creation?" Keynote Address of the Fifth Symposium of Religion, Science, and the Environment, Stockholm, Sweden, June 2. Accessed October 19, 2012. http://www.orthodoxytoday.org/articles2/MetJohnCreation.php.

IV

Our Catholic Faith in Action

TEN

Discernment of the Church and the Dynamics of the Climate Change Convention

John T. Brinkman, MM

The Church's long-standing critical regard and concerned guidance concerning humanity's most recent phase of industrial development is registered in no small part by the writings of the social teachings of the Church: *Rerum Novarum* and *Quadragesimo Anno* to *Caritas in Veritate*. This continuum may well be noted as part of a sequence in which concerns regarding economic and social equity have been progressively seen as intimately related to matters of ecological integrity. This is particularly true of the ecological vision presented by Pope Benedict XVI.

On July 24, 2007, Pope Benedict XVI addressed the clergy from the dioceses of Belluno-Feltre and Treviso in a wide ranging-conversation. With regard to environmental matters, the Holy Father said:

> Today we all see that man could destroy the foundations of his existence, the earth, and so we cannot just simply do with our earth, in reality entrusted to us, whatever we want and whatever appears useful and promising at a given moment. We must respect the internal laws of creation of this earth, learn these laws and also obey these laws, if we wish to survive. Therefore, this obedience to the voice of the earth, to life, is more important for our future happiness than the voices of the moment, the desires of the moment. To sum up, this is the first criterion to learn that life itself, our earth speaks to us and we must listen if we wish to survive and decipher this message of the earth. And if we must be obedient to the voice of the earth, this is even more so for the voice

> of human life. Not only must we heal the earth but we must respect each other, all others.

His Holiness again took up the theme of ecological concern intimate to the theme of human ecology in the 9 June 2009 *Caritas in Veritate*. His message for the celebration of the World Day of Peace, 1 January 2010 stated: "If you want to cultivate peace, protect creation.

Direct references to climate change and to the United Nations Framework Convention on Climate Change (UNFCCC) were evident in recent papal statements. In his 2010 Address to the diplomatic corps the Holy Father said:

> I share the growing concern caused by economic and political resistance to combating the degradation of the environment. This problem was evident even recently, during the XV Session of the Conference of the States Parties to the United Nations framework Convention on Climate Change held in Copenhagen from 7-18 December last. I trust that in the course of this year, first in Bonn and later in Mexico, it will be possible to reach an agreement for effectively dealing with this question. The issue is all the more important in that the very future of some nations is at stake particularly some island states.

At the Angelus on 28 November, the first day of the November 28-December 9, 2011 UNFCCC conference in Durban, South Africa, His Holiness made a direct reference to "the complex and troubling event of climate change." On the second day of the South African conference, the Holy Father addressed seven thousand Catholic young people whose Franciscan movement focused on ecological concern. The annual papal audience and address to the Vatican diplomatic corps on 9 January 2012 drew attention to "the obligation of nations to act as a family for the common good in their resolution of the climate change crisis."

The environmental concern of the church and the ecological vision of Pope Benedict XVI might best be summarized by the final paragraph in the May, 5, 2011 Pontifical Academy of Sciences report, *Fate of Mountain Glaciers in the Anthropocene* (2011, 1):

> We call on all people and nations to recognize the serious and potentially irreversible impact of global warming caused by the anthropogenic emissions of green house gases and other pollutants, and by changes in forests, wetlands, grasslands and other land uses. We are committed to ensuring that all inhabitants of this planet receive their daily bread, fresh air to breathe and clean water to drink, as we are aware that, if we want justice and peace, we must protect the habitat that sustains us.

However, we must be mindful that care for creation is not only to be considered an extension of the social doctrine of the Church, i.e., an application of principles derived from Church teachings presented to the wider human social forum. Environmental concern must be seen as coming from the very depths of our faith tradition. The intent of this writing is to state that environmental challenge is an issue that calls us to reflect on the most fundamental insights of our faith. For our faith response springs from the depth of our tradition's confident credence in the mystery of divine presence to creation. Nothing less would be adequate to address current human interference with the earth processes upon which all life depends. At this critical junction, the voice of the Church could well be a most timely and significant factor in focusing world and US leaders' and the public's attention on the sacred dimension of our imperiled environment, on the spiritual challenge and on the ethical and moral issues underlying the need to combat climate change and thereby protect the world's most vulnerable from its otherwise devastating impacts.

DISCERNMENT OF THE CHURCH

Climate change as defined in the UN FCCC discourse and in the Intergovernmental Panel on Climate Change (IPCC) Fourth Assessment Report (4AR) is unequivocal and unavoidable. In effect, it is an event that has occurred but the effects of all the GHG emissions expelled into the atmosphere to date have yet to be fully felt for the inertia of the atmospheric processes. It has become clear that to achieve a global warming limit of two degrees Celsius +2C above pre-industrial levels in the effort to avoid the most dangerous effects of climate change will still entail significant consequences for the earth and for the human community.[1]

The Intergovernmental Panel on Climate Change (IPCC) Fourth Assessment Report (4AR): "Climate Change Impacts, Adaptation and Vulnerability" informs us that "Unmitigated climate change would, in the long run be likely to exceed the capacity of natural, managed and human systems to adapt." Current documented actualities of the decreased capacities of the earth to naturally sequester carbon and of the present increase of Green House Gas (GHG) emissions at the higher end of the 2007 IPCCC 4AR projections state the conundrum and the challenge of climate change mitigation.

We are cautioned: "If humanity wishes to preserve a planet similar to that on which civilization developed and to which life on Earth has adapted, paleoclimate evidence and ongoing climate change suggest that CO2 will need to be reduced from its current 385 ppm to at most 350 ppm" (Hansen 2008, 1). At the time of the editing of this paper, the current measurement as of March 8, 2013, is 395 ppm.[2] "Carbon dioxide levels measured at the Mauna Loa observatory in Hawaii jumped by 2.67 parts per million (ppm) in 2012 to 395 ppm, said Pieter Tans who leads the greenhouse gas measurement team for the US National Oceanic and Atmospheric Administration (NOAA). The record was an increase of 2.93 ppm in 1998. . . .Tans told the Associated Press the major factor was an increase in fossil fuel use. 'It's just a testament to human influence being dominant,' he said. 'The prospects of keeping climate change below that [two-degree goal] is fading away'" (Vidal 2013). We are also informed by other current literature such as the UNEP Emissions Gap Reports that the world community has the technology and the means to accomplish the necessary mitigation measures albeit within a rapidly closing window of opportunity.

This endeavor of the Convention to avoid the most dangerous effects of climate change must be driven assuredly by facts and by values. Indeed it may be argued cogently that values are emerging as a central focus in the articulation of a timely and effective resolution to the climate change crisis. From the view of the contribution of the Church, the focus on values provide an opportunity for the Church to present tenets explicit in the perennial teachings of the Church and coincident with the deepest aspirations of humanity.

Before addressing climate change issues that would invite and even compel greater Church response, it would be best to present the sense of cosmological commitment that is a hallmark of our tradition of thought and faith and is the heart of its assured regard for the integrity of the earth. This theological disposition coherent with the ethical stance that is indispensable for environmental resolution is best presented under the rubric of Earth Ethics.

EARTH ETHICS

This focus on the need for an ethical stance toward the earth itself as the fundamental context for the resolution of our environmental crises was conceptually refined at the March 16-23, 2003, 3rd World Water Forum (WWF) in Kyoto, Japan. Participants focused solely on technological and economic response as primary in solving ecological challenges contended that economic efficiency, social equity and environmental sustainability were to be considered equal considerations in confronting global environmental crises. In this plenary round table, they were roundly informed that politics, economics and earth sustainability were not equal. Attention to earth sustainability was paramount and the essential substratum for all social and financial solutions. In effect, economics and politics are beyond the economical and political in their dynamism for they are rooted in the earth. In the same conference, at a UNEP and WWF sponsored panel, "Water, Nature and the Environment," Mr. Delmar Blasco, Secretary General of RAMSA-the Wetlands Convention concluded his remarks with a memorable insight. He stated that "unless and until humanity achieves an ethical stance that recognizes the intrinsic value of nature, all of our efforts may well end in failure."

This stance in which nature or the natural course of things is considered a central term of reference and a criterion for human action finds more than a resonance with the thought of St. Thomas Aquinas. In the *Summa Contra Gentiles* (SCG), Aquinas reflects on creation's seven day sequence in which each instance of creation is recognized as good. When all things were completed, created reality is noted to be "very good." By virtue of the complete order of things, the integral functioning that manifests the universe as universe, we participate in and are encompassed by what is most noble. Aquinas states (SCG 2:45.10, 139) "but all things together are very good, by reason of the order of the universe, which is the ultimate and noblest perfection in things." This integral functioning of the course of the cosmos comprises not merely one perfection among others but the ultimate and noblest perfection in all things. We may well add the monitum: "unless and until humanity achieves an ethical stance that recognizes the intrinsic value of nature, (i.e., the order of the universe) all of our efforts may well end in failure." It becomes increasingly apparent and urgent that the human community begin to function with something akin to a shared and integrated vision of the natural world.

For his own time, Aquinas presented in *Summa theologiae* (ST) a comprehensive vision of the phenomenal order of things, the interrelatedness of all things and the role of the human within it. In his metaphysical vision, the datum of the cosmos, i.e., "the whole universe together" manifests the primary meaning and mystery of things. For Thomas, the universe images forth in its multiplicity and diversity, the simple goodness of God. No single creature or creation, not the majesty of a mountain found in the earth's impulse to leap skyward, nor the mind of man found in creation's impulse to leap into consciousness can participate perfectly in or adequately represent the datum of divine goodness imaged forth by the universe in its totality of beings. Everything needs every other thing to manifest this divine attribute. In this, the cosmos itself constitutes the primary mode of revelation. "For goodness, which in God is simple and uniform, in creatures is manifold and divided; and hence the whole universe together participates in the divine goodness more perfectly and represents it better than any single creature whatever" (ST 1948, 1.47.1, 246).

In this scheme of things, humanity is not required in itself but by the differentiation of all things. Indeed, it is the "greatest perfection of the universe" that required that there be creatures of intellect and will. Hence it is by human capacity to know and the ability to will that humanity finds its designated place and capacity for reciprocity and fulfillment in the wider contexts of things. "Hence, the complete perfection of the universe required the existence of some creatures which return to God not only as regards likeness of nature, but also by their action. The greatest perfection of the universe therefore demanded the existence of some intellectual creatures" (SCG 2.45.10, 140).

In this dynamic cosmic order which constitutes a journey of return for all beings, there is simply no place for an anthropocentrism in a modern sense of human action seemingly privileged over-against the course of the universe. As Blanchette (1992, 15) remarks, "Human beings stand at the pinnacle of nature. But the human being is still only a part of the universe, and as such it is itself ordered, along with the other parts (though in its own peculiar way) to the good of the universe as a whole." It is by participating in a coherent and enhancing manner in this "return" by the unique capacities of seeking the truth and willing the good that the human finds its noble and privileged status. Clarke (2001, 304) affirms:

> In the emergence of creatures from their first source is revealed a kind of circular movement (*circulatio vel regratio*), in which all things return, to their end, back to the very place from which they had their origin in the first place (*Expositio in Libros Sententiarum*, Book I, Dist. 14, q. 2, art 2). St. Thomas actually uses the same structure of the emanation and return of all creatures from and to God as the basic organizing plan of his whole *Summa Theologiae*, his greatest work.

In the concluding chapter of *The One and the Many* entitled "The Great Circle of Being and Our Place in It, The Universe as Meaningful Journey," Clarke outlines the ancient concept of the Universe as a Journey which St. Thomas more directly took from Plotinus. Here is the *Exitus*: the Many pour out from the One. In the process of going out they find their differentia, they unfold their potentialities and fulfill their capacities as best they can in the course of life and in that very process are attracted to and are drawn back by the pull of the Good, toward the Source from which they came. Hence the journey is both an exiting from and a returning to (a *Reditus*) the road back.

If there is a particular "ecological turn" (i.e., the time when a certain clarity of thought is reached wherein the collaboration of a tradition's sense of transcendence and immanence form an essential and intimate rapprochement with the natural world), it would be found in Thomism. From the perspective of "Earth Ethics," such a conceptual turn of events can best be attributed to Thomas Aquinas. His synthesis of prior thought and his refined sense of creation in large measure accepted the cosmology inherited from the Greek Platonist, Dionysius. He does so with its refined articulation and shared sense of the numinous quality of the natural world. The visions of Dionysius and Aquinas alike are infused from the start and sustained throughout by a sense of sacred marvel before the divine origin of the world. Their works are characterized by an openness to what is unknown and undiscovered, inspired by the desire to comprehend, not necessarily to fully explain, but to contemplate and ponder, to accept and appreciate and to articulate as best possible according to the deficient modes at their disposal (O'Rourke 1992, 60-61).

However, Thomas radically transforms the Platonic texts and transcends its thought in the process. Succinctly put: "Thus, whereas for Dionysius it is a hindrance to our discovery of God that human knowledge is oriented towards finite beings, this for Aquinas is the very foundation of our natural disclosure of God" (O'Rourke 1992, 56). This is a radical disclosure. Humanity in this perspective is rooted in the earth and intimately and properly a part of the universe. Man is not ordained toward ecstatic escape from this realm. It is here by the very structure of his being that his orientation to God and his knowledge of the divine is founded in finite reality. In Thomist epistemology, the phantasm of sensory perception, the proximate impress of this finite world is the point of initiation for all concepts of the mind and the channel through which ultimate reality and truth are perceived. "Thomas asserted that all our knowledge, including the spiritual, and also our knowledge of God, took its starting point (and therefore always remained somehow dependent upon) sense perception" (Pieper 1966, 29).[3]

Although this is a radical disclosure, it is far from a radical departure in a tradition where nature is revelatory. There are certain themes in thought traditions and cultures that comprise so central and defining an attribute that if lost or diminished, they are reconstituted again in novel and renovated forms so as to preserve an identifiable continuum of cultural history or a recognizable continuity of thought. The remarkable renovation and reconstitution of Platonic cosmology in this Thomist revision does imply an "ecological turn" in the sense that it preserves and indeed recapitulates in an enhanced manner, the intimacy of the sacred to phenomena. This is an instance of exigent exposition driven by the sheer necessity to retain and deepen essential truths in the light of intellectual challenge and spiritual redress. It is important that we recognize this reaffirmation as part of the tradition that retained divine and phenomenal intimacy within the categories of thought and in accord with the philosophical concepts available through epochs of monumental change. Thomas drew Transcendent Presence and finite reality closer together.

INTIMACY OF THE SACRED AND PHENOMENA

Other traditions of thought, religious insights and schools of philosophy would find extraordinary if not incomprehensible the earlier affirmation of creation and incarnation extolled in the statement of Tertullian in his *De Ressurrect. Caris* 6 (Vagaggini 1968, 77):

> Think again about God, so concerned with and devoted to Adam's body: with his hand, his senses, his activity, his counsel, his wisdom, his providence, and above all his affection that guided the formation of his features. In fact, in everything that came to be expressed in the human body, there was the light of Christ, the future man, shining

> through. For he too was to be clay and flesh and word, and then was earth.

This is but a glimpse of Patristic sensitivity to human life as designed for an intimate relation to the phenomenal order of things in the very process of its divinization. It is significant that the varied and culturally diverse Fathers of the Church were determined in their efforts to discount dichotomies[4] which would have fractured spirit from matter, divinity from humanity and Word from cosmos. It is enough for our purposes to confirm that our tradition is one which consistently confirms Divine intimacy to the natural world, human integration in that world and above all the integrity of the order of creation as revelatory of God.

In this regard, it is important that we ourselves "think again" and recognize how germane is our tradition of thought and how unique is its potential to contribute to environmental reflection and resolve. It may well be said that the problematique of climate change, the alienation of the human community from the earth community, poses unprecedented challenges for the revision of thought in an age similar to the Patristic period plagued with dichotomies that distort the meaning and purpose of the universe and the role of the human within it.

Our time poses profound alienation. Never before has the human community been so alienated from the Earth community. It may be cogently argued that the resultant divisions within the human community itself offer the spectrum of a widening divide over the most fundamental issues of life and survival. These are most difficult times. Difference is not met by deference. Economic efficiency often disparages social equity. The artifice of mono-culture militates against the rich diversity of the earth and its differentiated cultures. These divisions contradict our Catholic credence at its core.

The most fundamental contradiction resides in the alienation of the human community from the earth community evidenced in the climate change crisis. Hence, the current and compelling context of climate change needs to be reviewed.

DYNAMICS OF THE CLIMATE CHANGE CONVENTION

The compelling context of climate change and the need for response are presented by the Three Maps first proposed by Hans-Martin Fussel[5] in "Climate Change Vulnerability and Responsibility" at the 10 December 2007 UN FCCC COP 13 Climate Change and Justice Panel. They provide a schema to appraise our present situation and the critical need for response. The precautionary principle, the principle of sustainability and the principle of equity are Convention references which have the potential to offer pathways that could reach out beyond relative perspectives that serve the interests of the few over the many. In effect, they are ethical principles in progress.

The Three Maps

- The Map of Carbon Intensive Development is a World Altering Map
- The Map of the Most Vulnerable to Climate Change is the Map of the Developing Countries
- The Map of Resource Competition in a Carbon Constrained World is an Outline for the Sundering of the Human Social Regime throughout the world

The Map of Carbon Intensive Development is a World-altering Map. The decreased capacities of the earth to naturally sequester carbon and the increase of GHG emissions at the higher end of the IPCC projections state the challenge of earth alterations. Arctic ice retreat and sea level rise are observed as escalated in their rate of change. Rising ocean acidification caused by seas absorbing more carbon dioxide is disrupting the ability of corals to build their calcium carbonate structures. Warmer seas stress corals still further.[6] These living sea walls are the sites of nutrition essential for marine ecosystems. Global warming is a factor in changes in terrestrial habitat and the plight of species migrating to higher altitudes and contributes to the current high rate of species extinction.[7] The warming of the oceans and its atmospheric effects are being linked to extreme events. Although weather variability has its cause-effect uncertainties, climate change scientists have drawn lines of probability between climate change and "extreme events" with observations published as early as 2005 (Emanuel 2005, 686-88). Hurricanes have been cited as an example of weather intensity linked to climate change (Union of Concerned Scientists 2012).[8] The earth and its life sustaining processes are in danger of perhaps irreversible alteration. In light of this, humanity is being challenged to live life on the bases on which life has been granted to it.

The Map of the Most Vulnerable to Climate Change is the Map of the Developing Countries. This map charts the global regions most open to climate-impact threats to basic human needs. In the IPCC AR4 extreme flood-risks to densely populated megadeltas are cited with particular reference to the Ganges-Brahmaputra and the Mekong. Water scarcities due to melting glaciers include the regions fed by Himalayan ranges and the corridor for the extreme events of tropical cyclones spans the latitudes from the Philippines to Bangladesh due in part to sea-surface temperature rise. Diminished fresh water availability, coastal area vulnerability and crop fluctuations are but a few of the projections cited in regard to Asia. The assessment report emphasized the vulnerability of Africa to the impacts of climate change even at the lowest concentrations of greenhouse gasses that have been modeled. An increase in average global temperature of two degrees Celsius is likely to raise the annual average in some Africans areas by four degrees or more and result in increased water stress for approximately a third of the continent's population. Drought-increases are charted and future drought projections are outstanding in Southeast Asia, Africa and Latin America. These continents now encompass the areas of the highest agro-economic vulnerability. Seeing the intricate interrelated intricacies of the natural world as a web of life and the dynamic that engenders and cradles life itself, we are mindful that in the Right to Life discourse within the Church; the rights of innocent life and the rights of the most vulnerable have taken on refined articulation. Within this second map, the least among us are the most vulnerable. It is quite clear that those least responsible for climate change are the most threatened by its consequences.

The Map of Resource Competition in a Climate Constrained World is an Outline for the Sundering of the Human Social Regime. The alternative to cooperative action would be to invest more heavily in national interests in the wake of which local and regional conflicts would be exasperated in the competitive demand for decreasing natural resources. This would raise the specter of an age of technologically refined competition and conflict. Remote sensing indicates the vastness of the natural gas reservoir in the China-Japan disputed area of East China Sea ("China Tells" 2010, 10).[9] There is competition for land in developing countries to secure water resources and food supplies against the event of shortages caused by climate change (Kajiwara 2010, 25).[10] According to recent reports the annual defense spending of Asian powers has grown to $224 billion in 2011 ("China Spurs" 2012, 1).[11] Navigable waters made accessible by Arctic melt have opened competitive claims to explore mineral resources in the disputed maritime regions ("China Tells" 2010, 10).[12]

The meaning of peace and the principles for attaining and sustaining peace have voluminous exposition in our social doctrine writings. Most recently the demise of peace and the "culture of death" have been insightfully linked. Yet references which have well served the discussion of homicide and genocide and human conflict as such are less effective for the consideration of biocide and geocide. There are theological references which focus on Divine Peace as the source of the integrity of each entity in the universe and of the integration of the cosmos, itself.

Some would look upon these three mappings as present indices of future inevitabilities. Such positions see current business-as-usual trends to be insurmountable for their grounding in a belief in unlimited growth and the "progress" and developmental benefits it offers. In contrast determined efforts to stabilized GHG in the atmosphere and thereby preserve the planet as we have come to know are supported by more comprehensive concerns and by principles inscribed in the Climate Change Convention.

The Three Principles

The precautionary principle, the principle of sustainability, and the principle of equity are Convention references which have the potential to offer pathways that could reach out beyond relative perspectives that serve the interests of the few over the many. In effect, they are ethical principles in progress.

The Sustainability Principle is inferred in the 1987 publication of the Brundtland Commission's Report, *Our Common Future*, in which sustainable development was described as "development that meets the needs of the present without compromising the ability of future generations to meet their own needs." Sustainability defines an equation when social development and environ-mental protection are seen as integral parts of human development.[13]

The Precautionary Principle takes on great importance not only because it admonishes us to do no further harm but also because it cautions us to be aware that present action may if left untended, determine negative effects not amendable at a later time when scientific data is completely conclusive.[14]

The Principle of Equity forms a "basis" for action.[15] Equity finds a primary reference in the disposition of reasonableness and moderation in the exercise of one's rights. It may be interpreted that equity calls us to a more ultimate term of reference in light of which legitimate rights are modified so that more fundamental rights are preserved. During the Bonn 16 May inter-session workshop: "Equitable Access to Sustainable Development," there were moments of exchange that aimed to define equity in terms of essential needs, e.g., "basic right to food, water and home" or the "fundamental right for life and development." The overall imperative is to start reducing emissions significantly and not to create new ways to increase them under an appeal to "rights" formulated for national advantage. Clearly it is inequitable to refrain from adequate emission reductions which threaten the very life, well-being and cultures of peoples. One intervention that stood out in its balance and equitable intention was that of Chile's Minister Counselor, Waldemar Coutts Smart:

> As emissions increase both in developed and developing countries, no progress will be made unless we all increase, in differentiated but real ways, our responsibility and capacity to act. We believe that the right to develop by some larger economies should not threaten the right to exist of the most vulnerable, particularly the small island developing states. Should not we rather promote a proactive case of development and climate protection, such as complementary objectives with an explicit narrative for cooperation that challenges the obvious North against

> South, or large against small rationales that usually tend to antagonize and hinder global consensus?

In the search for values that would inform action, conceptual advance in defining equity and sustainability and the precautionary is being sought to refine core principles, i.e., indicators to which one might appeal for guidance. Such norms are best compared to the Confucian sense of moral principle imaged by the polar star. It does not tell one where one must be but from where ever one finds oneself, it guides one. This outline of reflective effort points to the ethical dimension structured in the Convention process.

ETHICS, MORALS AND THE SPIRITUAL

Climate change is an issue that raises questions of values and questions of a properly considered ethical and moral and spiritual nature. Economic or scientific analysis cannot tell us what value to place on the lives of future generations, or how far the developed world should help the poorest nations to adapt to the effects of climate change, and develop low-carbon energy. We recognize that climate change raises profound moral and ethical questions and present essentially a spiritual challenge.

In the academic discussions concerning the environment and climate change, the terms "ethical" and "moral" have been used interchangeably and without any nuance or differentiation. In order to add some clarity to the matter, I propose that for our purposes, ethics will be defined by reason, morality by fulfillment, and the sacred by transcendence.

The ethical might well find an expression in social justice. However, guided by our theologically informed sense of Earth Ethic, the society of reference is the earth itself. Justice or the act of justification might best take on the analogy of type-setting in a printing press, i.e., the task is one of justifying the text so it says what it was meant to say from the beginning of its composition. Here, our text is the composition of the cosmos and its most proximate articulation, the earth. Our redaction in type-setting is to have this text say what it was meant to say from the beginning of time. The basic ethical norm is the well-being of the comprehensive community and the attainment of human well-being within that community.

The moral mind stands in relation to and is directed toward fulfillment. There is fulfillment within time and there are trans-temporal modes of fulfillment. For our present focus, the moral sphere of action might be recognized in the instance of stewardship which attends to and protects the matrix of all life. Everything needs every other thing in the universe to emerge with its own defined spontaneities in accord with which all entities find their fulfillment.

Beyond the imperatives of the ethical in our strivings for social justice and the moral in an engaged stewardship, there is the spiritual or more properly the sacred defined by the impulse toward the transcendent. The sacred as an ultimate term of reference and the ground of meaning and purpose finds particular reference to the order of creation itself with particular explication in the various traditions of faith. It is the contention of this writing that in the ecological challenge of climate change, only the sacred with save us.

In the preface to Kathleen Deignan's work (2003, 18-19) on Thomas Merton, *When the Trees Say Nothing: Writings on Nature*, we are informed:

> An absence of a sense of the sacred is the basic flaw in many of our efforts at ecologically or environmentally adjusting our human presence to the natural world. There is a certain futility in the efforts being made truly sincere, dedicated, and intelligent efforts to remedy our environmental devastation simply by activating renewable sources of energy and by reducing the deleterious impact of the industrial world. The difficulty is that the natural world is seen primarily for human use, not as a mode of sacred presence primarily to be communed with in wonder and beauty and intimacy. The deep psychic change needed to withdraw us from the fascination of the industrial world, and the deceptive gifts that it gives us, is too difficult for simply the avoidance of its difficulties or the attractions of its benefits. Eventually only our sense of the sacred will save us.

Today the primary task of the human is to become once again integral with the natural world in such way as to sustain the interrelated functioning of the natural order of things. In the historic record, it has been the sense of the sacred that has guided the human venture and has healed the deepest disorientations of the human mind and heart. Perhaps in the final analysis the critical "human capacity to respond" will hinge less on our political will and economic prowess and national determination and more on our recovery of a sense of the sacred that affirms the meaning of the universe and the role of the human within it.

THE CONTRIBUTION OF THE CHURCH

There is a recognized need for voices of conscience in the process of the UNFCCC discourse. The closing remarks of Mr. Yvo de Boer (2009, 8), Executive Secretary of the UN Framework Convention on Climate Change at the 13 December 2009 Forest Day 3 in Copenhagen were most significant. He pointed to the need for a continued articulation and guidance of "concerned parties" in the issues of climate change in the near future. His actual wording appealed to those who must continue to be the "conscience" of this on-going process "to ensure the social and environmental integrity of the architecture of the Copenhagen outcome."[16]

There is a need for multiple voices of conscience to emerge from within these proceedings. There is also a need for religious articulation to be present to and be informed by the dynamics of the framework processes themselves. It is within this process and in the details of the negotiation that the necessity for a spiritual vision for humankind will become most evident. When in negotiations, a "tree" or a "forest" is reduced to the status of a sink or carbon-sequester apart from any consideration of habitat, bio-diversity, indigenous land tenure, aesthetic place or sacred presence which evoke the ecstatic and confirm the true role of the human in the wider community of being; the religious voice must affirm a more integral view of reality. In the editing brackets and in the margins of protocol formulations where modes of assistance to developing nations become loopholes in emission reductions, where mechanism intended to assist become incentives for mere profit; there is a need for a presence engaged in the interpretation of such occurrences as diminishments of the universe and as denigration of the sacred as sacred.

The engaged Church must be a "concerned party" that views its "option for the poor" coherent with climate change convention goals. The particular bishops' conference and the area and local church are in unique positions to aid the voice of the most vulnerable to be heard throughout its region and through the network of established conferences of bishops in the Church universal. To make known the suffering of the marginal and their specific adaptation needs is a function best presented from the experience of the local church and amplified through wider conferences of concern which would include ecumenical and interreligious outreach.

CONCLUSION

Care for the phenomenal order of things is an emerging context for the mutually enhancing presence of each faith tradition to every other faith tradition. In this very process of articulation and reflection, traditions of faith grasp anew their essential teachings and reaffirm their perennial message in terms inclusive of the effort to restore the integrity of the Earth. In this effort each tradition recognizes itself within a new context. The spiritual challenge of the environmental crisis draws each tradition to reflect on the natural world in its most profound sense of mystery:

> As part of the long cosmic process it can be said that the varied spiritual traditions scattered across the globe are not of yesterday, nor are they simply of the earth. In some manner they were born when the galaxies appeared in the limitless swirl of space; the dynamic at work in all this has found unique expression in the formation of the green earth with it myriad forms of life and their completion in man. In man the galaxies become aware of themselves; the globe comes to its most

intense form of life, matter reaches its high transformation in those interior spiritual experiences wherein the human comes to itself in both its personal identity and identity with the universe. This process takes place in the presence of that divine mystery in which everything finds its peace and its perfection (Berry 1974, 173-74).

NOTES

1. Should humanity manage to stay below +2 C, there will still be significant suffering for the poorest regions of our world. There is therefore a growing consensus that the effort to keep global warming below +1.5 C is an ethical imperative. A 1.5 C global warming limit would correspond to a 350 ppm atmospheric condition.

2. The most reputable source for current atmospheric CO2 concentration comes from the National Oceanic and Atmospheric Administration in the U.S. NOAA has been recording monthly atmospheric CO2 as measured at Mauna Loa Observatory, Hawaii since 1959 and all their data and analyses are available at http://www.esrl.noaa.gov/gmd/ccgg/trends/.

3. Pieper refers to Aquinas' *Summa Theologica* 1.12.12 and *Commentaria in Librum Boethii DeTrinitate* 6.3.

4. For our purpose, there are five polemics that have particular importance. Each combat a misunderstanding of corporeity either in man or in Christ with reference to the way of salvation willed by God: (1) The polemic of the Fathers against dualism, either in its Docetist form as with Ignatius of Antioch, its developed Gnostic form as with Irenaeus or Tertullian, or in an Encratist form as with Clement of Alexandria. (2) The polemic against those who denied the resurrection which we find in the Apologetes and the anti-Gnostic Fathers. All the Fathers who sought to defend the Christian position on this point developed the theme that man is neither solely a soul nor solely a body. (3) The anti-Arian polemic. When the Fathers found themselves obliged to explain the relations between the divinity and humanity in Christ they clarified the notion that in the work of salvation, Christ's humanity including his body was the instrument through which divinity operated. (4) The polemic against Apollinarianism. To show that Christ is perfect man and that he assumed an integral human nature, the Fathers explained what an integral human nature was. In this context, particularly, we see the relevance of the axiom: the Word did not save what it did not assume. (5) The polemic on the hypostatic union, especially that of Cyril of Alexandria against Nestorius. Consult Cipriano Vagaggini, O.S.B., *The Flesh, Instrument of Salvation*, 66-67.

5. Hans-Martin Fussel is affiliated with the Center for Environmental Science and Policy, Stanford University, Stanford, CA 94306, U.S.A., and the Potsdam Institute for Climate Impact Research, Telegrafenberg, 14473 Potsdam, Germany. He is co-author of "Climate Change Vulnerability Assessments: An Evolution of Conceptual Thinking."

6. According to a report issued by the International Union for Conservation of Nature at the October 2012 Convention on Biological Diversity (CBD), 33 percent of reef-building corals are now at risk of extinction. "Researchers from the Australian Institute of Marine Science (AIMS) say the Great Barrier Reef has lost half of its coral in little more than a generation. And the pace of damage has picked up since 2006."

7. In "Biodiversity Funds to Double" that appeared in *The Japan Times* on October 21, 2012, p. 3, the International Union for Conservation of Nature was reported as stating: "A quarter of the world's mammals, 13 percent of birds, 41 percent of amphibians and 33 percent of reef-building corals are now at risk of extinction."

8. In an October 30, 2012, press release, the Union of Concerned Scientists stated: "As Hurricane Sandy dissipates and recovery efforts begin, people are asking what role climate change plays in influencing such storms. Oceans have absorbed much more of the excess heat from global warming than land and scientists understand that

when hurricanes form, higher water temperatures can energize them and make them more powerful. Warming is also causing the atmosphere to hold more moisture and concentrate precipitation in stronger storms, including hurricanes. In the case of Hurricane Sandy, it retained much of its strength as it tracked across ocean water that was 9 degrees (F) warmer than average for this time of year. . . . More broadly climate change is increasing sea levels globally, which affects all coastal storms, including hurricanes. Locally, sea level rise along the Mid-Atlantic and New England coasts has been among the highest in the world."

9. China claims indisputable sovereign rights on South China Sea Islands and their adjacent waters.

10. According to Kajiwara, arable land leased for agricultural production and export without proper social safeguards has been labeled "Neo-colonialism" by Jacques Diouf, the director-general of the Food and Agriculture Organization (FAO).

11. According to this article in *The Japan Times*, David Berteau, the director of the Washington-based Center for Strategic and International Studies, was reported as stating: "While troop numbers have remained constant, overall annual spending has grown to $224 billion in 2011, according to a report released Monday [15 October 2012] by the Center for Strategic and International Studies think tank. Spending particularly accelerated in the second half of the decade . . . . The levels of increases and concentration of spending is nothing like we saw in the 1950's and 1960's, or even the 1970's and 1980's, between East and West."

12. "At the September 2010 Arctic Forum meeting, Arctic Council member states presented competing territorial claims for the control of polar fossil fuel reserves," Kajiwara reported in this article.

13. This formulation stated in the 5 June 1997 Newcastle, Australia UNEP conference presentation: "How We are Travelling Along the Road to Sustainability" by K.A. Edwards of UNEP Nairobi conference presaged what has become in the on-going discourse, "the three pillars of sustainability:" economic feasibility, social equity and environmental integrity.

14. Article 3, paragraph 3: "The Parties should take precautionary measures to anticipate, prevent or minimize the causes of climate change and mitigate its adverse effects. Where there are threats of serious or irreversible damage, lack of full scientific certainty should not be used as a reason for postponing such measures, taking into account that policies and measures to deal with climate change should be cost-effective so as to ensure global benefits at the lowest possible cost."

15. Article 3, paragraph 1: "The Parties should protect the climate system for the benefit of present and future generations of humankind on the basis of equity and in accordance with their common but differentiated responsibilities and respective capabilities. Accordingly, the developed country Parties should take the lead in combating climate change and the adverse effects thereof."

16. Yvo de Boer, UNFCCC Executive Secretary, called for capitalizing on the present historical moment, with 43,000 participants registered at COP 15 and 120 Heads of State and government scheduled to attend the high-level segment. He noted that with the adoption of the Kyoto Protocol, attention to climate change had shifted away from a broader agenda on sustainable development, biodiversity and food security, and added that the Bali Roadmap provides an opportunity to address other issues than industrialized country emission reduction targets. He suggested that, while politicians will focus on targets, finance and MRV, other relevant constituencies should ensure that attention is not diverted from the four building blocks (mitigation, adaptation, technology and finance), capacity building and REDD+, even when the spotlight shifts elsewhere. He thus called on Forest Day 3 participants to contribute to ensuring the social and environmental integrity of the architecture of the Copenhagen outcome."

REFERENCES

Aquinas, Thomas. 1955. *Summa Contra Gentiles: Book Two: Creation.* Translated by James F. Anderson. Garden City: Hanover House.

———. 1948. *Summa Theologica.* Translated by Fathers of the English Dominican Province. 3 volumes. New York: Benzinger Brothers.

Benedict XVI, Pope. 2010. "To Cultivate Peace, One Must Protect Creation." Papal Address to Diplomatic Corps on January 11. *Zenit* January 11, 2010. Accessed March 20, 2013. http://www.zenit.org/en/articles/papal-address-to-diplomatic-corps.

Berry, Thomas. 1974. "Contemporary Spirituality: The Journey of the Human Community." *Cross Currents* 24 (Summer/Fall): 172-83.

"Biodiversity Funds to Double." 2012. *The Japan Times*, October 21.

Blanchette, Oliva. 1992. *The Perfection of the Universe According to Aquinas; A Teleological Cosmology.* State College: Pennsylvania State University Press.

"China Spurs Asia Defense Spending." 2012. *The Japan Times.* October 17.

"China Tells U.S. to Stay Out of South China Sea Dispute." 2010. *The Daily Yomiuri*, Beijing (AFP-JiJi), September 22.

Clarke, W. Norris, S.J. 2001. *The One and the Many: A Contemporary Thomistic Metaphysics.* South Bend: University of Notre Dame Press.

de Boer, Yvo. 2009. *IISD Forest Day Bulletin* 148.3. December 15.

Deignan, Kathleen. 2003. *When the Trees Say Nothing: Writings on Nature.* Notre Dame: Sorin Books.

Emanuel, Kerry. 2005. "Increasing Destructiveness of Tropical Cyclones Over the Past 30 Years." *Nature* 436 (August 4): 686-88.

Fussel, Hans-Martin. 2007. "Climate Change Vulnerability and Responsibility." Climate Change and Justice Panel, United Nations FCCC COP 13, Bali, Indonesia, December 10

Hansen, James, Mikiko Sato et al. 2008. "Target Atmospheric CO2; Where Should Humanity Aim?" GISS-7, April 2008, Cornell University Library.

Kajiwara, Mizuho. 2010. "Critics Cry Foul Over Neocolonialism' Disguised as Investments." *The Asahi Shimbun GLOBE*, September 27.

O'Rourke, Fran 1992. *Pseudo-Dionysius and the Metaphysics of Aquinas.* Leiden: Brill.

Pieper, Josef. 1966. *The Silence of Saint Thomas.* Chicago: Regency Logos Edition. Pontifical Academy of Sciences. 2011. *"Declaration by the Working Group." In Fate of Mountain Glaciers in the Anthropocene: A Report by the Working Group.* Accessed March 15, 2013. http://www.casinapioiv.va/content/dam/accademia/pdf/glaciers.pdf.

Union of Concerned Scientists. 2012. "Hurricane Sandy Underscores Climate Change Threat to Coasts." October 30.

Vagaggini, Cipriano O.S.B. 1968. *The Flesh Instrument of Salvation.* Staten Island NY: Society of St. Paul.

Vidal, John. 2013. "Large Rise in CO2 Emissions Sounds Climate Change Alarm." *The Guardian*, March 9.

ELEVEN

American Lifestyles and Structures of Sin

The Practical Implications of Pope Benedict XVI's Ecological Vision for the American Church

David Cloutier

Few issues are as abstract as climate change. The issue is so abstract, in fact, that much ink gets spilled simply establishing that it is a genuine problem. And then, once the problem is made clear, the myriad sources and causes of emissions and the complex effects quickly lead to discussions of national policies and international treaties. James Garvey laments in his book *The Ethics of Climate Change* (2008, 137): "There is a lot of talk in the philosophical literature and elsewhere about the moral demands of climate change, but almost none of it has to do with the moral demands placed on us just insofar as we are individual people."

As a moral theologian, I am interested in remedying that problem by developing the concrete implications of Pope Benedict XVI's environmental vision. In *Caritas in Veritate,* he offers the first-ever extended treatment of environmental issues in a papal encyclical, presenting an overall framework and naming two major problems that must be addressed to live out this vision. I will describe these two problems, and then turn to the circumstances of the American Church to develop specific sites for living out this vision. I contend that the pope's general vision requires us to challenge the "suburban way of life" as a kind of "structure of sin" that stands in the way of implementing the pope's vision, and in particular, that our desire for luxury is at the center of that way of life. As Garvey

(2008, 143) puts it, "One of the conclusions we are avoiding, perhaps above all the others, is a personal conclusion: I ought to change my comfy life." *Luxuria* is a vice frequently critiqued by the Church Fathers, and its contemporary instantiations are a significant barrier to the application of pope's vision in American parishes. Yet, even as we recognize the necessity of making different *personal* choices, we as a Church need to recognize and challenge the *structures* which incline many well-intentioned persons to continue living in unsustainable ways within this dominant cultural pattern. Thus, this lifestyle is not simply a matter of sinful choices, but constitutes a structure of sin.

How might we go about determining that a lifestyle is a structure of sin? I will proceed in three parts. First, I will analyze Pope Benedict's environment vision, with attention to what he names as "sinful." Secondly, I will identify and use data to show how something called "the suburban lifestyle" contributes to the sinful use of resources. Thirdly, I will name both the personal and structural "conversions" that are necessary to confront this structure of sin.

UNDERSTANDING THE POPE'S VISION

In *Caritas in Veritate,* Pope Benedict XVI teaches that the natural environment has "a grammar which sets forth ends and criteria for its wise use" (2009, #48). This idea, that "nature expresses a design of love and truth," is the foundation of the pope's ecological vision. He rejects the idea that nature is simply "a heap of scattered refuse" (2009, #48, quoting the ancient Greek philosopher Heraclitus), which we may use in any way we wish. Benedict develops a vision that insists on ecology, on the idea that nature has a design "prior to us," with which we should cooperate. If we do not cooperate, we "abuse" nature, rather than use it. As I have previously argued, this notion of the "grammar of creation" is central to developing principles for wise use (Cloutier 2010).

From the outset, we must recognize that Benedict also ties environmental concern directly into more traditional Catholic moral issues. In speaking to the United Nations in 2008 on the subject of climate change, he notes that addressing hunger "demands far more than a mere scientific study to confront climate change." Rather, "it is necessary first of all to rediscover the meaning of the human person" (Benedict XVI 2012, 86). In *Caritas in Veritate,* the connection is fully made: "The book of nature is one and indivisible: it takes in not only the environment but also life, sexuality, marriage, the family, social relations . . ." (2009, #51) Just as Pope John Paul II developed his "theology of the body" in order to draw tightly together sexuality and human dignity, so Benedict is developing a "theology of the earth" in order to show that our treatment of nature just as much reflects our respect for human dignity, and thus for God who is its Creator and Source.

Benedict's environmentalism, however, goes beyond an anthropocentric to a more eco-centric approach, to borrow terms from Michael Northcott's (1996) classic study. Northcott's project insists that any real approach to the problem of climate change must begin with a real reconnection of creation and redemption, embodied in the actual (changed) lives of local communities. For Benedict XVI, the secular environmental movement is a promising site of evangelization specifically because it implicitly is committed to a notion of *logos*, of the world and reality as an ordered whole which must guide embodied community life. This notion of *logos* is at the center of Ratzinger's understanding of early Christianity, and its rejection of pagan gods in favor of pagan philosophy (Ratzinger 2004, 137-43), as well as his vision of how a rejection of a prior order is at the root of the modern road to unbelief (Ratzinger 2004, 57-58). At the heart of Benedict's environmental theology is not simply a concern over actual harms, but of a pattern of life that violates the order inherent in creation. In short, care for creation is just as much a theological "life" or "dignity" issue as are other prominent Catholic moral concerns. Call it a truly "consistent ethic of life"—or a "consistent ethic of God's creation."

What does this abstraction mean for you and me, or for the average parish? It is a nice idea, one that is unlikely to encounter much resistance . . . until its *implications* are examined. Pope Benedict identifies two concrete problems: one, "the technologically advanced societies can and must lower their energy consumption" (Benedict XVI 2009, #49). The data make it clear, effectively and compactly summarized by Sir John Houghton (2010) within a Christian frame, that there is really no alternative to a substantial reduction in energy consumption if global temperatures are to be held to a few degrees Celsius increase. In his 2008 World Day of Peace message (2012, 59), he insists on the "stewardship of the earth's energy resources," requiring the rich countries to "reassess their high levels of consumption" so countries may gain basic goods. The pope is well aware that the per capita carbon emissions in wealthy countries are 20 to 50 times that of the poorest nations.[1] Such a stewardship will require international cooperation, which is particularly resisted *because* it would force recognition that our ordinary lifestyle in America is egregiously dependent on environmental destruction. Any international plan would require us to be "the biggest loser" on an energy diet.

Thus, the first of the Pope's action steps is dependent upon the second, the lifestyle question. In the WDP message (2012, 59), he urges a universal solidarity "without extravagance and without waste" in which we "prefer the common good of all to the luxury of the few and the poverty of the many." Right now, wealthy countries are living "extravagant" and "wasteful" lives, a lifestyle, Benedict says, of "hedonism and consumerism" pursued "regardless of [its] harmful consequences" (2009,

#51). In an informal question and answer session with Italian priests early in his papacy, the Pope was even more blunt: "It's not just a question of finding techniques that can prevent environmental harms, even if it's important to find alternative sources of energy and so on. All this won't be enough if we ourselves don't find a new style of life, a discipline which is made up in part of renunciations" (quoted in Allen 2008).

SUBURBIA AS OUR STRUCTURE OF SIN

Given the concrete character of the two steps proposed, why hasn't the American Church taken up these claims? In part, there is simply a failure of catechesis: a large number of Catholic Americans hold attitudes toward private property (very pro) and toward international organization (rather anti) that do not reflect the long-standing teaching on the universal destination of goods and of subsidiarity as an aspect of solidarity that the Pope presumes. The US Bishops' 1991 pastoral statement, "Renewing the Earth," highlights themes of "global interdependence," "solidarity," and "the universal purpose of created things" as key piece of Catholic social teaching that must be understood in order to respond properly to environmental issues (USCCB 1991, 230).

However (and this is the main argument of the paper), the resistance is more than theoretical. Catholics concerned about effectively communicating the Pope's ecological vision run into a clear "structure of sin"—which must ultimately be traced back to personal sins. In this case, generalizations about "hedonism" and "excessive energy use" must be traced back to the particular choices which reinforce these problems. There is unavoidable discomfort in this "tracing back," since so much of our lives are wrapped up in patterns that seem counter to a sustainable life within creation's grammar, a pattern which I want to identify here as "suburban."[2]

What Is "Suburbia"? The Dream of House and Car

Thus, my claim: suburbia is a structure of sin which implicates us in the sinful overuse of energy resources and "hedonistic" lifestyle choices. In order to establish this claim, I have to define what "suburbia" is, how it constitutes a particular pattern of life. So, what is "suburbia"? As Tom Martinson (2000, 137) writes, "To many Americans of all races and incomes, the suburbs simply represent that amorphous quality called the American Dream." For many in early America, the American Dream was really about small proprietorship—farmers, shopkeepers, and the like, being able to establish and prosper in business, independent of an aristocracy. In the post-WW2 era, after decades of war and depression, this dream for a broad swath of America dream shifted away from independent ownership and became oriented toward the consumer lifestyle, the "goods life," as Brad Gregory (2012, 235) calls it. At the center of this movement was a remarkable geographical and lifestyle shift: the rise of what we now call "the suburbs."

Characterizing this shift is not easy, but its magnitude is easily seen in US Census figures for central cities from 1950 to 2010, as reported in the 2012 *World Almanac*. Detroit lost over a million residents, and Chicago nearly a million. Proportionally, cities like Cleveland (lost fifty-seven percent of its population) and St. Louis (sixty-three percent) were hit even harder. The suburbs came to represent both a particular geographical pattern of settlement—contrasted with "urban" and "rural"—and also a set of quality-of-life aspirations. Even in technically urban or rural areas, suburban quality-of-life aspirations came to dominate the sort of lifestyle people sought.

What distinguished that lifestyle was above all the rise of the personal automobile, and the concomitant rise of interstate highways, especially in and out of central cities. The subsidized highway system meant that "distance from the city was now measured in time rather than mileage" (Palen 1995, 60; also Hayden 2003, 165-68). Car ownership then dictated the further "look" of many suburbs, which especially emphasized separation of industrial, commercial, and residential functions, discouraging walkability and encouraging auto commuting. Such strong distinctions were further enhanced by accelerated tax depreciation for commercial real estate, which made (cheap) new buildings attractive purely as an investment proposition, ones that were best left behind (for more new construction) after the tax write-off was complete (Hayden 2003, 162-64; Mitchell 2006, 6).

Besides the car, the suburban lifestyle was most especially distinguished by the single-family dwelling, surrounded by a lawn. Such a desire appealed to latent romantic, "country" images of American innocence, in particular "the Jeffersonian self-sufficient yeoman," with "Jefferson's views as to the corrupting effect of urban life" (Palen 1995, 93). Suburbanites were not really seeking space for food self-sufficiency; instead, they sought the ideal of what Robert Kirkman (2010, 64) calls "the pastoral middle landscape" of "houses in a park . . . away from the city with its grid of streets, its noise and its filth," an image that goes back to Roman villa life. The idea was to be near the city, while also being able to escape it. These ideals were further subsidized by FHA lending programs that favored certain style of homes, lot designs, and areas "protected from adverse influences" (Gillham 2002, 32-46). Often enough, communities were built with extensive design rules and "restrictive covenants," to fend off "bourgeois nightmares" of the city, including the walkability of a neighborhood that combined residential and commercial uses, and lower and upper-income residences.[3] All of this was deeply idealized above all for children—suburban "safety" and an assurance that schools would be populated with others "like" one's own family seemed better than the strife and "adverse influences" in central cities. The disastrous difficulties of central cities, especially after the mid-1960s —often exaggerated by their transmission into people's living rooms via the new technology of television—became both effect and cause of suburbanization.

While some of this description can seem like a stereotype, "the stereotype is real," as Kenneth Jackson states in his classic study. Suburbia is "both a planning type and a state of mind," and he argues that suburbia

> is perhaps more representative of its culture than big cars, tall buildings or professional football. Suburbia symbolizes the fullest, most unadulterated embodiment of contemporary culture; it is a manifestation of such fundamental characteristics of America society as conspicuous consumption, a reliance upon the private automobile, upward mobility, the separation of the family into nuclear units, the widening division between work and leisure, and a tendency toward racial and economic exclusiveness (Jackson 1985, 4).

Jackson's description does not really enter into some of the fiercest "suburbia debates" in the literature, about its effects on "community." Martinson (2000, 36), a suburb defender, suggests that suburbanites have a different understanding of "sense of community." Whereas city dwellers abstract intellectually and look for signs of cultural grandeur, he says, suburbanites focus much more on warm and intense interpersonal connections. On the other hand, Robert Putnam's well-known *Bowling Alone* suggests that even these personal ties have weakened considerably in the last 50 years. Putnam devoted an entire chapter to "mobility and sprawl" in his analysis, and notes that extended sprawl has particularly negative effects because of time lost commuting and because of a loss of a sense of "well-defined and bounded" community. We need not enter judgments about vague categories like "community"; Jackson's description above simply states a set of lifestyle expectations and standards. For the purposes of this paper, I only want to recognize the standards and ask how they are related to the problem of climate change.

How do we measure the impact of this lifestyle on climate change? First and foremost, we must see the peculiarly energy-intensive character of American life.

Overall US energy consumption *per capita* increases 40.2 percent from 1960 to 2000, with by far the biggest proportional jump in the 1960s, and a flattening out but still increasing use after that.[4] Also, as Table 11.1 shows, US energy use outpaces other developed countries, often by a significant amount. Comparison of actual energy use—rather than carbon emissions—makes clear that regardless of the source of energy, we use a disproportionate amount.

Table 11.1. Comparison of overall energy use per capita, 2007, in tonnes of oil-equivalent

Italy	3.18
UK	3.73
Germany	4.06
Japan	4.09
France	4.26
Sweden	5.47
Norway	5.49
US	7.82

Source: OECD 2009.

Relative energy consumption can depend on many variables. Since we want to pay attention to lifestyle in this paper, let's take Germany, a country with a similar energy mix, climate, and emphasis on industry. In 2006, prior to the Great Recession, German fossil-fuel CO_2 emission were 10.0 tonnes per person, compared to 19.07 tonnes per person for the U.S. (OECD 2009). Note also that the trend line for German emissions is downward for decades, even accounting for greater population growth in the U.S.[5]

How do we account for such differences? I cite three interrelated factors, all of which stem from the suburban aspirations cited above: overall car usage, housing density, and housing size. One factor– car use. A 1997 FHA comparison (U.S. Department of Transportation 2011) shows that Germans drive 3,967 miles per year, compared with 5,701 miles per year in the US.[6] Why do Americans drive so much more? In 1960, sixty-four percent of workers commuted by car, while by 2009, ninety-two percent of American workers used a car (Union of Concerned Scientists 2012, 61-62). This is the result of the suburban evolution of geography, as is the fact that over sixty percent of car trips made in the US were six miles or less round-trip (Mitchell 2006, xv). What are these trips? Well, if you think about American suburbia, you will realize that the nearest supermarket, discount store, multiplex, or regional park is probably a couple of miles away from your house. Just far enough to drive!

This difference is driven by the second factor—housing density. Typical *low-rise* but pre-1945 American central cities, like Cleveland, Minneapolis, and Pittsburgh—not the largest ones like New York or Chicago—have a population density in the range of 5000-7000 persons per square mile.[7] Compare this to the density of post-1945 mega-suburbs like Aurora, CO (2102) or Henderson, NV (2392), or newer cities, like Charlotte, NC (2457).[8] The spatial pattern of development has both driven and been driven by the "dream of house and car." Less density both makes a car required and is driven by cars. A remarkable comparison can be seen in looking at a recent study of metro Washington, DC (Metropolitan Washington Council of Governments 2011), where eighty percent of all trips away from home were by car, with only two densely populated urban neighborhoods showing a major difference (nineteen percent drive/ fifty-six percent walk in one, forty-four percent drive/twenty-one percent transit in another). My small city/exurb of Frederick, Maryland, with a walkable downtown center surrounded by a lot of suburban-style subdivisions, was more typical: eighty-two percent car/two percent transit/0.9 percent bike/eleven percent walk. Compare this with the report on usage from Freiburg, Germany, a city of 220,000, where the mix went from thirty-eight percent car/eleven percent transit/fifteen percent bike/thirty-five percent walk in 1982 to thirty-two percent car/eighteen percent transit/ twenty-seven percent bike/twenty-three percent walk in 2007 (Union of Concerned Scientists 2012, 79-80). This is no accident: planned development in European cities emphasizes dense centers that can contain (within walking distance) typical daily services, and can serve as a public transportation hub to the central city.[9] Such density can even be achieved with single-family units—my own neighborhood in 1920s-era bungalow-belt Chicago featured blocks of 20-22 single-family houses on each side of a 1/8-mile block.

Yes, the houses were small and close together. Low density development is also driven by problem #3, expanding house size. As Tim Flannery notes, humans have always "engaged in a constant battle to maintain thermal comfort." Today, in the US, that effort at heating and cooling takes fifty-five percent of total household energy use (Flannery 2005, 202). So perhaps the largest chunk of American extra use must be chalked up to our preference for large, detached single-family dwellings. The simple size of these dwelling has increased by more than double since the early 1950s (Polter 2007, U.S. Energy Information Administration 2009).[10] The American average housing size is over 1000 sq. ft. larger than the standard in most European countries, and many more European units are semi-detached (Yunghans 2011).[11]

Roughly speaking, if we drive like Germans and live in well-insulated 1300-sq-ft townhouses, we could cut around 6-7 tons off of average emissions (and much more for some). If we attempt to cut our emissions in half—and this is *less* than what the majority of scientists (McKibben 2011; Northcott 2007, 51-52; Houghton 2010) believe is necessary for heavy-use developed countries, if warming is to be limited to 2°C (3.6°F), we will be looking very clearly at questions of transportation choice, housing size, and housing location. Mike Berners-Lee (2011, 22-23) estimates that a standard plastic grocery bag has a carbon footprint of 10 grams—five of them every week for a year accounts for 5.5 pounds of emissions, equivalent to a *single* cheeseburger, or a *quarter of a gallon* of gasoline! Now, I am all for bringing bags to the grocery store, and there are other environmental problems created by all those bags. However, given that 60 percent of personal emissions come from housing and personal transportation (Union of Concerned Scientists 2012, 16), serious cuts in energy use mean attention to the "lifestyle choices" of house and car.

How is Suburbia a Structure of Sin?

It is here where concern for climate change runs into the aforementioned "built in" aspirations of the suburban life; hence we try to ignore their significance. Wendell Berry (2010, 41) writes that, facing the end of cheap fossil fuels as well as climate change, we have first of all sought "to delay any sort of reckoning," maintaining "a dogged belief that what we call 'the American way of life' will prove somehow indestructible."

Hence, we need to introduce the idea of "structures of sin." Pope John Paul II (1987, #36) defines structures of sin as existing social structures which "are rooted in personal sin, and thus always linked to the concrete acts of individuals who introduce these structures, consolidate them and make them difficult to remove. And thus they grow stronger, spread, and become the source of other sins, and so influence people's behavior." This definition is helpfully complex. A "structure of sin" *influences* people's behavior. Elsewhere, the pope (1987, #16) mentions how such structures "often function almost automatically," as if they are merely natural. Hence, the perception of status associated with particular housing choices in our society "influences" what housing people buy, as does the existence of a structure such as the home mortgage interest tax credit, as does the difficulty or ease of selling the home, as does the existence of various types of zoning strategies. The difficulty in switching to cleaner alternatives is blocked by powerful and wealthy lobbies for existing coal and oil corporations, many of which are among the largest in our economy.[12]

However, as the Pope (1984, #16) also notes, these structures are always "rooted in personal sin." Influence is not total causation. Structures of sin should not be understood "in a way that leads more or less to a watering down and almost abolition of personal sin . . . ." Why not buy a very solid 1920's 3-bedroom duplex a few blocks from where I live in Frederick—a nice street, near the hospital— except . . . no big lawn, extra bathrooms, and vaulted-ceiling entrance . . . and the feeling that Frederick is just a little too "gritty". . . and you'd have to share an entry porch with the other occupant of the duplex. We need to see that choice, as well as transportation choices, particularly ones where you are "locked in," as *genuine moral choices*, made more difficult by "structures of sin" which incline us in the wrong direction. To use Benedict XVI's language, suburbia's understanding of ideal housing and transportation patterns runs counter to the grammar of creation itself, as evidenced by the disproportionate energy use required to fulfill the ideals.

Objections

Is this development pattern really a "structure of sin"? And aren't there efficiency steps and technological advances that can be chosen, instead of questioning people's personal choices? Two very important questions. First, let's consider whether this pattern merits the language "structure of sin." It's typically been used for large problems like international development issues (the Pope's usage) and racism. The term is important for two reasons: one, the term generally attempts to bring to the surface deeper and more intractable problems that hamper straightforward efforts to confront evil. It points to a crucial disjunction between people's "good intentions" and the intractability of the problem, because changes in individual attitude don't help (and can even hurt) if the mediating structures involved are not addressed. In the case of suburbia and climate change, the above arguments need to help us recognize that a combination of good will and recycling, with a few light bulb changes, *cannot address the problem*. One writer makes the correct point that confronting climate change requires "a change in attitude to complete the journey," one that rejects monetary success and rugged individualism (Gunshinan 2012, 17-19)—but such attitude changes can only go so far when your job requires a commute or housing prices drive you away from walkability and transit. Mark O'Keefe (1990, 69) defines social sin as "the embodiment of sin in unjust structures which exist relatively independent of human agency and conscious willing." Many Americans support action to confront climate change, just as they may want to solve world hunger and end racial discrimination—yet they often oppose changes that would cost them money or require lifestyle changes.[13] The sticking point comes in the disconnection of this good "wish" and the actual day-to-day choices made or avoided. Our choices in relation to the environment are mostly not direct—they don't involve pouring motor oil down the sewer; they are mediated by existing structures. The genuinely correct choices cannot be made if the mediating structures do not receive attention. In this case, the "mediating structure" is the form of development and lifestyle aspiration called "suburban."

Secondly, the term is used specifically by Pope John Paul II in order to "theologize" social structures. That is, the problem is not merely a *technical* one; it is a moral and spiritual one. As the pope (1991, #36) notes elsewhere, "a given culture reveals its overall understanding of life through the choices it makes in production and consumption." These are precisely the choices made every day to sustain the suburban consumption-driven economy. Tim Gorringe (2002, 242) rightly argues that "the problems associated with the built environment are not primarily technical but spiritual, that is to say they are fundamentally a question of values, of our understanding of the whole human project." Like art, literature, and any other cultural endeavor, it both manifests and further reinforces meanings. Suburbia is not just a sloppy structure, it is a structure of *sin*. Suburbia is a moral and spiritual problem, not because of some elite "bias" against ordinary people, but because the structure means that we are forced to depend on fossil fuels to such a large extent.

Acknowledging the seriousness of the pattern is one thing; on the other hand, many suggest the solution lies not in lifestyle critique, but in technological and efficiency gains. Maybe Pope Benedict is right about reducing energy use, but are suburban aspirations really an example of the "hedonistic" lifestyles he criticizes? After all, studies do show that people tend to focus disproportionately on "curtailment efforts," rather than "efficiency efforts"—they turn off lights rather than replace windows—even though efficiency efforts can often net larger gains (Union of Concerned Scientists 2012, 18-21). The focus on efficiency efforts is an attempt to confront the problem of climate change in a "lifestyle-neutral" fashion—without Pope Benedict's second critique. True, smaller cars are not exactly lifestyle-neutral, but they require much less sacrifice than walking, biking, or carpooling. One author writes that any "realistic" change must be one carried out "without altering the way people drive or buy vehicles. In North America, the issue of an individual's right to buy the car or truck of his choice is about as contentious as gun control. And a lot more people own vehicles than guns" (Tertzakian 2006, 189).

There are three difficulties with putting a lot of hope in lifestyle neutrality. First, efficiency gains can sometimes have inverse psychological effects. We have seen enormous efficiency gains since the 1970s, especially for heating and cooling (Union of Concern Scientists 2012, 100-105), yet these have been largely "consumed" by increased usage—sometimes people think, oh, the lights are energy efficient . . . so I can leave them on! If our cars are more efficient, it won't help if we drive them twice as much. Second, the magnitude of the problem surpasses purely technological solutions. In reference to "greening the grid," James Garvey (2008, 103) notes that to achieve even a 1/7 reduction in emissions would require a fiftyfold increase in current wind capacity or 7,700 square *miles* of solar panels, 700 times current solar capacity. Electric cars aren't (much) greener if we just have more coal plants. In short, there are ceilings to efficiency gains. Third, even theoretically "lifestyle-neutral" choices are not really lifestyle-neutral. Scientist Tim Flannery (2005, 303) maintains that "you can, in a few months" make changes that cut your emissions by seventy percent with only "a few changes to your personal life, none of which requires serious sacrifices." Solar hot water, a hybrid car, some solar panels, and some retrofitting on the house—you are set. In fact, Flannery is really talking about the same kinds of choices I am, and he admits that even with these changes (none of which, it should be noted, require new or innovative technologies) "our family is vigilant about energy use, and we cook with gas. And I'm fitter than before because I use hand tools rather than the electrical variety to make and fix things" (Flannery 2005, 304). So, if one is willing to accept that giving up power tools, leaf blowers, absolutely unlimited hot water, and large cars are not really "serious sacrifices," then good! But is driving a different, smaller car "lifestyle-neutral"? Furthermore, for many people, genuine investments in efficiency gains DO mean lifestyle tradeoffs: money spent on new windows, solar panels, or a hybrid vehicle is money that can't be spent on other aspects of the suburban lifestyle. Thus, Flannery's admirable proposals may look like "serious sacrifices" to many caught in suburban structures and aspirations.

CONFRONTING THE STRUCTURE OF SIN: LETTING GO OF LUXURY

Appeals to efficiency gains through technological advance are ways of dodging the moral questions about lifestyle that Benedict rightly raises, a lifestyle that is simply unsustainable. Even outside a Christian context, "lifestyle change is fast becoming a kind of 'holy grail' for sustainable consumption policy" (Jackson 2008, 352). The author goes on to note that rapid lifestyle change is something people are used to in this era, citing the rapid diffusion of cell phones, the Internet, SUVs, flat-screen TVs, and even air conditioning. Unfortunately, these changes often tend toward a more carbon-intense direction; surely Catholics can and should respond to the pope's call for lifestyle changes in a different direction. A structure of sin approach to this problem allows for the recognition of both the personal moral choices involved in this lifestyle, as well as the structural issues that tend to reinforce these bad choices.

Thus, a response to the structures of suburban life must be both personal and social. We must start with the personal. As we saw Benedict propose, a new lifestyle must include disciplined "renunciations." Elsewhere, I have focused on recovering the traditional critique of luxury as a way of reintroducing morality into people's everyday economic choices (Cloutier 2012). Combatting luxury begins with the question, "How can I do without this?" The ancient critiques of luxury apply well to what is sought in suburban housing and transportation choices. Indeed, what emerged in the post-WW2 era was a lifestyle previously reserved for the very wealthy now available to all. Historian Peter Brown (2012, 197) outlines the life of the Roman wealthy of the fourth century. At the center of this life of luxury was the country villa, "presented as a place of unproblematic abundance." These villas were distinguished not only by their size, but also by the elaborate furniture and decoration they featured. The decoration often emphasized "innocent abundance," portraying scenes of nature full of "limitless resources" (and notably, omitting any image of the workers providing the resources or the state of the land). Particular emphasis was put on *private baths* and on *food preparation*, since both contributed to the image of comfortable abundance and innocent happiness. Luxury baths and kitchens–how things don't change! Brown also notes the importance of the body in these households, the need for "exuberant, wellgroomed hair" and "glowing complexion." Moreover, the country villas themselves were not usually occupied year-round, since business took their owners into cities, and so another mark of luxury was the ability to have easy access to one's own transportation, so that one could "escape" from the city in order to enjoy this "clean" nature.

As is well-known, many Church Fathers were none too kind to these displays of wealth. Basil proclaims, "What will you tell the judge, you who dress up your walls and leave humans naked? You who groom and adorn your horses and will not look at your naked brother? You whose wheat rots, and yet do not feed the hungry?" (Gonzalez 1990, 178). Ambrose likewise says: "You give coverings to walls and bring men to nakedness. The naked cries out before your house unheeded; your fellowman is there, naked and crying, while you are perplexed the choice of marble to clothe your floor" (Phan 1984, 175). He goes on to criticize gold bits for horses and jewelry. This kind of evidently unnecessary ornamentation is worthy of condemnation, not praise. The Fathers emphasize how it signals a neglect of the poor; now, with so many people living this lifestyle, we can add to their indictment neglect of the entire Creation, while retaining in our sights how suburban housing and private vehicles are a way to keep us "protected from the poor." As Michael Northcott (2007, 209) sums it up, "The still-growing material and spatial differentiation between luxury and social housing, between gated community and ghetto, is indicative of a culture of luxury that drives the profligate use of energy, and which is at the same time a culture of injustice."

To be blunt, there cannot be sustainability in the United States without abandoning our commitment to luxury. "Luxury" in this case should indicate our constant temptation to preferring private comfort, status, ease, and convenience over the common good of reducing energy use. As William Leach (1993, 295, 290) notes in his history of the rise of consumer society in America, the period 1890-1930 saw "luxury's complete redefinition" in order to promote an economy which, in the words of a 1920s booster, "had shifted from a 'needs' to a 'desires' culture." This shift often was conveyed by the language "standard of living." It became assumed that social progress meant above all a "rising standard of living," by which was meant more and better devices, and more and more comfort—one manual notes especially the new ideal of "a standard of one room per person exclusive of bath" (Leach 1993, 296) Today, perhaps, we think, one room AND one bath! Such an ideal was often surrounded by "traditional" aspirations, such that a car gave you "access to nature" or a large, up-to-date home was reflective of a fulsome concern for family.

Issues of luxury are crucial for climate change. Even as far back as the 1970s, Barbara Ward (1979, 12) noted that the years since World War II "with oil at less than two dollars a barrel" meant that "yesterday's luxuries became today's necessities." Table 11.2 lists the carbon impact of some evidently excesses "wants" that we have.

The activities in the left-hand column are among the highest-impact, most extravagant consumer activities. Yet here again, even when compared to these egregious examples, the *much* bigger impact lies with housing and transportation choices.

The most obvious luxuries here is our desire for excessively large and over-powered private vehicles, and our construction of housing and housing areas entirely dependent on lengthy driving of such vehicles. This sort of discourse needs to return to our churches, especially in areas where sufficient wealth means both that people can afford to make changes in their lives and that people can properly set trends. Outside my parish one day, a woman, very fashionably dressed, was dropping off her daughter at the Catholic school from her Chevy Suburban; behind them, a granola-looking mom in denim overalls and a t-shirt was pulling in to do her drop-off . . . from a Prius. (The Prius, notably, had a prominent rosary dangling from the rear-view mirror.) The real concern here is whether the parish and school are doing anything to educate about these choices as a genuine part of Catholic faith, just like issues of sexuality and life.

The Prius-versus-Suburban face-off here is easy to cast off as a kind of social stereotype. Unfortunately, when one realizes that the Prius burns

Table 11.2. Carbon impact of various activities, in kg of CO_2e

1 lb of air-freighted strawberries/wk, for a year	171.6	20 extra miles (10 each way) on your commute, average US car	4080
3 bottles of distantly-produced beer/day, for a year	328.5	Driving a 40-mpg car as opposed to a standard 22-mpg car, for 10000 mi/yr	5000
1 cheeseburger/day, for a year	910	Shaving 33% off national-average heating and cooling use, electric	3134
A dozen roses/wk, for a year, air-shipped from Colombia	218.4	Family of four 3-hour driving vacation (in SUV) vs. flying 1000 miles	2600

Sources: Berners-Lee (2011), except heating and cooling savings, derived from average heating and cooling energy use for all households (US Energy Information Administration 2009) and the EPA's (2012) carbon emission average for electricity in the United States , with a BtU to KwH conversion (US Environmental Protection Agency 2004). (All estimates from Berners-Lee's numbers for driving are somewhat higher than the usual, because he also takes into account the energy needed to produce/replace the car, much as the typical mileage rate includes depreciation. This seems reasonable: driving less also means the car is likely to last longer.)

five times less fuel than the Suburban, it is a real issue. I can guess (from the few times I've ridden in an SUV) why their owners are reluctant to give them up. They are so comfortable. They give you a commanding view of everything. You feel as though you are in some sort of protected tank. This feeling is, in some sense, illusory, since SUVs are no safer than other cars (although they are much more destructive to others). Illusory or not, the feeling is really one of luxury, of an experience of being in a car in a certain kind of way, with a certain kind of power and separation. As with luxuries, once you get used to this, it must be extremely hard to get into a subcompact car. Are we willing at least to challenge this as hedonism and luxury, given its effects?

The housing question is more difficult, for two reasons. One, parishes are territorial, and increasingly in the suburbs, that means they are relatively homogenous in terms of housing. Two, housing itself is a much more permanent choice. The aforementioned Catholic school mom could be in a situation to get a new vehicle in the next five years, but she is less likely to be able to give up her housing.

However, there are still crucial issues here. The most obvious is encouraging people with existing houses to do two things to counteract luxury. One, we should spend our money on retrofitting our houses, *instead of* on marble countertops, designer bedspreads, ever-larger TV sets, and other vanity purchases. This has the additional benefit of detaching us and our children from consumer culture itself, which is an explicitly-stated objective in papal encyclicals . . . and it helps reduce a not-insignificant aspect of their contribution to climate change. (The other two big contributors would be to change our diet to less meat, especially less beef, and to cut out most long-distance flying for leisure.) Two, we should keep our heat lower and our cooling higher—sixty-five and seventy-eight degrees would be a good start.[14] Once again, people will howl; as illustrated in Table 11.2, the difference in usage is significant. Moreover, consider the advice of Seneca (1965, 89), the ancient moralist, who writes: "Avoid luxury, . . . If a man always has always been protected from the wind by glass windows, if his feet have been kept warm by constant relays of poultices, if the temperature of his dining room has been maintained by hot air circulating under the floor and through the walls, he will be dangerously susceptible to a slight breeze." So, let's put on those slippers and sweaters, or bring out those window fans.

On the level of housing, though, addressing structures is important in the medium to long-term. In many cases, such as *development planning*, decisions are often strongly driven at the local level, a level where large parishes can exercise substantial weight. Local and regional planning are often contentious efforts that take place among a rather confusing set of local players. In my exurban Frederick County, the growth fight is fierce. Largely rural until twenty years, Frederick has expanded rapidly, and the last few years have seen the county government engulfed in contentious and shifting plans for future development—smart growth or sprawl growth? Becoming a DC suburb or retain our "quality of life"? A recent BOCC developed a comprehensive twenty-year plan for more concentrated, denser development connected to existing towns. However, the plan was resisted fiercely by developers and by many landowners who recognized that their land would not be open for development. So the present BOCC overturned the past plan, in favor of much wider latitude in land use away from towns. In the midst of all the lengthy debate over this issue, I never once saw any Catholic parish—or even any Christian community—even sponsor education programs on this issue, much less lobby the (very Christian) conservatives on the current BOCC. The other area in which explicit church efforts would be helpful *structurally* would be in advocating for carbon taxation, which I will just indicate here as an obvious structural point. As Michael Northcott (2007, 141) comprehensively argues, taxation should return to its origins, on actual material goods, in order to provide a limit on their use.

Parishes advocating on zoning and carbon taxes may seem a stretch to some. Again, the necessity is to recognize the environmental issue as both personal and social. Our personal choices are affected by social structures which incentivize and disincentivize behavior. Tim Jackson (2008, 353-54) notes two important structures: the problem of "lock-in" and the problem of bad price signals. "Lock-in" explains that consumer choices and trends are often responses to initiatives begun elsewhere, or to expectations and patterns started elsewhere. Thus, parents who want their children (understandably) to go to good schools often choose a relatively new and well-off suburban area . . . because the area has good schools. Because it is well-off (and wants to stay that way), the homes are pricey and large. And once people sink a lot of their money into such housing, they are "locked in" to a significant portion of their auto and home consumption costs. Bad price signals happen when the most environmental choice is not the cheapest choice. If public transit is more expensive than driving, how many people will use the transit just because they are limiting carbon emissions? Yet the cost of driving is artificially low—the "negative externality" of carbon emissions needs to be somehow added to the cost of the gas for the car. Thus, churches can and should take stands on important structural issues like zoning and carbon taxation, as a necessary and logical part of confronting climate change.

I recognize that there may be significant objection to and distaste for the above points. Yet what is being suggested is hardly St. Francis; it's much easier! In light of the Pope's real challenge, if we comfortable Americans can't sacrifice some luxury, and can't even talk about it, then we should recognize how deeply attached—*religiously* attached—we are to this idea of material progress, embodied in suburbia. Even in the 1920s, social critic Samuel Strauss criticized "consumptionism" and a focus on "luxury and security and comfort," on "the standard of living above all other values." He writes, "No minister in any pulpit offers any cure which requires that the nation's 'standard of living' sags back" (quoted in Leach 1993, 266-268). The fact is, middle and upper-middle class Americans are the single most important group of people to "convert" on these issues, in the entire world, not least because they "probably have much more economic power than the vast majority of people on the planet" (Garvey 2008, 141). If they lived their lives differently in this way, not only would there be a concrete difference in carbon emissions, but also perhaps the largest political roadblock to international cooperation on this issue would be removed.

CONCLUSION: LIVING THE REALISTIC LIFE

An unwillingness to confront this challenge to our conventional lifestyle, a lifestyle reinforced over and over again, every day, by the people sitting right there in the pews of Catholic parishes from coast to coast, should not be surprising. That is characteristic of structures of sin. It is true that "no one wants to hear that they have to cut back on life's taken-for-granted luxuries" (Tertzakian 2006, 208). Social structures that seem "just the way things are" make sin invisible—as we know in terms of slavery. These choices may seem "unrealistic"—what is the single most hopeful thing to realize is that *they are not*. The good news is we do not have to give up the advances in things like medicine, water treatment, intelligent building, and common transportation systems that mark the greatest advances of our culture. These advances simply *do not require* the kind of profligate energy use we now have. In one sense, climate change is an easy problem to confront, because as I've illustrated, serious progress can be made with existing technologies and simple (though serious) lifestyle change. We need no "miracle" scientific development to make a ton of progress.

However, we do need conversion, since what is "unrealistic" is the life we live now, and we are in practice deeply attached to the aspects of this suburban paradise I've described here. Maybe converting hearts is actually harder than developing new technologies, for it requires us to reorder our loves. Pope Benedict maintains that this same loss of reality infects many aspects of our postwar American lives—our illusions about sex and the unborn, as well as our financial and environmental illusions. None of these illusions are easy to confront. Yet Benedict invites us to see that God's dream is simply not the postwar American dream.

NOTES

1. Michael Northcott (2007, 49-50) uses the figure 0.2 tons for the annual emissions of the least-developed countries, a figure at least 30 times less than even the lowest U.S. per capita estimates.

2. Debates over the environment, drawing on reservoirs of compassion, often focus on harm done to distant parts of the globe. However, such images then get contrasted to the progress out of poverty made by nations like Brazil and China, who are doing so partly through large increases in their use of environmental resources, including fossil fuels.

3. On this idea, which began with enclaves for the very wealthy but then spread, see Fogelson 2005.

4. The number decreases from 2000 to 2010, though one of the questions is the extent to which this is because of the economic slump and because of big increases in the outsourcing of production. Given that the number also dropped 1980-1990 (with the first advent of smaller cars), but bounced back 1990-2000 (years of economic expansion but also the rise of the SUV), both the economic issues and the cars we drive are big contributors.

5. These numbers do not include all emissions, but only emissions from the actual use of fossil-fuels in transportation, industry, and domestic consumption. The EPA (2013) estimates that globally, fifty-seven percent of emissions are caused by carbon from burning fossil-fuels, with the other major contributors being deforestation/biomass decay (seventeen percent), and methane (forteen percent), approximately a third of the latter also being related to fossil-fuel production and use, specifically natural gas. Mike Berners-Lee (2011, 11), who has tried to incorporate every possible detail for other activities beyond burning fossil fuels, makes an estimate of 28 tonnes CO2e for an average North American, compared 15 for the UK. These estimates show that the further impact is roughly proportional to the "direct" carbon impact measured by the OECD numbers.

6. If we assume that Germany also has a 10 mpg more efficient overall car fleet, this difference in miles driven would account for a ton to a ton and a half of the difference in emissions.

7. Some examples include Milwaukee (6,188/sq. mi.), Cleveland (5,107), Buffalo (6,471), Minneapolis (7,088), and Pittsburgh (5,521). Compare this density to larger cities like Boston (12,793), Chicago (11,842), and New York City (27,012). All figures from the 2012 *World Almanac*.

8. For more comparisons on density, see Jackson (1985, 5, 11).

9. For a direct comparison of American and Swedish developments that illustrate the difference, see Popenoe (1977).

10. The EIA survey is fascinating. According to the data, the average size of all types of US dwelling units combined was 1,971 square feet, but the average single-family detached house was 2,483 square feet. Units constructed prior to 1990 averaged between 1,600 and 1,900 square feet, but the average jumped over 2,200 in the 1990s and then over 2,400 in the 2000's.

11. In my neighborhood in Frederick, a development of new, super-insulated, geothermally-heated 1300 square-foot attached townhouses net a carbon reduction of almost 5 *tons* per year over the "average" (even without their nifty added solar panels). That's a savings that covers over 15,000 miles of driving in a forty mpg car.

12. In the 2011 Fortune 500, 3 of the top 10 were oil companies, 2 others were car manufacturers, and another was the top mortgage issuer, Fannie Mae, a company with a deep interest in maintaining the value of current housing stock. Indeed, a look at the top 100 largest corporations—from aircraft manufacturers to shippers to agro-chemical conglomerates to massive nationwide retailers—reveals how much of our economy depends on continued, unquestioned petroleum supplies, especially. Northcott (2007, 7) notes that imagined technological fixes are meant to avoid recognizing global warming as "the earth's judgment on the global market empire, and on the heedless

consumption it fosters." He rather devastatingly shows how so many supposed technical or economic fixes are in fact schemes to insure that large, corporate actors and wealthy nations can continue to reap their accustomed economic benefits. James Kunstler (2005, 222-223) observes, not incorrectly, that the recent decades of the American economy were "no longer about anything except the creation of suburban sprawl and the furnishing, accessorizing, and financing of it . . . . Nothing else really mattered except building suburban houses, trading away the mortgages, selling the multiple cars needed by the inhabitants, upgrading the roads into commercial strip highways with all the necessary shopping infrastructure, and moving vast supplies of merchandise made in China for next to nothing to fill up those houses." It should also be mentioned that perhaps the most important determinant of "sprawl" in this era is the increasing *unaffordability* of housing, particularly in the fastest-growing metros. Just down the street, in the far-from desirable Brookland neighborhood, brand-new (dense! Near Metro!) townhomes are going in . . . starting in the 600s! Urging people to buy housing near the Metro in such a situation is perhaps not the greatest "influence" on their ultimate choice.

13. A 2012 Stanford/Washington Post poll (Eilperin 2012) showed two-thirds of Americans want the U.S. to be a world leader in confronting climate change, and over half think it is a very important problem. Yet strikingly, "People don't see a lot of downside for taking action to stop global warming. Only 12 percent say that the things people would do to help stop it would make their own lives worse." Moreover, more than seventy percent oppose direct tax increases on electricity and gasoline.

14. Programmable thermostats are a technical solution that is also very helpful. According to one source Union of Concerned Scientists (2012, 93), simply turning down the heat and AC when asleep and at work can cut fifteen percent off your heating and cooling emissions. Yet only about one-third of households have these, and fewer use them.

REFERENCES

Allen, John A. 2008. "A Round of Questions for the 'Shepherd-in-chief'," *National Catholic Reporter*, August 15. Accessed May 27, 2013. http://ncronline.org/blogs/all-things-catholic/round-questions-shepherd-chief .

Benedict XVI. 2012. *The Environment*. Huntington, IN: Our Sunday Visitor.

———. 2009. *Caritas in Veritate*. Accessed May 27, 2013. http://www.vatican.va/holy_father/benedict_xvi/encyclicals/documents/hf_ben-xvi_enc_20090629_caritas-in-veritate_en.html .

Berners-Lee, Mike. 2011. *How Bad Are Bananas? The Carbon Footprint of Everything*. Vancouver: Greystone Books.

Berry, Wendell. 2010. "Faustian Economics." In *What Matters: Economics for a Renewed Commonwealth*, 41-54. Berkeley: Counterpoint.

Brown, Peter. 2012. *Through the Eye of a Needle: Wealth, The Fall of Rome, and the Making of the Christian West, 350-550 AD*. Princeton: Princeton University Press.

Cloutier, David. 2010. "Working with the Grammar of Creation: Benedict XVI, Wendell Berry, and the Unity of the Catholic Moral Vision." *Communio* 37: 606-33.

———. 2012. "The Problem of Luxury in the Christian Life." *Journal of the Society of Christian Ethics* 32: 3-20.

Eilperin, Juliet, and Peyton M. Craighill. 2012. "Temperatures climbing, weather more unstable, a majority says in poll." *Washington Post*, July 13. Accessed May 27, 2013. http://www.washingtonpost.com/national/health-science/post-stanford-poll-finds-more-americans-believe-climate-change-is-happening/2012/07/12/gJQAh92wgW_story.html .

Flannery, Tim. 2005. *The Weather Makers*. New York: Grove Press.

Fogelson, Robert M, 2005. *Bourgeois Nightmares*. New Haven: Yale University Press.

Garvey, James. 2008. *The Ethics of Climate Change*. New York: Continuum.

Gillham, Oliver. 2002. *The Limitless City: A Primer on the Urban Sprawl Debate*. Washington: Island Press.

Gonzalez, Justo. 1990. *Faith and Wealth: A History of Early Christian Ideas on the Origin, Significance, and Use of Money*. San Francisco: Harper & Row.

Gorringe, Tim J., 2002. *A Theology of the Built Environment: Justice, Empowerment, Redemption*. Cambridge: Cambridge University Press.

Gregory, Brad. 2012. *The Unintended Reformation*. Cambridge: Harvard University Press.

Gunshinan, Jim. 2012. "Power Surge: How Faith Helps Answer Our Energy Challenges." *America*, May 7, 17-19.

Hayden, Dolores. 2003. *Building Suburbia: Green Fields and Urban Growth, 1820-2000*. New York: Random House.

Houghton, Sir John. 2010. "The Changing Global Climate: Evidence, Impacts, Adaptation and Abatement." In *Keeping God's Earth: The Global Environment in Biblical Perspective*, edited by Noah J. Toly and Daniel I. Block, 187-215. Downers Grove IL: InterVarsity Press Academic.

Jackson, Kenneth T., 1985. *Crabgrass Frontier: The Suburbanization of the United States*. New York: Oxford University Press.

Jackson, Tim. 2008. "Sustainable Consumption and Lifestyle Change." In *The Cambridge Handbook of Psychology and Economic Behaviour*, edited by Alan Lewis, 335-62. New York: Cambridge University Press.

John Paul II. 1984. *Reconciliatio et Paenitentia*. Accessed May 27, 2013. http://www.vatican.va/holy_father/john_paul_ii/apost_exhortations/documents/hf_jp-ii_exh_02121984_reconciliatio-et-paenitentia_en.html.

———. 1987. *Solicitudo Rei Socialis*. Accessed May 27, 2013. http://www.vatican.va/holy_father/john_paul_ii/apost_exhortations/documents/hf_jp-ii_exh_02121984_reconciliatio-et-paenitentia_en.html .

Kirkman, Robert. 2010. *The Ethics of Metropolitan Growth*. New York: Continuum.

Kunstler, James. 2005. *The Long Emergency*. New York: Grove Press.

Leach, William R. 1993. *Land of Desire: Merchants, Power, and the Rise of a New American Culture*. New York: Random House.

Martinson, Tom. 2000. *American Dreamscape: The Pursuit of Happiness in Postwar Suburbia*. New York: Carroll & Graf.

McKibben, Bill. 2011. "Resisting Climate Reality." *New York Review of Books*, April 7, 60-64.

Metropolitan Washington Council of Governments. National Capital Region Transportation Planning Board Meeting. 2011. "2011 TPB Geographically-Focused Household Travel Surveys Initial Results." Accessed May 27, 2013. http://www.mwcog.org/uploads/committee-documents/k11dXlle20120517145044.pdf .

Mitchell, Stacy. 2006. *Big-Box Swindle*. Boston: Beacon Press.

Northcott, Michael. 1996. *The Environment and Christian Ethics*. Cambridge: Cambridge University Press.

———. 2007. *A Moral Climate: The Ethics of Global Warming*. Maryknoll NY: Orbis.

O'Keefe, Mark. 1990. *What Are They Saying About Social Sin*. New York: Paulist Press.

OECD. 2009. *OECD Factbook 2009: Economic, Environmental, and Social Statistics*. Paris: OECD.

Palen, John J. 1995. *The Suburbs*. New York: McGraw-Hill.

Phan, Peter C. 1984. *Social Thought: Message of the Fathers of the Church*. Wilmington DE: Michael Glazier.

Polter, Julie. 2007. "Attack of the Monster Houses." *Sojourners* 36 (3): 38-42.

Popenoe, David. 1977. *The Suburban Environment*. Chicago: University of Chicago Press.

Putnam, Robert. 2000. *Bowling Alone: The Decline and Revival of American Community*. New York: Simon and Schuster.

Ratzinger, Joseph. 2004. *Introduction to Christianity*. Translated by J. R. Foster & Michael J. Miller. San Francisco: Ignatius.

Seneca. 1965. "On Providence." In *Hellenistic Philosophy*. Edited by Herman Shapiro and Edwin M. Curley, 79-95. New York: Modern Library.

Strauss, Samuel. 1924. "Things Are in the Saddle." *Atlantic Monthly*, November, 577-88. Quoted in Leach 1993.

Tertzakian, Peter. 2006. *A Thousand Barrels a Second*. New York: McGraw Hill.

Union of Concerned Scientists. 2012. *Cooler Smarter: Practical Steps for Low-Carbon Living*. Washington DC: Island Press.

USCCB, 1991. "Renewing the Earth." In *And God Saw It was Good: Catholic Teaching and the Environment*, edited by Drew Christiansen and Walter Grazier, 223-43. Washington DC: USCCB.

U.S. Department of Transportation. Federal Highway Administration. Office of Highway Policy Information. 2011. "Annual Automobile Vehicle Miles of Travel (VMT) per Capita and Number of Automobiles per Capita 1997." Accessed May 27, 2013. http://www.fhwa.dot.gov/ohim/onh00/bar4.htm .

U.S. Energy Information Administration. 2009. "2009 Residential Energy Consumption Survey." Accessed May 27, 2013. http://www.eia.gov/consumption/residential/data/2009/#undefined .

U.S. Environmental Protection Agency. 2004. "Unit Conversions, Emissions Factors, and Other Reference Data." Accessed May 27, 2013. http://www.epa.gov/cpd/pdf/brochure.pdf .

———. 2012. "Clean Energy: Calculations and References." Accessed May 27, 2013. http://www.epa.gov/cleanenergy/energy-resources/refs.html .

———. 2013. "Global Emissions." Accessed May 27. http://www.epa.gov/climatechange/ghgemissions/global.html .

Ward, Barbara. 1979. *Progress for a Small Planet*. New York: W.W. Norton.

Yunghans, Regina. 2011. "Average Home Sizes Around the World." *Apartment Therapy*. July 20. Accessed May 27, 2013. http://www.apartmenttherapy.com/average-home-sizes-around-the-151738 .

TWELVE

American Nature Writing As a Critically-Appropriated Resource for Catholic Ecological Ethics

Anselma Dolcich-Ashley

In *Caritas in Veritate* (2009) and other writings, Pope Benedict XVI joyfully celebrates God's creative wisdom and steadfast love that are evident in the nature environment, and likewise prophetically denounces the ways in which humans subvert this and other gifts to their own sinful ends and purposes. However, according to the American Psychological Association (2010), Americans appear to be stuck in a pattern of denial regarding responsibility for climate change. I suggest that our incapacity to see and accept our culpability for environmental destruction also mirrors a similar incapacity to see and experience creation as a gratuitous gift of God in Christ, the basis of conversion, loving self-sacrifice and communion. In this paper I explore how the distinctive genre of American nature writing might provide interdisciplinary cultural resources in an American idiom to respond positively to the gratuitousness of creation's beauty and goodness, to identify and repent of culpable sinful actions, and to develop a capacity to grasp the truth which liberates.

AMERICAN NATURE WRITING—A RESOURCE FOR MORAL DISCERNMENT

Recent studies in genre and ethics point towards promising ways in which narrative and other literary forms mold ethical conscience and assist in ethical decision-making (McKenny, 2005). The genre of

American nature writing can assist Church responses to the ecological crisis because nature writing represents a type of rhetoric in the classic sense—a genre aimed at persuasion by means of inviting the reader into the place and space of the writer, and of enabling the reader not only to see as the writer sees but also to respond likewise. Non-academic literary genres have the capacity to examine levels of moral response otherwise difficult to probe via rational argument, to "push back moral horizons, contribute to the creation of social conscience and expose the complexity . . . of moral discernment" (Yeager 2005, 445). Analogous to the ways in which the stories, parables, events and persons of Scripture place contemporary readers in positions to make choices, take action and exercise virtue (Spohn 1999), classic pieces of American nature writing confront readers with the possibility of "renew[ing] and strengthen[ing] that covenant between human beings and the environment, which should mirror the creative love of God, from whom we come and towards whom we are journeying" (Benedict XVI 2010).

Examining three short but influential pieces from three prominent American nature writers whose writings spanned the twentieth century, this investigation correlates these with resources from the Catholic tradition.[1] The selected nature writers expose classic human reactions to the natural environment, disclose a process of conversion of heart, and raise fundamental questions regarding the human place in the natural world, thus providing faithful seekers to with key markers on the path towards communion with God and others in the context of the natural environment. The pieces to be examined here are "The Hetch Hetchy Valley" by John Muir (1912) from his book-length work on the Yosemite region, Aldo Leopold's "Thinking Like a Mountain" (1989) from his celebrated *The Sand County Almanac: And Sketches Here and There*, and Annie Dillard's meditation on 3 consecutive days on an island in Puget Sound in her short book *Holy the Firm* (1977).

These writers enable a "reading of the signs of the times" in regards to the modern ecological crisis not by offering blueprints for programs and policy, or even with making clear a general course of action—such as was obvious in 1962 with the publication of Rachel Carson's *Silent Spring*—but rather by summoning readers into active participation in a real-life ecological narrative, placing readers within view of their destiny for metanoic transformation (of which the concrete endings may be variable and only dimly perceived), and confronting persons with the challenges of living in communion with others identified both as human and non-human creation. Indeed in the case of Dillard, a virtuous disciple with an already wide-awake sensitivity to creation's tangible signs of God is only the starting point for the further journey into the terrifying and incomprehensible mystery of nature and natural events—of which the ultimate outcome, I suggest, is contemplation of and communion with God. By interpreting these writings[2] from the perspective of faith I seek to fill in some of the details of a course charted by Pope Benedict XVI (2010), whereby a commitment to the environment and the natural world neither devolves into a neo-paganism or pantheistic nature-worship—wrongly seen by many as the only possible spiritual response, nor takes cues solely from secular and scientific analysis—these are necessary but not sufficient. Rather, such a commitment holds the promise of reconnecting our experience of beauty with an encounter of God's goodness and liberating truth, and our sinful destruction of creation's goodness with the truth of judgment and the divine offer of reconciliation (Benedict XVI 2008).[3] By articulating the truth and rejecting oversimplified and sentimental responses to the natural world, these writers help us see the gospel unfolding in the human relationship with each other and in nature, a point fittingly raised by Benedict XVI in *Caritas in Veritate* (2009 #3):

> Without truth, charity degenerates into sentimentality. Love becomes an empty shell, to be filled in an arbitrary way. In a culture without truth, this is the fatal risk facing love. It falls prey to contingent subjective emotions and opinions, the word "love" is abused and distorted, to the point where it comes to mean the opposite. Truth frees charity from the constraints of an emotionalism that deprives it of relational and social content, and of a fideism that deprives it of human and universal breathing-space.

From this point, Christian communities may then, in an ongoing act of discipleship, engage in a "liturgical *askesis*" (Pfeil 2007) concretely reconnecting their own lives and their environment with God's creative power and love.

JOHN MUIR AND THE HETCH HETCHY VALLEY: THE HUMAN RESPONSE TO NATURE'S BEAUTY

With skillful combination of writing elegance and persuasive rhetoric akin to preaching, rather than a philosophical-ethical argument focused on rights and wrongs, Muir (1912, 255) succeeds in identifying California's Hetch Hetchy Valley as "our land," not in the possessive sense of human ownership (which led it to its being "harvested" for water resources) but rather in the sense that the land itself makes a claim upon us, the human community. Possession, if any, is reversed: by its very beauty the land places us in a location of responsibility and love—"this most precious and sublime feature of the Yosemite National Park, one of the greatest of all our natural resources for the uplifting joy and peace and health of the people" —and challenges us to include it in our account of community. In order successfully to retrieve Muir's essay as a resource for the Catholic moral imagination to apply to the ecological crisis, however, a critical appropriation attends to the characteristics and the performance of Muir's writing, while rejecting pantheistic interpretations by means of an authentic Christian theology of creation and doctrine of God.

The controversy over Hetch Hetchy represents perhaps "the first time in the American experience a national audience considered the competing claims of wilderness versus development" (National Park Service 2007). Located within the bounds of Yosemite National Park, the valley is evidence that breathtaking beauty in creation is no rare event. Utilizing the language of creation but with reference to "Nature" (capital N) rather than to God, John Muir observed (249-50):

> Yosemite is so wonderful that we are apt to regard it as an exceptional creation, the only valley of its kind in the world; but Nature is not so poor as to have only one of anything. Several other yosemites have been discovered in the Sierra that occupy the same relative positions on the Range and were formed by the same forces in the same kind of granite. One of these, the Hetch Hetchy Valley. . . . I have always called the "Tuolumne Yosemite," for it is a wonderfully exact counterpart of the Merced Yosemite, not only in its sublime rocks and waterfalls but in the gardens, groves and meadows its flowery park-like floor."

Indeed, the valley was a unique, and uniquely beautiful, ecosystem within a granite canyon, sustaining hundreds of animal species as well as two of the highest waterfalls in the US. Yet after the 1906 San Francisco earthquake and the devastating fires in its aftermath, public alarm for reliable supplies of water set in motion a process resulting in US Congressional passage of the 1913 Raker Act authorizing construction of the O'Shaughnessy Dam in the valley (within the boundaries of Yosemite National Park). The burgeoning human population of the area also caused increased demand for a reliable water supply. Since 1923 the Hetch Hetchy valley has been submerged (National Park Service 2007).

Muir's success in advocating for and preserving large swaths of land—he was instrumental in changing the idea of national parks into what it is today, a *system* of lands throughout the country and accessible to all, rather than isolating Yellowstone as "the" national park (Worster 2011)—lay largely in his capacity to render into words that transcendent human response to the sheer beauty of nature. To be sure, conveying this experience is the bread-and-butter of all nature writers. Yet Muir not only was able to articulate more keenly than others what goes on in the human spirit when confronted by nature's glories, but also was encountering some of the most spectacular natural scenery on earth. His prose invites us directly into this experience, thus for example, his descriptions of Hetch Hetchy's two great waterfalls place us immediately in their presence. Of Tueeulala Falls he writes (251-52):

> Imagine yourself in Hetch Hetchy on a sunny day in June, standing waist-deep in grass and flowers (as I have often stood), while the great pines sway dreamily with scarcely perceptible motion. Looking north-

> ward across the Valley you see a plain, gray granite cliff rising abruptly out of the gardens and groves to a height of 1800 feet, and in front of it Tueeulala's silvery scarf burning with irised sun-fire. In the first white outburst at the head there is abundance of visible energy, but it is speedily hushed and concealed in divine repose, and its tranquil progress to the base of the cliff is like that of a downy feather in a still room. By comparison to the Tueeulala's ethereal descent, the nearby Wapama Fall is a roaring river of water, 'pounding its way like an earthquake avalanche' in carrying more than twice the volume of water.

As to what goes on inside human persons confronted with such beauty, Muir is quite clear: ultimately we respond with love, and we are drawn into communion. Nature's beauty itself compels, calls and attracts us; its sheer loveliness gets through internal prejudices that see only inert "stuff" via the senses of sight, sound and touch, and render the overall experience as one of transcendence. The focus on the "subjectivity of the writer as the bearer of certain perceptions and feelings in response to the non-human world" (Clark 2011, 28) brings to the fore the relationship of humans to non-human nature, and the inner human response to beauty. Thus, writes Muir (255):

> [T]he Hetch Hetchy Valley, far from being a plain, common, rock-bound meadow, as many who have not seen it seem to suppose, is a grand landscape garden, one of Nature's rarest and most precious mountain temples. As in Yosemite, the sublime rocks of its walls seem to glow with life, whether leaning back in repose or standing erect in thoughtful attitudes, giving welcome to storms and calms alike, their brows in the sky, their feet set in the groves and gay flowery meadows, while birds, bees, and butterflies help the river and waterfalls to stir all the air into music—things frail and fleeting and types of permanence meeting here and blending, just as they do in Yosemite, to draw her lovers into close and confiding communion with her.

Muir's purpose is not merely to convey the beauty of northern California to the nature-deprived masses of America's burgeoning cities in the early twentieth century. After escalating to heights where nature's beauty climaxes in a human person's conscious recognition of love and communion, Muir shocks his readers with the contrast between such sublime beauty and its crass destruction as a city's reservoir.[4] His handling of the matter does not facilely condemn the modern world's need for electricity and water utilities—he was no neo-Luddite—but rather keeps the reader rooted in beauty, an experience which now brings grief: "Everybody needs beauty as well as bread, places to play in and pray in, where Nature may heal and cheer and give strength to body and soul alike" (256). For Muir the issue is not one of the rights of nature, or even of trail-hiking naturalists, but rather a matter of the heart and of basic principle. As the essay progresses, Muir ups the ante, moving into polemic and invoking Biblical narratives in an attempt to make sense of the depth of the tragedy. Comparing to situation to the stories of Jesus and the money-changers in the Temple and our parents' first sin in Genesis 3, Muir argues not from a philosophical-ethical system of right and wrong but rather from the more profound contrast of beauty and goodness threatened by evil made manifest in human hubris, where evil frequently masquerades as good. Hetch Hetchy only reveals a more widespread greed in the United States "by despoiling gainseekers and mischief-makers of every degree from Satan to Senators, eagerly trying to make everything immediately and selfishly commercial, with schemes disguised in smug-smiling philanthropy . . . Thus long ago a few enterprising merchants utilized the Jerusalem temple as a place of business instead of a place of prayer; . . . and earlier still the first forest reservation, including only one tree, was likewise despoiled" (257).

How can Muir's discourse assist contemporary Catholic communities in their responses to global climate change? First, it is worthwhile to note echoes in recent papal messages, especially Pope Benedict's emphasis on the beauty of creation as worthwhile in itself. Such experiences not only bring sought-after rest and relaxation; they also assist in our vocational and moral discernment wherein our choices and actions for our lives remain impressed by natural beauty and reflective of our desire for its preservation. People loved Hetch Hetchy for its sublime beauty and, because of that love, committed themselves through legal structures to preserve it forever as part of Yosemite National Park. Pope Benedict XVI (2009, #2, 13) echoes the understanding that creation's beauty alone is an object of contemplation, bringing an understanding of our proper human "vocation and worth" as well as recognizing the "love of the Creator" which brought forth and sustains such beauty. It is a very significant fact that many people experience peace and tranquility, renewal and reinvigoration, when they come into close contact with the beauty and harmony of nature.

Secondly, contemplation of nature ushers Christians once again into the experience of Biblical authors, especially those of the Wisdom tradition, for whom the natural world points to and partially discloses a loving God, both author of creation yet wholly other from it. The distinctions articulated by this tradition find resonance in Benedict XVI's caution against eliding creation and Creator in contemporary movements of pantheism and neo-paganism.[5] Muir's prose—especially his use of the term "Nature" (capital N), sometimes described metaphorically as expressive of human perception, sometimes as a quasi-conscious reality—alarmed some of his strict Protestant contemporaries who could not but see nature-worship in such language. While an investigation into Muir's personal faith and theological interpretation of nature is beyond the scope of this paper, it is well known that he memorized most of the Bible during his childhood in a devout Protestant family, and also sympathized with American Transcendentalist writers such as Henry David Thoreau. A recent biographer noted that by his sixties, Muir's "earlier pantheistic tendencies, which celebrated every nodding flower, every zephyr, as divine it itself, became more muted"; as Muir himself said in a 1903 interview, "some scientists think that because they know how a thing is made, that therefore the Lord had nothing to do with it. They have proved the chain of development, but the Lord made the chain and is making it" (Worster 2011, 374-75).

As Muir himself indicates, a theological understanding of creation yields a far more satisfactory explanation of the Hetch Hetchy Valley's beauty than simply a psychological feel-good experience in response to its beauty[6] or a positing of nature as divine. For Christians, the natural world is not an end in itself, but rather reveals the Creator: As Pope Benedict notes (2010, #6), "is it not true that what we call "nature" in a cosmic sense has its origin in a "plan of love and truth?" Indeed, nature's revelation of the Creator connects us not only with God's Wisdom inherent in the natural order, but also with God's plan of grace and mercy revealed in Christ and secured for eternity in the new heaven and new earth.[7] Since the dynamic of rhetoric depends not only upon the writer, but also upon the interactive reception of the written piece within the readers, it seems that Muir's very use of the term "creation" will steer biblically-informed Catholic communities not in the direction of pantheistic nature-worship, but rather towards their own scriptures which grappled with the very same issue. The author of the book of Wisdom understood natural beauty and the human response to it, yet also articulated a theology of real *creation,* that is, a difference between the created world and the God who stands beyond the created world as incomprehensible mystery and wholly-other Creator. This author critiques those who too closely blur creator and creation; such persons "were ignorant of God [and] foolish by nature; and they were unable from the good things that are seen to know the one who exists, nor did they recognize the artisan while paying heed to his works; but they supposed that either fire or wind or swift air, or the circle of the stars, or turbulent water, or the luminaries of heaven"—or perhaps in Muir's case "Nature"—"were the gods that ruled the world" (Wis. 13:1-2). With reason informed by faith, the sacred author sees beyond beauty in itself to the living God who stands over and against matter and history: "If through delight in the beauty of these things people assumed them to be gods, let them know how much better than these is their Lord, for the author of beauty created them" (Wis. 13: 3). Thus a sound Christian understanding of creation inexorably supports a fundamental Christian doctrine of God.

If this is so, then a Christian theology of creation also places our experience of creation's beauty within an eschatological context of salvation. Rather than negating Muir's praises of nature, such a theology gives it a context and necessary horizon. Encounters with nature and natural beauty provide us with a glimpse of a God's own life as a gift and invitation; the gratuitousness and sheer abundance of natural beauty in creation reveal not only that God is the wholly-other creator, but also that God is "relational, ecstatic, fecund, alive as passionate love" (LaCugna 1991, 1), offering us the gifts of creation for our free use that we might freely respond in love and gratitude. Encounter with natural beauty prepares us for that true re-creation undertaken continuously in God's providence, definitively in the Paschal Mystery, and finally at the eschaton. Thus, reflection upon natural beauty and creation leads Christian faithful inexorably to the triune God whom we encounter in the economy of salvation; and so a sound grasp of Trinitarian theology prevents a slide into pantheism.

Thirdly, Muir's evocation of the Biblical stories of the first sin and of the moneychangers in the temple as analogues to the Hetch Hetchy affair places the matter within the context of human concupiscence and the mystery of evil. Muir's writing exposes the simultaneously disordered and well intentioned attitudes and practices of persons which have led to environmental catastrophe. Again, a caution is in order, as Muir's goal of social activism to counter the incipient destruction of Hetch Hetchy is itself subject—as are all human efforts—to the mystery of evil itself in human hubris, personal sin and structural sin. For faithful Christians, however, the *telos* of creation does not ultimately rely upon limited if well intentioned human efforts prone to sin. Rather, ongoing efforts to protect creation arise from our response to God's gratuitous gift of creation, and are oriented to an eschatological horizon. In this context, Christian efforts towards environmental preservation make sense as actions of discipleship. For example, since 1923 a small group of concerned citizens has kept alive as a concrete goal the natural restoration of the Hetch Hetchy Valley (Restore Hetch Hetchy 2012). Action plans have included the long and slow work of conducting studies to demonstrate alternative means (and other reservoirs) to provide San Francisco with water and electricity, and pursuing the issue via referendum ballot.[8] Even though this project is not itself religiously organized, a Christian interpretation—those with eyes to see and ears to hear—might note that those who really did fall in love with Hetch Hetchy became seriously involved in a practice—an *askesis*—of (limited) re-creation as a way to express their gratitude for God's gift of the valley. Christians themselves, on the basis of their own encounter with God in the beauty of creation, can thus join forces with others seeking to restore wilderness areas such as the Hetch Hetchy Valley.

ALDO LEOPOLD: GOODNESS, CONVERSION, AND "THINKING LIKE A MOUNTAIN"

Aldo Leopold's essay "Thinking Like a Mountain" moves us away from Muir's grand and joyful vistas to a very particular mountain which will challenge us, as St Benedict of Nursia (1981, 157) urges in the prologue to his famous *Rule,* to "listen with the ear of the heart." Such a challenge is potentially transformative but not easy; indeed, so difficult that Leopold knows we will meet it with deepest resistance. In a single metaphorical phrase—"thinking like a mountain"—Leopold unseats the conventional thinking of his day which exalted the human thought processes behind the conquering and closing of the American frontier. With these four words Leopold puts us on a quest to encounter and understand the goodness of creation—if only we have the humility to accept that goodness. On the mountain lives a wolf and her pack. Leopold (1949, 120) challenges the reader to decode the wolf's language: the emphasis is not on the animal's verbal sounds, but on what is communicated by its wolf's very being. All living things, even the mountain's pines, "pay heed to [the] call" of the wolf ostensibly lest they court peril, but "behind these obvious and immediate hopes and fears there lies a deeper meaning known only to the mountain itself. Only the mountain has lived long enough to listen objectively to the howl of a wolf." Leopold only knows that he does not possess such objective knowledge himself, but, after decades of life experience and reflection, the little he does know represents a paradigm shift, indeed, a *metanoia,* from his prior worldview.

One of three reflections in his famous *Sand County Almanac* upon his young adult days (1909-1924) doing survey work for the United States Forest Service in Arizona and New Mexico, Leopold's essay conveys its truth in the form of an extended metaphor or parable based upon a true event. The value of metaphor lies not in mathematical or philosophical precision, but rather in its attuning our knowledge, awareness and responses to the truth, often in unexpected, surprising or ironic ways. Metaphor ingeniously conveys complex or paradoxical truths, and captures our curiosity, by a cognitive dissonance for which the human brain naturally seeks resolution (Lakoff and Johnson 1980). Yet by challenging us with a metaphor, neither does Leopold give us marching orders but rather a "vision of reality upon which we must base our ethical decisions and actions" and a pattern we need to detect and follow in different, but analogous, circumstances (Spohn 1999, 54-55). Rather than lapsing into rigid moralism or fanciful sentimentality, "thinking like a mountain" inexorably leads to a profound internal conversion, achievable only by the humbling experience of contrition for sin.

Christians familiar with Jesus' favorite means of preaching the Kingdom of God through the use of parables[9] understand that such analogical forms of speech both "hook" us and surprise us in order to encourage further thought and personal engagement. To interpret a parable or metaphor literally strips it of meaning. In thinking like a mountain Leopold is not denigrating human rationality any more than Jesus commands us to become housecleaners in search of coins or farmers tossing seeds onto random patches of soil. In so doing, Leopold's essay moves us past our comfortable constructs and forces us to examine our fundamental conceptions undergirding our relationship with the natural world.

What happened that made Leopold finally grasp and become convinced that he did not think like a mountain, and for what did he come to feel contrition? He was working as part of a survey team in the mountains of New Mexico, and heard "a deep chesty bawl echo[ing] from rimrock to rimrock, roll[ing] down the mountain and fad[ing] into the far dark blackness of night." The sound was so unsettling that he could describe it only as a kind of "wild defiant sorrow" and "contempt for all the adversities of this world"—sort of a hard-edged "blues" of the wild (Leopold 1949, 129). Some days later he and his team observed from afar a wolf family's crossing the river below. The wolves were so large, and so far away, that the men thought they were deer but recognized them as wolves when the pack emerged upon the bank, "all joined in a welcoming melee of wagging tails and playful maulings" (130). Then the men opened fire and the carnage began. Leopold writes, "We reached the old wolf in time to watch a fierce green fire dying in her eyes. I realized then, and have known ever since, that there was something new to me in those eyes—something known only to her and to the mountain."[10]

Just as the Lord called to the man and woman in the cool of the evening in Genesis 3, the dying wolf called out to Leopold, and his eyes were opened. The remainder of the essay expresses a sorrow he held for the rest of his life: ecosystemic destruction for both deer and for humans spiraled out of control as every state exterminated its wolves, a keystone predator. "I have seen every edible bush and seedling browsed, first to anaemic desuetude, and then to death. I have seen every edible tree defoliated . . . In the end the starved bones of the hoped-for deer herd, dead of its own too-much, bleach with bones of the dead sage" (131-132). In our hubristic quest for security from nature's power, "too much safety seems to yield only danger in the long run" (133). The goodness of God's natural creation, from the mountain itself housing all kinds of living things to the wolf family's fulfilling God's command in Genesis 1 to produce offspring according to their own kind,[11] has taken a back seat to humanity's shadow creation of technological dominance, which now threatens to destroy even human life.[12]

Catholics informed by the sacraments understand contrition as sorrow for sins; without confronting the painful truth of our willful separation from God, penance cannot heal. Thus St. Thomas Aquinas (1948, Suppl 1.1) sees contrition as a virtue and a grace given by God, which destroys pride understood as "the beginning of all sin because we cling to our own judgments." Indeed, Aquinas (Suppl. 2.2) deploys his own metaphor for contrition, that of a force capable of breaking up a rigid and hard material, analogous to a "state of continuity and solidity in our mind." Contrition pulverizes and subdues pride, which interferes with the human acceptance of God's Lordship. A feature of Leopold's real-life parable is his own contrition, which via a theologically and sacramentally informed conscience can be seen as not simply "changing his mind" in regards to factual right or wrong but indeed a movement of grace enabling detestation for and future avoidance of sin, as well as a commitment to a more prudential and loving relationship with all things wild.

How do we know that what is conveyed here really is contrition, and not attrition, which arises from a motive other than the love of God? It's easy to think here of criticisms of wildlife rescue efforts—wolves and other large mammals are, to humans, inspiring and noble animals (and awfully cute as babies), moving persons to work on their behalf but not, say, on behalf of some ugly endangered mollusk (*Scientific American* 2007). But Leopold himself displays the corrective here. His message is not a direct call to action—"save the wolves"—but rather a change of heart applicable in many places and many ways, and responsive to the goodness of the natural world broadly considered— "thinking like a mountain." He is not promoting a cause or an ideology, or a precise moral response for any given situation. Rather, he invites us, rather compellingly, through his own experience to a change of heart expressed by and in "thinking like a mountain." According to Aquinas (Suppl. 2.1), the "sorrow of contrition is based on the love of charity, the greatest love." Leopold's "thinking like a mountain" expresses a great love for the land, itself now understood as living and nonliving things occupying key niches (such as the wolf's essential role as predator), and offering Leopold the possibility of coexistence. In his guilt and regret he experiences a love whose first expression is the sorrow that he personally is culpable for contributing to the land's devastation. Finally, his love for the land clearly encompasses the human place within it: as an outdoor sportsman, Leopold (133-39) saw a place for hunting regulated by prudence.

The value of Leopold's essay lies in its capacity to draw us into his experience and find ourselves confronted by the challenge to divest ourselves of conventional thought and instead to "think like a mountain." While changes of heart in regard to environmental practices may be difficult to bring about, I suggest that Catholics may engage this essay from their understanding of contrition, penance and humility as necessary aspects of discipleship. In continually striving to "think like a mountain," communities may evolve more just and loving responses to the environment, as they allow contrition to substitute a new heart for the old one.

ANNIE DILLARD: GOD, NATURE AND TRUTH

In the late twentieth century Annie Dillard wrote not from encounters with pristine ecosystems untrammeled by human development but rather from within the "rambunctious garden" (Marris 2011) which the planet as a whole has become. Dillard additionally differs from Muir and Leopold because she intentionally questions and makes observations about her Christian faith quite synchronously with her reflections on nature. Her journey of faith takes place within encounters with nature, and leads her to difficult questions regarding creation's overpowering beauty and destructive power alongside human fragility and seeming insignificance. "All flesh is grass" observed Deutero-Isaiah (Isa. 40:6), but few since Job have paused to ponder the implications of this truth. Dillard is one who does.

Holy the Firm began as a self-imposed challenge from her home on an inhabited island in Puget Sound, to write an account of whatever transpired during three sequential days. On the first day, Dillard (1977, 25) explores an experience of which she summarizes, "[a]ll day long I feel created." As she gazes at the great Cascade mountains to the east she quotes the Psalms and recalls Chesterton on the Eucharist; as she salts her eggs she references Roman and Armenian Catholic baptismal rituals; as she glances out the window she sees the local Congregationalist church where on Sundays she joins a small group of worshippers. The Jesuit poet Gerard Manley Hopkins wrote that the earth is charged with God's grandeur, and Dillard goes one further, naming things—the sea, the air, the birds—as gods, not in a pantheistic way but rather as Psalm 8 envisions humanity—so wondrous, so beautiful, so full of life—that words referencing anything other than the divine fall short.

Her memory flashes back to a candle in Virginia's Blue Ridge region. At that time Dillard looked up from her reading just in time to see a moth trapped by the candle fire, its wings ablaze. As the evening wore on, the moth's body became beautifully and gruesomely transformed into a candle wick, giving forth light. "The wax rose in the moth's body from her soaking abdomen to her thorax to the jagged hole where her head should be, and widened into flame, a saffron-yellow flame that robed her to the ground like any immolating monk" (16-17). Only the naïve, or those in denial, can shake off the sense of foreboding as Dillard then launches into the day, delighting us in her delight in and sensitivity to God's creation all around her (19-30).

On the second day, she hears the sound of a light plane taking off, then, impossibly, the silence of a failed engine. The plane falls to the ground; its two occupants escape with their lives, but "the fuel exploded; and little Julie Norwich seven years old burnt off her face" (36). The catastrophe visited upon an acquaintance of Dillard's, whom she remembered as the child trying to dress up Dillard's cat in a black doll's dress and learning how to whistle while the adults gathered for a neighborhood afternoon of apple-cider-making, sends Dillard into a dizzying free-fall of Job-like questioning of God's goodness and justice, and the meaning of a creation where without warning the face of an innocent child can be burned off. In a manner reminiscent of the psalms of lament, Dillard confronts the truth of the moment: God is "abandoning us to time, to necessity and the engine of matter unhinged." It does not take a leap of faith to understand the matter, and in a reversal of hope's definition in Romans 8"[13]—this is evidence of things seen"—she conveys the hopelessness of reality: "one Julie, one sorrow, one sensation bewildering the heart, and enraging the mind, and causing me to look at the world stuff appalled" (46). Her rational, more settled mind knows "that God is all good. And I take it also as given that whatever he touches has meaning, if only in his mysterious terms . . . The question is, then, whether God touches anything. Is anything firm, or is time on the loose?" (47). In other words does God *really* hold us in the palm of his hand (Psalm 91), or are such lovely words only the voice of the desperate, unwilling to face reality, unable to see their existence as a cruel joke? In grief and rage she reflects (61-62):

> We do need reminding, not of what God can do, but of what he cannot do, or will not, which is to catch time in its free fall and stick a nickel's worth of sense into our days. And we need reminding of what time can do, must only; churn out enormity at random and beat it, with God's blessing, into our heads: that we are created, *created*, sojourners in a land we did not make, a land with no meaning of itself and no meaning we can make for it alone. Who are we to demand explanations of God?

The second half of this short book takes on a delusional cast as Dillard struggles to make sense of her ordinary, daily life now entirely upended by Julie Norwich's terrible accident. In contrast to Muir's prose, where creation's beauty manifests love, bounty and communion, Dillard's story confronts and probes the harsh alternative. Nature has presented undeniable evidence that maybe God isn't so gratuitous after all—maybe God is more like a capricious tyrant, overwhelming our small world with sheer power and enormity. Rejecting easy answers and exposing her own inner questioning of this event, she does her readers a service by confronting them with the truth that here there are no "deceptive consolations" (Yeager 2005, 456). In a world of gravity and temperature, of oxygen and hydrocarbons and sparks of flame—those very essential components on earth which give life and which, if we pass by the Julie Norwiches of this world like the priest and Levite hurrying past the man fallen in with robbers, bring praise to our lips—in this world where Julie's lips no longer exist to whistle—how can we claim that God is just and loving and providential? Dillard's wrestling with God as manifest in God's great work of creation is profound, daunting—and haunting.

Just as with Job's story, facile answers miss the point and dishonor the dignity of Julie and her family. Some, of course, find in the glorious account of creation in Job chapters 38-41 a God only of overpowering and incomprehensible greatness, in response to which Job's task is to find his place and apologize. The ending, wherein Job's life is restored tenfold, may reinforce such an interpretation. However, Dillard knows there is no Disney-esque happily-ever-after, and her prose—one moment sublimely considering the Seraphim who themselves burst into flame around God's throne and another moment conveying the unbearable violence and pain of this-world experience—will not allow the reader to engage in such fancies. With no face, no lips, no similarly-faced companions or family, Julie's future must unfold like that of a "nun," but the description of this "nun" is more like that of a prisoner. Has Julie been condemned to a faceless life, isolated as a victim by a God with "bladelike arms" in whose service she may sing or serve, because without her lips she will never whistle or kiss? (74-75).

Dillard's ending is as abrupt as that of the Book of Job. The final, repentant paragraph is revealing, and reversing. "Julie Norwich; I know. Surgeons will fix your face. This will all be a dream, an anecdote, something to tell your husband one night . . . People love the good not much less than the beautiful, and the happy as well, or even just the living, for the world of it all, and heart's home." Dillard then willingly places herself, in a Christ-like sacrifice, in Julie's stead, as she writes: "So live. I'll be the nun for you. I am now" (76). Like Job when confronting God's awesome and mysterious power in nature, Dillard encounters not an answer to her questions, not a justice which evens out the score, but communion—and this not from an exalted, incomprehensible God living above the clouds, but emerging from within her own heart as God's own grace enables her to make a gift of self to the other. Within the context of a nature so dangerous that a child's face can be burned off, sometimes it seems the only morally responsible course of action is to tame nature and control it. But Dillard does not give into the temptation to play God: rather she ends up encountering God and voluntarily, out of love, taking upon herself Julie's suffering. Two key images from the final chapter are a bottle of wine, purchased by Dillard for a church service, and the medieval concept of the basest of matter, "holy the firm." The communion wine, "Christ with a cork" (63), puzzles and almost amuses her—how can such a thing as "communion" be bottled, how can she possibly carry it? Yet this is what communion is—encountering God in the stuff of everyday existence. "Holy the firm" engages her in a mystical way—matter at the bottom of the *scala naturae* which, rather than being most removed from the divine realms, paradoxically curves around and touches God. We encounter God in the basest of things. From the perspective of faith, notes Gustavo Gutiérrez (2007, 87) in his spiritual commentary on the book of Job, Job's own wrestling and lamenting keeps him tied to the awful truth of his situation, but this truth eventually leads him to another truth, the truth of contemplation and communion in love. "Only when we have come to realize that God's love is freely bestowed do we enter fully and definitively into the presence of the God of faith . . . God's love, like all true love, operates in a world not of cause and effect but of freedom and gratuitousness."

Our calling as Christians is indeed to be the moth in the candle and the Seraph before God's throne. Because and through the natural world, we have been given a glorious if frightening opportunity to be that part of creation which is reconciled to God through the cross and enters lovingly into a relationship of communion with the Julie Norwiches of this world. The poor, the suffering, the vulnerable can meaningfully be located right at the center of Christian encounters with the environment and Christian environmental ethics. Hence, Pope Benedict XVI's (2009) connecting sustainability with efforts on behalf of the human poor and vulnerable is no ordinary moral duty defined merely as a distinctive "Catholic" mark in environmental ethics, but rather expresses "charity in truth" a charity and a truth whose utter unbearableness, apart from grace Dillard ably conveys.

CONCLUSION—AMERICAN NATURE WRITERS AND LITURGICAL *ASKESIS*

American nature writing can be a useful resource in Catholic responses to the environmental crisis because of the genre's force of moral persuasion and invitation to spiritual conversion. The American nature writings presented here not only enable an understanding of the human relationship to the natural environment but more importantly draw the reader into a journey through the contours of interior movements of response to the natural world. The sounds of Muir's evident delight in the beauty of Hetch Hetchy Valley fall on to Christian ears like a psalm of praise to God about creation. Leopold and Dillard remind us that the negative way of contrition, sorrow and sacrifice also has sacramental value leading to real change of heart and reconciliation with God and creation. Unlike a moral-philosophical treatise, the genre of nature writing draws us vicariously into the created world presented by the writer, challenges our defenses, helps us to identify with people and the land, and has the result of "altering the landscape of perception and response at levels of sensibility and supposition . . . often immune to rational arguments" (Yeager 2005, 477).

More practically, however, from this point Catholic communities then can move into a liturgical and sacramentally expressive *askesis* often perceived in positive terms of creation's material expression of God's love and glory, within which negative experiences of sorrow, diminishment and humility may be located. "[T]hrough the created gifts of nature, men and women encounter their Creator. The Christian vision of a sacramental universe—a world that discloses the Creator's presence by visible and tangible signs—can contribute to making the earth a home for the human family once again" (Christiansen and Grazer 1996, 231). Unlike ascetical practices understood as conditioning the isolated individual to approximate a philosophical ideal—say, turning off the lights to promote sustainability—liturgical asceticism arises within and among members of a community responding to God's gifts and moving towards an eschatological purpose of fullness of life in God. As described by moral theologian Margaret Pfeil (2007, 127):

> [L]iturgical asceticism encompasses not only the disciplines of liturgical worship but also those sustained practices of daily life undertaken with a conscious awareness on the part of the moral agent of the way in which these disciplines express one's baptismal commitment rooted within a particular faith community. Thus, a liturgically rooted *askesis* suggests a broadened conception of sacramentality beyond formal ritual workshop to include intentional acts of asceticism such as virtuous habits of energy conservation.

Internal assent and conviction are a first step in liturgical *askesis*, since they align the will with the intellect, often require the mutual and loving support of others in the community, and help to dislodge the kind of denial, distancing or apathy evident in Americans' incapacity to take responsibility for global climate change. Here, in the movement of will, is where nature writers exert their effect. The Christian community then can enter into practices of discernment, social justice and the corporal and spiritual works of mercy recognizing that their actual place in the material world is the basis from which they are called to participate in God's own life.

NOTES

1. A Catholic response to climate change may draw from relevant secular resources and aspects of American culture to formulate responses both American and Catholic. At the same time the correlation works in the opposite direction as well: secular nature writings, when viewed through the lens of a Biblical and sacramental imagination, take on additional richness and urgency not available outside a faith context, and indeed help us place our responses to the ecological crisis within the life of faith rather than as an extraneous addition on to it.

2. Identifying the depth and effect of the authors' own religious formations upon their nature writing remains beyond the scope of this investigation. While Muir and Leopold both received extensive Christian formation in their upbringing, their purpose in these works is not, as in Dillard, to make explicit connections between Christian faith/community and experiences in the natural world; but nonetheless the power of their prose cannot help but speak (like Dillard's as well) to persons of faith. Regarding Muir, see Donald Worster, 2011, *A Passion for Nature: The Life of John Muir* (New

York: Oxford University Press). Regarding Leopold, see Curt D. Meine, 2010, *Aldo Leopold: His Life and Work* (Madison WI: University of Wisconsin Press).

3. "[I]t is necessary to again understand the close connection that binds the search for beauty with the search for truth and goodness… It should be noted that the Greek text [of Matthew's gospel] speaks of *kalà erga,*of works that are good and beautiful at the same time, because the beauty of works manifests and expresses, in an excellent synthesis, the goodness and profound truth of the action, as well as the coherence and holiness of those who perform it" (Benedict XVI 2008).

4. "Sad to say, this most precious and sublime feature . . . one of the greatest of all our natural resources for the uplifting joy and peace and health of the people, is in danger of being dammed and made into a reservoir to help supply San Francisco with water and light, thus flooding it from wall to wall and burying its gardens and groves one or two hundred feet deep" (Muir 1912, 255-56).

5. "On the other hand, a correct understanding of the relationship between man and the environment will not end by absolutizing nature or by considering it more important than the human person. If the Church's magisterium expresses grave misgivings about notions of the environment inspired by ecocentrism and biocentrism, it is because such notions eliminate the difference of identity and worth between the human person and other living things. In the name of a supposedly egalitarian vision of the 'dignity' of all living creatures, such notions end up abolishing the distinctiveness and superior role of human beings. They also open the way to a new pantheism tinged with neo-paganism, which would see the source of man's salvation in nature alone, understood in purely naturalistic terms" (Benedict XVI 2010, #13).

6. For example, one critic of the genre of nature writing accuses it of having a "misleadingly cosy feel" (Clark 2005, 5).

7. Christians "contemplate the cosmos and its marvels in light of the creative work of the Faith and the redemptive work of Christ, who by his death and resurrection has reconciled with God "all things, whether on earth or in heaven" (Col 1:20). Christ, crucified and risen, has bestowed his Spirit of holiness upon mankind, to guide the course of history in anticipation of that day when, with the glorious return of the Saviour, there will be "new heavens and a new earth" (2 Pt 3:13) (Benedict XVI 2010, #14).

8. A ballot referendum is only one strategy currently being deployed by the "Restore Hetch Hetchy" movement. According to the "Restore Hetch Hetchy" group, studies conducted in recent years indicate that with proper conservation and reclamation efforts, the San Francisco water district could free itself from dependence upon the Hetch Hetchy reservoir. Financial and other considerations necessary to remove the dam are, however, another matter. On Nov. 6, 2012 San Francisco voters rejected a ballot measure to restore the valley (Aiello, 2012).

9. "Jesus told the crowds all these things in parables; without a parable he told them nothing" (Mt. 13:34, NRSV).

10. "In those days we had never heard of passing up a chance to kill a wolf. In a second we were pumping lead into the pack, but with more excitement than accuracy: how to aim a steep downhill shot is always confusing. When our rifles were empty, the old wolf was down, and a pup was dragging a leg into impassable slide-rocks" (Leopold 1949, 130).

11. Besides Genesis 1, other biblical sources call attention to God's care of and delight in even the wild animals, useless and possibly threatening to humans but beloved of God. Cf., Psalm 104:10-13, 20-30; Job 38-39.

12. Occupying a "keystone" place in the ecosystem, the American gray wolf has been extensively studied, thanks to the reintroduction in 1996 of wolves to Yellowstone National Park, and their subsequent spread throughout the region. After wolf extermination in the first half of the twentieth century, Yellowstone suffered a variety of ecosystemic disorders, including overpopulation of grazing species and destruction of stream banks and forest understory due to the foraging of herbivores. Rodent and small mammal populations grew out of control (Smith, Peterson and Houston 2003).

13. "For in hope we were saved. Now hope that is seen is not hope. For who hopes for what is seen?" (Rom. 8:24, NRSV).

REFERENCES

Aiello, Dan. 2012. "'Complicit' San Francisco Voters Reject Plan to Restore Yosemite's Hetch Hetchy." *San Francisco Examiner,* November 7. Accessed March 15, 2013. http://www.examiner.com/article/complicit-san-francisco-voters-reject-plan-to-restore-yosemite-s-hetch-hetchy .

American Psychological Association. 2010. "Psychology and Global Climate Change: Addressing a Multifacted Phenomenon and Set of Challenges." Accessed March 15, 2013. http://www.apa.org/science/about/publications/climate-change.aspx .

Aquinas, Thomas, Saint. 1948. *Summa Theologiae.* Supplement. Translated by Fathers of the English Dominican Province. New York: Benziger.

Benedict of Nursia, Saint. 1981. *The Rule of St. Benedict.* Edited by Timothy Fry, OSB. Collegeville MN: The Liturgical Press.

Benedict XVI, Pope. 2008. Message to the President of the Pontifical Council for Culture on the Occasion of the 13th Public Conference of the Pontifical Academies on the Theme: "The Universality of Beauty: A Comparison between Aesthetics and Ethics." Accessed March 15, 2013. http://www.vatican.va/holy_father/benedict_xvi/messages/pont-messages/2008/documents/hf_ben-xvi_mes_20081124_ravasi_en.html .

———. 2009. *Caritas in Veritate.* Encyclical Letter. June 29. Accessed March 15, 2013. http://www.vatican.va/holy_father/benedict_xvi/encyclicals/documents/hf_ben-xvi_enc_20090629_caritas-in-veritate_en.html .

———. 2010. "If You Want to Cultivate Peace, Protect Creation." Accessed March 15, 2013. http://www.vatican.va/holy_father/benedict_xvi/messages/peace/documents/hf_ben-xvi_mes_20091208_xliii-world-day-peace_en.html .

Christiansen, Drew, SJ and Walter Grazer. 1996. *"And God Saw That It Was Good." Catholic Theology and the Environment.* Washington DC: United States Catholic Conference.

Clark, Timothy. 2011. *The Cambridge Introduction to Literature and the Environment.* Cambridge: Cambridge University Press.

Dillard, Annie. 1977. *Holy the Firm.* New York: Harper & Row.

Gustavo Gutiérrez. 2007. *On Job: God-Talk and the Suffering of the Innocent.* Maryknoll NY: Orbis.

LaCugna, Catherine Mowry. 1991. *God For Us: The Trinity and Christian Life.* New York: HarperCollins.

Lakoff, George and Mark. Johnson. 1980. *Metaphors We Live By.* Chicago: University of Chicago Press.

Leopold, Aldo. 1989. "Thinking Like a Mountain." In *A Sand County Almanac: And Sketches Here and There,* 129-33. New York: Oxford University Press.

Marris, Emma. 2011. *Rambunctious Garden: Saving Nature in a Post-Wild World.* New York: Bloomsbury.

McKenny, Gerald. 2005. "Focus on Genre and Persuasion in Christian Ethics." *Journal of Religious Ethics* 33: 397-535.

Muir, John. 1912. "Hetch Hetchy Valley." In *The Yosemite,* 249-62. New York: The Century Company.

National Park Service, United States Department of the Interior 2007. "Hetch Hetchy Valley site bulletin." Accessed March 15, 2013. http://www.nps.gov/yose/planyourvisit/upload/hetchhetchy-sitebull.pdf .

Pfeil, Margaret. 2007. "Liturgy and Ethics: The Liturgical Asceticism of Energy Conservation." *Journal of the Society of Christian Ethics* 27: 127-149.

Restore Hetch Hetchy. 2012. "The Plan." Accessed March 15, 2013.http://www.hetchhetchy.org/theplan. "Ballot Initiative." Accessed March 15, 2013. http://www.hetchhetchy.org/images/Reports/Ballot_Initiative.pdf .

Scientific American. 2007. "Podcast: Could Humans Cause Survival of the Cutest?" February 1. Accessed March 15, 2013. http://www.scientificamerican.com/podcast/episode.cfm?id=7A0DAD53-E7F2-99DF-39390424CB742FC.

Smith, Douglas, Rolf Peterson, and Douglas Houston. 2003. "Yellowstone After Wolves." *BioScience* 53: 330-340. Accessed March 15, 2013. https://wildlife.state.co.us/SiteCollectionDocuments/DOW/WildlifeSpecies/SpeciesOfConcern/Wolf/YellowstoneAfterWolves.pdf .

Spohn, William C. 1999. *Go and Do Likewise: Jesus and Ethics*. New York: Continuum.

Worster, Donald. 2011. *A Passion for Nature: The Life of John Muir*. New York: Oxford University Press.

Yeager, D.M. 2005. "'Art for Humanity's Sake': The Social Novel as Mode of Moral Discourse." *Journal of Religious Ethics* 33:445-483.

Appendix A

Keynote Address at the Catholic Consultation on Environmental Justice and Climate Change November 7, 2012

Bishop Bernard Unabali, Diocese of Bougainville, Papua New Guinea

Thank you for the opportunity to be with you this weekend—it is truly an honor and a privilege. Tonight I will speak with you about the impacts of climate change in my part of the world, specifically on the Carteret Islands. You have asked me to give the keynote address, but I was joking with [Catholic Coalition on Climate Change Director] Dan Misleh that, really, "it is you who has the keys; I have only a small note. I can speak about my experience with those impacted by climate change, but the real key to solving this crisis lies with each of you here tonight."

As was mentioned, I am Bishop Bernard Unabali from the Diocese of Bougainville, Papua New Guinea. I was ordained in 1985, and have served as a priest for the last twenty seven years. I was made an auxiliary bishop of the Diocese in 2006, and became Bishop of Bougainville in 2010. The theme of my coat of arms is "serving the people of God today." This is the idea that we must try, as much as possible, to make faith *relevant* to those we serve; the meaning and values of our faith must be translated to the times, cultures, and lives of people today. In addition, we must find ways to hand on the faith to those who come after us and to teach and share this gift of faith with future generations. We have an especially wonderful opportunity to do that now during this Year of Faith.

I would like to first give you a bit of background about the context of Bougainville and Papua New Guinea. Today, Papua New Guinea is called by this single name. Prior to this, however, it was known by two distinct names: Papua and New Guinea. Historically, this island was cut into two: the eastern half was British, and the Western half, including Bougainville, was German New Guinea. After World War I, the League of Nations was created and granted a mandate to Australia for administering the External Territory of Papua and New Guinea. This arrangement was maintained until September 16, 1975, when Papua New Guinea became one nation.

Bougainville is largely Catholic, mostly due to Marist Missionaries from Germany. The population is 200,000, and the Catholic population is 160,000—so roughly between sixty and seventy percent of Bougainville is Catholic. The total population of Papua New Guinea is about seven million, and Catholics are around sixty to seventy percent as well. There are nineteen dioceses in Papua New Guinea, and ninety percent of the bishops are European. I am the second local bishop in Bougainville. In my Diocese we have a few orders of men and women religious, around 150 primary schools, fifteen clinics and health centres, and thirty local priests. The idea of a local church is very strong in Bougainville.

By local, indigenous tradition there is a direct relationship between people and the environment; it is total, and includes peoples' spirituality. As such, many of the things that are now being talked about in the Catholic Church—for example, the fact that we are intimately related to the natural world and must care for creation in order to care for human life and dignity—already existed in the local traditions and cultures in my part of the world. Given this, part of my role has been to help the people in my Diocese recapture this part of their tradition, a part that has been lost, but is now also being reclaimed within the Christian context.

The geography of Bougainville is very mountainous, with many rivers and a large amount of pristine forests. In the 1990s, we had a civil war for independence in Bougainville. The war was partly caused by latent tensions from the colonial exchange between Britain and Germany and partly fueled by hostilities surrounding the Panguna copper mine in Bougainville. This was due to the fact that a large amount of the revenue generated by the mine was going to the mainland Government of Papua New Guinea, while a comparatively small amount of the revenue stayed in Bougainville despite the fact that the mine caused large amounts of local environmental degradation on our island. Although the fighting has subsided, Bougainville remains an Autonomous Region of Papua New Guinea. Bougainvilleans were promised a referendum on independence from Papua New Guinea in 2001 peace accords, and that referendum is to be held between 2015 and 2020.

The war caused a tremendous amount of damage to the people, land, and infrastructure on the island of Bougainville. In order to respond to this widespread destruction, the Autonomous Bougainville Government has sought a comprehensive reconstruction program that prioritizes projects and progressively tailors efforts to meet the needs of the various Bougainvillean populations, including women and youth. Likewise, in my role as Bishop of Bougainville, within a comprehensive pastoral vision, I have identified five priorities for the people in my Diocese as we seek to recover from the lasting effects of the war:

Prayer and spiritual life;
Fiscal education and formation—teaching people about making money, using money, and saving money;
Development—focusing on what is good development, what is bad development, and what needs to be done now;
Media—given its immense power today, and we must learn how to use media to tell our story and pursue our goals; and
Social conscience—education, citizenship and proper patriotism, right relationships with God, one another, and the environment.

Fostering right relationships and fixing broken relationships seems to be of particular importance in addressing many of our modern challenges, environmental and otherwise. Only when we establish right relationships can we enjoy true peace and justice. Closely related to this is the development of a proper social conscience that finds the right balance between rights and obligations. This is particularly important today since rights are often overemphasized to the exclusion of the obligations that we owe one another.

Now that you have a better idea of the context in Bougainville, we can begin to talk more specifically about the issue of climate change. Through the education efforts of the Catholic Coalition on Climate Change, I believe that most of you are familiar with the film *Sun Come Up* and the situation faced by the inhabitants of the Carteret Islands. Although there have been several documentaries made about this situation, the Carteret people are still faced with the very difficult situation of relocating away from the consequences of climate change.

The Carterets are around four hours from Bougainville using outboard motor. In the 1980s, rising sea levels divided the Huene Island of the Carterets into two separate parts. In addition to this, more frequent severe storms and high tides have caused damage to many homes and harmed vegetation, arable land, and fresh water sources. In the longer term, the Islands are threatened by the coral bleaching caused by warmer sea waters. These are incredibly difficult conditions, and those of us on the mainland are very aware of these challenges.

Although I was familiar with the difficulties faced in the Carterets, I had never seen the situation in person for myself prior to 2006. Then, within a month of being made a bishop, I was invited to visit the Carterets and see the plight of these people for myself. I stayed for two weeks, and it is difficult for me to express how moved I was by their struggle. First, let me say that I visited at an especially windy time, and I remember praying the Rosary every day that my return trip to the mainland would be safe!

In addition to experiencing the severe weather that is impacting the Islands, I also saw and heard very moving testimonials about how the local people are being harmed by the effects of climate change. They showed me how the salt water from storm surges and higher tides is killing some of the coconut trees that supply their staple food. They told me about how the salt water is also resulting in smaller coconuts that taste differently than in years past. They showed me the places where the salt water has created saline marshes that have killed taro and tapio crops and made the land un-farmable. They shared with me how they must now travel to the mainland in order to buy food using their handmade shell money or wait for sporadic shipments of food to the Islands from the government.

In response to these conditions, several attempts to relocate the people from the Carteret Islands to the mainland have been made. Originally, in the 1960s, Bishop Leo Lemay, SM, the second American bishop, offered the islanders land at Mabiri but somehow it was not seen as a need by them? And so the first formal relocation occurred in the 1980s and 1990s, at Kuviria near Mabiri in central Bougainville under the provincial government but this was unsuccessful due to the Bougainville civil war, hostility towards the Islanders, and shortage of proper facilities. In 2003, the government of Papua New Guinea approved a plan to relocate all of the Islanders to the mainland by 2007. However, this never happened due mainly to a lack of government funding and their ongoing bigger holistic plan and land traditional owner problem.

Given these failed attempts, the Carteret Islanders decided that relocation would not successfully occur unless it was led by the people themselves. In order to do this, the Council of Elders of the Carteret Islands established an NGO in 2007 called Tulele Peisa, which means "Sailing the Waves on our own." Tulele Peisa is now working to plan and implement a voluntary relocation program to the island of Bougainville, and the organization focuses on three key program areas: research and advocacy, fostering relationships, and training and capacity-building. In doing so, the Carteret Islanders are acting on the realization that any successful relocation effort must be comprehensive in nature and must attend to the dynamics of both the Islands and the mainland.

Although Tulele Peisa has been coordinating the relocation, we in the Diocese of Bougainville have been working with them to try to assist their efforts. When I visited the Islands in 2006, I first met with the young people. I did this for two reasons: The first is that they are the future of the Islands, and so must be involved in the process that will impact the rest of their lives. Second, I met with them because I wanted to begin fostering dialogue between them and young people from Bougainville. Young people listen to other young people and knowing this I brought two young islanders to the mainland Bougainville to meet with the young there and begin to form them on leadership and establish relationships too.

This proved to be a valuable step in both directions. On the one hand, the young people from mainland part of the Diocese were able to hear first-hand the struggles that their peers were experiencing on the Islands. Moved by these stories, a group of young people went to the island and began to experience what they had merely heard and then went back to the mainland and began to gather sympathetic supporters. On the other hand, the young people on the Islands, especially the two who had been exposed and experienced life on the mainland started to share with their island peers. They began to think more about what they needed to do in order to be accepted when if they relocated to the mainland. For these reasons alone, this was a very positive interaction.

In addition to these positive outcomes, these conversations between young people laid the groundwork on which we began to build in 2008 and 2009 through a series of four meetings on the mainland. The Diocese of Bougainville has title to land in Tinputz, and our Diocese decided to offer this land as a site on which some families could relocate from the Carteret Islands. In order to begin this process, the young people from the Islands with whom I had met came and visited with their counterparts on the mainland. Since they had met previously on the Islands, they were able to engage quickly in productive dialogue and move forward with the relocation process. Over the next several months, leaders and young people from Tinputz visited the Carteret Islands to better understand their situation and discuss logistics, and leaders and young people from the Islands continued to visit with their counterparts on the mainland.

While these meetings were going on, I tried to assist in the relocation effort by writing a circular to the Diocese explaining the situation and why we were called to respond. I pointed out that we are all brothers and sisters in Christ and that we had a moral obligation to come to the aid of the Islanders who were now in need of our help due to the harmful consequences of climate change. We created an Awareness Team that has been traveling around the Diocese for the last two years explaining the situation to people and educating them on why and how we are responding. Finally, we established December 8 as a Diocesan-wide a day to remember the Carteret Islanders, and took a special collection to assist the relocation program.

This process of involving everybody in open dialogue and Christian charity paved the way to helping five families move from the Carteret Islands to Tinputz. Our organizing plan is to integrally spread out the Islanders at various "transit points," as we are calling them, across the Diocese rather than to simply place four thousand people in one area. There is a parish in the western part of the Diocese that has more land to which we hope to relocate more of the Islanders, and I recently traveled there with the Awareness Team to begin the discussion of establishing a second transit point there. This is still in the making.

For a long time, climate change has been an abstract idea about something that will happen sometime in the future. Although more frequent severe weather events around the world are perhaps starting to increase awareness of this issue, the world still lacks the sense of urgency necessary to mitigate climate change and prevent runaway global warming. My hope is that hearing the story of the Carteret Islanders will help you and others see the human impacts of climate change that are happening right now.

My Diocese has done its best to live the Gospel by responding to the plight of those who are being impacted by climate change—despite the fact that they have done virtually nothing to cause the problem. With this in mind, I hope that the conference this weekend and the discussions that follow will help Christians everywhere to recognize climate change as a moral issue, to draw on the richness of Christian tradition to address the root causes of climate change, to better care for the natural world of which humanity is part.

Appendix B

Homily: Catholic Consultation on Environmental Justice and Climate Change, November 8, 2012

Bishop Donald Kettler, Diocese of Fairbanks

Reflecting on the proceedings of this Consultation, one thing has become very evident to me. God's message regarding the created world is clear and simple. We are not owners of our world; we are to care for and work creatively with it. Even though we continue God's creative work today, we are not owners even of our own creative work. We are cooperators, associates, taking creation, using it for the common good, and then passing it on in better condition—to our children. If we truly believe this and live by this understanding of our role, the way we are called to respond to issues regarding our natural environment becomes clearer.

Climate change is a fact. Its causes are complex, and they certainly include human activities. The solutions are equally complex. They will require changes in how we produce and use energy, and that should cause us to examine our lifestyles.

In Alaska, the way the impacts of climate change are being felt throughout the state are a study of extremes. For example, this past winter we had record snow falls and -65 degrees Fahrenheit in Fairbanks. In other parts of the state, there are more sobering and long-lasting impacts with which people are being forced to deal. I invite you to reflect with me on two climate change-related situations facing us in Alaska.

The first one is similar to what we heard last night from Bishop Unabali about islands in the Pacific. The Yup'ik Eskimo village of Newtok, home to 300 people, will be covered by the rising ocean waters within ten to fifteen years. Fifty feet of protective coastland washes away annually around Newtok, and I heard recently that less shore ice in the fall and winter is also diminishing a protective land boundary around the village.

Because of this, the village has begun a six-mile move inland. Despite their best efforts, whether or not the villagers will be able to make the move in time is uncertain, mainly because of the cost of moving entire households ($250,000 per home) and concerns about the lack of availability of public resources to assist the move. Unfortunately, our native peoples do not have much of a voice with government or even with many of their fellow Alaskans. I am worried that, if funding determines the future of a village, they might soon be without a home. Do they have a right to stay there, or will they have to leave most everything and move to an urban area?

Even if they are able to move their village inland before it is engulfed by the rising sea, I am still concerned about whether our native peoples will be able to continue the subsistence lifestyle of hunting, fishing, and gathering that they established eight or nine thousand years ago. What is needed at this point, it seems to me, is a deeper development and understanding of a theology of human rights which endorses the right to a homeland, preserves native culture and, in this case, supports the subsistence lifestyle. Perhaps the theologians here today can help with this.

The second climate change-related situation currently faced in Alaska concerns the state's vast amount of natural resources. Besides its people, Alaska is abundantly blessed with fish, birds, animals, and many mineral sources, including oil, natural gas, gold, copper, and coal. Oil, gas, and coal are not renewable and often not clean enough sources of energy to sustain a life-supportive climate. When exploited improperly, the extraction of these sources of energy can generate contamination that harms the lives and dignity of local people and the local environment. We need support to assure that our environment and people are protected, even while these sources are farmed and mined. Many native Alaskans' livelihoods are closely linked to the environment. They practice subsistence living, or supplement their food supplies by hunting and fishing. Environmental degradation has a direct impact on the lives and dignity of local peoples.

In spite of these social and environmental costs, the financial structure of Alaska, its schools, government, and human services depend heavily on the revenue generated by the extraction of these energy sources. When extraction of natural sources and subsistence hunting and fishing are considered together, we find ourselves in a bind. We rely heavily on the income generated by extraction, but we also need to protect our environment and the health of our people. Our environment is increasingly compromised by the activities of energy source extraction and, in the case of climate change, the burning of the fossil fuels.

Given this delicate balance, it is not sufficient for those of us concerned about God's creation—which should include all people of faith and goodwill—to merely say "no" to all resource extraction activities. We need "bridge builders" who will work civilly with government and society to protect the environment while also assuring that our people can secure their basic needs. This can and should be a role for the faith community, and Catholics can use the gift of our social teachings to offer guidance through these dilemmas. Using the Church's rich social tradition, we have the opportunity to bring people together, raise ethical questions, foster understanding, and help forge a way forward so that all life—human and non-human—is respected and protected, and so that the common goods of the earth and future generations are safeguarded.

Your presence here as we celebrate the Word and Eucharist is important. Equally important is our recognition of what God has given to each of us and how we are to use it. The message is clear, and the path ahead is challenging. As we seek to learn from the ecological vision of Pope Benedict XVI and maintain the integrity of God's good gift of creation, we pray: "O God, direct and help us to do your will."

Index

abortion, 99, 176, 190n3
agrarian reform, 107, 120n6
agriculture, xxvii–xxviii, 89, 103, 104–105, 105, 107, 108, 109–111, 111–112, 116, 117, 118, 119, 120n6, 121n20, 134, 135, 137, 140, 142, 143, 145, 161, 212n10
alien good, 68
Ambrose of Milan, St., 106, 227
American dream, 218, 231
analogy of faith, 186
angel, 63, 64–65, 77, 78n3
animals, xxx, 4, 20, 22, 35, 51, 62, 74, 79n10, 80n16, 106, 117, 129, 136, 174, 176, 247, 253n11, 264
Anthropocene,. *See also Fate of Mountain Glaciers in the Anthropocene* 83, 85, 86, 87, 91, 98, 99, 140–141, 142, 143
anthropocentric, 63, 71, 72, 190n5, 217
apokatastasis, 178
apotheosis, 181
Aquinas, St. Thomas, xxvii, 7, 13, 38n5, 46, 49, 49–51, 54, 55, 61, 62, 62–63, 64, 66–67, 67–68, 69, 73, 74, 78, 78n2–78n3, 79n10–79n15, 79n7–79n8, 91, 106, 130, 131, 132, 142, 201, 202, 211n3, 246–247
Aristotle, 7, 13, 62, 64, 79n11, 79n5, 113
asceticism, 252
askesis, 107, 239, 244, 251, 251–252
Augustine of Hippo, St., 4, 7, 8, 10, 11, 12, 15, 26, 38n5, 46, 47, 47–48, 49, 49–51, 55, 104, 109, 181
Austin, John, 50
automobile, 137, 219, 220

Bacon, Francis, 113, 114, 118, 120n13–120n16
Basil the Great, St., 106–107, 227
beauty, x, xv, xvi, xxvii, xxx, 76, 131, 132, 140, 142, 143, 144, 145, 150, 153, 154, 155, 166, 184, 209, 237, 239, 239–240, 240, 241, 241–242, 243–244, 248, 249, 251, 253n3
Berry, Wendell, 76, 118, 223
biomimicry, 134, 137, 138, 139, 146
Bonaventure, St., xxv, 3–5, 6–9, 9, 10, 10–11, 11, 12–13, 13, 14, 15, 16, 16n4–16n5, 16n7, 131, 144
bonum alienum. *See* alien good
bonum proprium. *See* proper good
bonum suum. *See* proper good
Brennan, Andrew, 45

carbon: emissions, 84, 127, 134, 139, 141, 217, 220, 228, 230, 231; taxation, 230
Caritas in Veritate, ix, xvi, xxvii–xxviii, 30, 38n6, 43, 56, 72, 90, 94, 95, 96, 97, 99, 103, 104, 107, 108, 110–111, 115, 116, 118, 197, 198, 215, 216, 237, 239
Carteret Islands, xxiii, xxviii–xxix, 87, 127, 257, 259, 260, 261, 262
catechesis, xxvii, 173–174, 175, 177, 185–186, 187, 187–188, 188–189, 218
Centesimus Annus, xvii, xviii, 3, 29–30, 61, 70, 71
charity, xiii–xv, 65, 68, 78n2, 90, 92, 120n10, 159, 164, 239, 247, 251, 253n4, 261
Christ. *See* Jesus Christ
Christocentric spirituality, 6
Christology, 156, 175, 179, 183
Church: domestic, 76, 77; Fathers, 90, 153, 204, 211n4, 216, 227
circulatio vel regratio, 202
civic virtues, 69
civil society, xv, 77

climate change : adverse effects of, xxi, xxiii, xxix, xxv, xxviii–xxix, 84, 85, 87, 89, 99, 100n2, 141, 199, 200, 205, 207, 208, 212n14–212n15, 215, 259; mitigation, 199, 200, 212n15–212n16; refugees of, xxiii, 100n2
climate crisis, xxiii, xxiv, xxvii–xxviii, xxxi, 127, 134, 146, 200
Coakley, Sarah, 133–134, 145
common good, xiii–xiv, xxvii, 16, 16n6, 50, 51, 61, 62, 62–63, 64, 65–66, 67, 68, 69, 70, 72, 73, 75–76, 78, 78n2, 79n10, 91, 93, 97, 145, 150, 162, 174, 191n9, 198, 217, 227, 263, 265; political, 62, 63; pre-political, 62
communion, xv, 12, 24, 27, 28, 35, 110, 155, 156, 180, 181, 237, 238, 241, 249, 250–251
Compendium of the Social Doctrine of the Church, 21, 96, 162
consummation, 8, 12, 185
contemplation, 128–129, 143, 153, 238, 242, 251
contemplative spirit. *See* contemplation
contrition, xxviii, 245, 246, 246–247, 251
conversion, xxviii, 29, 145, 146, 191n9, 216, 228, 231, 237, 238, 245, 251
convertibility: of being and goodness, xxvii, 44, 46–47, 47, 49, 55; of being and order,. *See also* created order; creation; disorder; dynamic cosmic order; disorder xxvii, 44, 47, 49, 55
cooperatives, 109
cosmic soteriology, xxvii, 174, 177, 178, 179, 180, 183, 185–186, 187, 188–189, 191n11
cosmocentric, 72, 75
cosmology, xxvii, 6, 15, 62, 70, 79n9, 203
covenant, ix, xv, 33, 116, 130, 154, 157, 158, 159, 160, 179–180, 219, 238
created order, x, 4, 8, 10, 12, 45, 49, 52, 53–54, 56, 77, 103, 105, 108, 114, 115, 115–116, 127, 131, 134, 141, 142. *See also* creation; convertibility; disorder; dynamic cosmic order; order
Creation: as gift, 105, 108, 244, 265; care for, xvi, 4, 6, 159, 174, 200, 217, 258; doctrine of, 104, 112, 115, 120n15; dominion over, xi, 114, 175; goods of, xii, xxvii, 83, 88, 90, 91, 96, 97, 98, 100n6, 160; grammar of, xxvii, 103, 104, 115, 116, 119, 216, 224; guardians of, 26; new, 160, 182; order of, 105, 204, 209; priests of, 181, 189; protectors of, xxx; respect for, ix, xiii, 25; stewards of, xi, xvi. *See also* stewardship; threats to, xii; convertibility; created order; disorder; dynamic cosmic order
Creator, x, xi–xii, xv, xvi, 9, 11, 26, 30, 47, 51, 56, 64–65, 66, 69, 99, 103, 114, 116, 128, 131, 140, 142, 153, 159, 165, 174, 175, 217, 242–243, 244, 252
cross, 130, 144–145, 179, 182, 189–190
Crutzen, Paul, 85–86, 87, 88, 141

Darwin, Charles, 86, 131
Dawkins, Richard, 132, 133
Day of Atonement, 179, 179–180
Deanne-Drummond, Celia, 133
deification, 189
Dei Verbum , 186
Delio, Ilia, xxiii, 4, 10, 16n5, 27, 38n9, 133
de Lubac, Henri, 120n9, 183
deontological. *See* deontology
deontology, xxvii, 44, 45, 48, 52, 55–56
Democrat, 104, 176. *See also* Democratic Party
Democratic Party, 190n2. *See also* Democrat
development, xi, xiii–xiv, 4, 5, 28, 30–31, 37n2, 46, 56, 62, 90, 91, 109, 110, 112, 113, 114, 131, 133, 133–134, 146, 173, 197, 205, 206, 207, 221–222, 224, 229–230, 231, 232n11, 232n9, 240, 243, 248, 259, 264; authentic, 56; ecologically sound, 153; integral human, ix, xiii–xv, xxv, 20, 30, 31, 37, 38n7, 72, 91, 94, 105, 207, 212n16; sustainable, 27, 207

dialogue between theology and science, 14
dignity, xvi, xxv, 23, 30, 31, 32, 38n6, 43–44, 52, 78n1, 105, 108, 120n10,

181, 189, 216–217, 250, 253n5, 258, 264
Dillard, Annie, xxviii, 133, 238, 248–251, 251, 252n2
discernment, xi, 6, 13, 70, 88, 106, 112, 113, 114, 116, 131, 205, 237, 242, 252
disorder, 48, 92, 174, 244, 254. *See also* convertibility; created order; creation; dynamic cosmic order; order
divinization, 180, 181, 182, 183, 184, 188, 204
dynamic cosmic order, 202. *See also* convertibility; created order; creation; disorder; order

Earth community, xxx, 204
eco-centric, 217
ecological, xi, xx, xxi–xxii, xxiii, xxiv, 20, 21, 27, 62, 69, 76, 77, 78, 92, 119, 131, 144, 158, 167n3, 173, 175, 187, 189, 190n6, 191n9, 209, 238; awareness, x, 173; challenges, xxvii, 71, 200, 209; concern, 61, 69, 71, 154, 198; crisis, x–xi, xiii, xxi, xxv, xxvii, 99, 107, 115, 135, 149, 153, 154, 167, 167n3, 175, 190, 191n12, 191n9, 238, 239, 252n1; degradation, xxi, xxiv, xxx, xxxi, 92, 127, 144, 168n12, 190; design, 134, 135, 139, 146; ethics, xxviii, 69, 79n15, 167n3; integrity, xxviii, 22, 197; issues, xx, xxiv–xxv, xxxi, 153, 190; problems, xiv, xix–xx, xxiii, xxiv–xxv, 29; responsibility, xv, xvi; systems, xix, xx, xxix, xxxi, 20, 21, 30; turn, 202, 203; vision, xx, xxii, xxviii, xxxi, 76, 197, 198, 216, 218, 265
ecology, xxii, xxix, xxv–xxvii, 4, 7, 14, 70, 78n3, 116, 118, 150, 152, 153, 167n2, 169n19, 177, 189, 191n11, 216; deep, 75; environmental, 29–30, 30, 32, 116; human, xv–xvi, xxv–xxvii, 3, 4, 13, 14, 14–15, 16, 20, 29–30, 30, 32–33, 35, 36, 37, 38n7, 61, 70, 70–71, 72, 75, 76, 77, 116, 154, 155, 198; natural, xxv–xxvii, 61, 70, 71, 75, 76
economy, xi, 87, 88, 91, 93, 94, 96, 97, 105, 108, 111, 117, 223, 224, 227, 232n12, 244
ecumenical, xii, 150, 167n3, 189, 210
Ecumenical Patriarch, xx, 167n3
efficiency gains, 225–226
emanation, 8, 202
emergence of creatures, 202
energy, xiii–xiv, xxix, xxviii, 85, 89, 94, 96, 107, 117, 135, 136, 137, 138, 139, 140, 142, 145, 184, 208, 209, 217, 218, 220, 221, 222, 223, 225–226, 227, 228, 231, 241, 252, 263, 264; non-renewable, xii, xxix, 96, 264; renewable, xiii, 96, 100n8, 264
environment: as gift, ix, 94; concern for the, 188; human, 93; imperiled; natural, ix, xx, xxviii, 27, 44–45, 46, 56, 92, 94, 115, 237, 238, 263, 264; protection of, 189, 190n5, 264; value of the, 95
environmental: action, 19, 33; catastrophe, 244; challenge, 175, 198, 259; changes, 87; concerns, xxvii, 150, 157, 173, 174, 198; consciousness, 167; costs, 95, 264; crisis, xxvii–xxviii, 28, 149, 154, 166, 173, 175, 176, 190n2, 200, 210, 251; degradation, xii, xiv, xv, 21, 27, 28, 83, 87, 91, 95, 96, 98, 134, 217, 237, 258, 264; education, 33, 137; ethic, 9, 28, 35, 38n9, 44, 47, 158, 160, 251; exploitation, xii; goods, 88, 90, 100n6; harms, 107, 218; holism/holist, 19–20, 21, 27; impact, 96; integrity, 209, 212n13, 212n16; issues, xxii, 218, 230; justice, xxxi, 54; measures/policies, xiii; movement, 19, 20, 37, 37n1, 176, 217; natural, xxvii, 44; philosophy, xxvii; preservation, 244; protection, 96, 153, 190n2; refugees, x; resources, xii; sustainability, 200; theology, 158, 160, 217; vision, xxvii, 47, 54, 55, 215–216
environmentalism, 19–20, 20, 21, 22, 37, 176, 190n2, 217; personalist, 19–20, 21, 22, 37, 38n9
equity, 95, 97, 207, 208; principle of, 204, 207; social, xxviii, 197, 200, 204, 212n13, 212n15

equivocal agent, 67
eschatology, xxvii, 12, 133, 165, 166, 167, 169n19, 174, 177, 178, 181, 182, 183, 184, 185, 187, 188, 189, 190n6–190n7, 243–244, 252
eschaton, 184, 185, 186, 188, 191n11, 244
eternal reason, 73–74, 74
Eucharist, xxiii, xxix, 5, 153, 154, 158–159, 163, 167n2, 184, 185, 188, 191n11, 248, 265
evolution, ix, 7, 13, 14, 28, 85, 87, 100n9, 131–133, 133, 137, 139, 140, 142, 144, 145–146, 184, 185, 211n5, 221
extinct, 70, 127, 132, 141, 142, 143–144, 205

family, x, xii, xvi, 27, 28, 29, 32, 33, 35, 36, 43, 70, 72, 76, 90, 91, 98, 105, 110, 128, 135, 162, 198, 216, 219–220, 222, 226, 227, 228, 232n10, 242, 246, 250, 252
Fate of Mountain Glaciers in the Anthropocene,. *See also* anthropocene 198
First Vatican Council, 177
fossil fuels, xxi, xxxi, 85, 139, 145, 199, 212n12, 216, 223, 225, 232n2, 232n5, 264. *See also* carbon; climate change; climate crisis; energy; greenhouse gas; The Map of Resource Competition in a Carbon Constrained World; The Map of the Most Vulnerable to Climate Change; non-renewable; renenewable
Francis of Assisi, St.,. *See also* St. Francis Pledge to Care for Creation and the Poor xxiv, xxix, 4, 9, 27, 35, 189, 230
Francis, Pope, xx, xxix–xxx, xxx, 189
Franciscan intellectual tradition, 5–6, 16
Frings, Joseph Cardinal, 177
future generations, ix, xii–xiii, xix, xxi, xxix, xxv, 90, 92, 94, 95, 98, 99, 175, 177, 207, 208, 212n15, 257, 265. *See also* solidarity, principle of

genre, xxviii, 143, 151, 152, 168n6, 237, 237–238, 251, 253n6
gift, 153, 161, 162, 163, 182, 191n9, 209, 237, 244, 250, 252, 257, 264; creation as.
God the Father, xvii, 7, 8, 10, 11, 187. *See also* Creator
good intentions, 224
goodness, xi, xv, xxvii, xxx, 9, 11, 23, 24, 25, 26, 29, 36, 38n5, 44, 46, 47, 48, 49, 52, 53–54, 56, 66, 67, 70, 74, 131, 142, 146, 150, 153, 155, 157, 160, 174, 175, 201, 237, 239, 242, 245, 247, 249, 253n3
greenhouse gas, xxxi, 84, 141, 199, 205. *See also* carbon; climate change; climate crisis; energy; fossil fuels; The Map of Resource Competition in a Carbon Constrained World; The Map of the Most Vulnerable to Climate Change; non-renewable; renewable
Gustafson, James M., 23, 28, 37n4

Haught, John, 71–72, 133, 177, 190n6
Hetch Hetchy Valley, 238, 239–240, 241, 242, 243, 244, 251, 253n8
Holy Spirit, xvii, 11, 26, 131, 155, 165, 182, 185, 187, 189, 253n7
hope, xx, xxix, xxx–xxxi, xxxi, 28, 61, 114, 115, 118, 152, 177, 187, 189, 190n6, 225, 245, 249, 254n13, 261, 262
housing, density, 221
Howard, Albert, xxvii, 116–117, 118, 121n20
human rights, ix, xi, xxvii, 54–55, 62, 83, 90, 93, 94, 96, 97, 98, 99, 154, 155, 162, 206, 207, 259, 264
hunger, 34, 104–105, 105, 108, 109, 110, 216, 224

Incarnation, xxvii, 5, 8–9, 15, 130, 149, 181, 203
inequities, xxi, 97
injustice, 46, 55, 87, 106, 107
interreligious dialogue, 189
intrinsic value, xxxi, 23, 128, 200–201
irreversible alteration, 205

Jackson, Wes, xxvii, 116, 118, 119, 134–135, 135, 140, 142, 143
Jesus Christ, xxx, 7, 8–9, 9, 15, 24, 29, 104, 108–109, 120n10, 130, 131, 152, 153, 165, 169n22, 178, 179, 179–180, 181, 183, 187, 188, 189, 190, 242, 245–246, 253n10
Job, 128, 129, 248, 249, 250, 253n11
John Paul II, Pope, x, xiv, xix, xx, xxi, xxv, 3–4, 20, 61, 70, 71, 72, 78, 85, 91, 92, 93, 96, 97, 99, 105, 150, 152, 153, 191n9, 216, 223, 224
John XXIII, Pope, 99, 177
Jung, L. Shannon, 28
justice, xiv, xvii, xxi, xxiii, xxvii, 21, 27, 36, 46, 55, 75, 87, 90, 92, 93, 96, 97, 99, 100n7, 109, 119, 120n10, 120n5, 129, 158, 162, 168n13, 169n28, 198, 204, 208, 249, 250; ecological/ environmental, xx, xxxi, 54; social, xxvii, 159, 162–163, 168n11, 169n25, 175, 176, 188, 208, 209, 252. *See also* injustice

Kant, Immanuel, 139
Kettler, Bishop Donald J., xxiii, xxix
keystone predator, 246
kinship care, 20, 34, 36, 37, 38n8

law, 46, 49, 49–52, 73, 74, 91, 94, 106, 112, 130, 131, 132, 158, 176, 197, 209; eternal, 51, 52, 62–73, 74, 131; humanitarian, 46; moral, xvi, 46, 132; natural, xxvii, 44, 46, 49, 50, 51, 54, 55, 74, 77, 78, 79n15, 106, 131, 133

Leopold, Aldo, xxviii, 238, 245, 246, 247, 248, 251, 252n2, 253n10
luxury, xxviii, 215, 217, 226, 226–227, 228, 229, 230

Magisterium, xix–xx, xvi, xxi, xxiii, xxiv–xxv, xxix, xxvii, 25, 27, 28, 62, 70, 72, 75, 83, 90, 93, 99, 120n3, 150, 151, 152, 153, 168n5–168n6, 186, 253n6
McFague, Sallie 28, 71, 79n9
metanoia, 245
metaphysics, xxvii, 4, 8–9, 10, 11, 13, 14, 15, 43, 44, 45–46, 46, 47, 49, 55, 183, 191n9, 201
metaphor, 9, 10, 11, 13, 14, 15, 28, 128, 157, 242, 245, 246, 247
Methodist-Catholic Dialogue, xx, 191n11
modernity, 20, 24, 104, 112, 113, 114, 115, 143, 144
moral sphere of action, 208
Muir, John, xxviii, 238, 239–240, 240, 241, 241–244, 245, 248, 249, 251, 252n2, 253n4
Myers, Gene, 22

natural inclination, 63, 64, 65, 67, 68, 73, 78n3
natural law. *See* law
natural location, 75
nature: book of, xvi, xxv, 15, 72, 73, 146, 216; degradation/deterioration/ exploitation of, x, xv, 33, 91, 92, 93; dominion over, 56; human, 30, 45, 51, 63, 71, 72, 74, 75, 77, 78n3, 79n11; intellectual, xxvii, 44, 52, 53, 55; nonhuman, xxviii, 28; protection of, 30, 33; relational, 11, 21; respect for, x, xvi, 71, 72, 76; social, 62
Nowak, Martin, 133, 142
natural world
nonhuman, 26, 36
numinous quality of, 202
non-renewable, xiii, xxix, 96, 100n8, 264. *See also* climate change; climate crisis; energy; fossil fuels; greenhouse gas; renewable

omega, 184–185, 190n7
Omega Point. See omega
Origen of Alexandria, 178
Orthodox, xx, 150, 167n3, 169n20, 189

pantheism, xvi, 56, 87, 157, 239, 240, 242–243, 244, 248, 253n6
parable, 238, 245–246, 247, 253n10
parochial school, 77
Pascal, Blaise, 52–53
Paschal Mystery, 155, 158, 163, 174, 178, 182, 183, 186, 188, 244

Paul VI, Pope, x, 30–31, 43, 91, 95, 108, 169n21
perfection of the universe, 54, 201
plants, 22, 45, 74, 117, 129, 134, 135, 136, 137, 138, 145, 174, 225
Pontifical Academy of Sciences, 88, 198
poor,. *See also* poverty; solidarity, principle of ix, xii, xix, xxi, xxiii, xxix, xxv, xxx, xxxi, 6, 87, 90, 93, 94, 97, 106, 130, 154, 161, 162, 163, 175, 208, 210, 211n1, 217, 227, 240, 251
Porter, Jean, 38n5, 62, 79n4, 133
poverty,. *See also* poor; solidarity, principle of xiv, xxi, xxvii, 87, 107, 109, 217, 232n2
preamble to the faith, 65
precautionary principle, 204, 207
priest, xxiv, 179, 181, 182, 218, 249, 257, 258
proper good, 66–67, 73
property, xiii, 34, 69, 90–91, 92, 93, 95, 98, 99, 106, 162, 218

rapprochement, 202
Ratzinger, Josef, xxvii, 3–4, 13, 15, 16n2, 16n6, 47, 77, 80n17, 112–113, 114, 114–115, 120n15, 120n3, 120n7, 152, 168n16, 177, 178, 179, 180, 181, 182, 183, 184, 185, 190n7, 217
redaction criticism, 94, 95
redemption, 130, 143, 144, 145, 150, 154, 157, 161, 164, 165, 174, 176, 178, 179, 181, 182, 185, 187, 188, 189, 217
Republican Party, 176, 190n2
Ressourcement, 183
Resurrection, xvii, xxvii, 144, 165, 177, 178, 186, 211n4, 253n8
rhetoric, 90, 99, 238, 239, 243
right to life, xi, xxvii, 83, 90, 98, 99, 206
Rolston III, Holmes, 87, 133

sacrament, xxi, 149, 150, 155, 156, 157, 162–163, 166, 167, 169n19–169n20, 246
sacramental, xxi, xxvii, 24, 26, 76, 127, 146, 157, 162–163, 166, 167, 169n18–169n19, 247, 251–252, 252n1
sacramentality, xxvii, 149, 150, 157, 158, 167, 169n19, 169n27, 252
sacrifice, 69, 144, 145, 168n11, 179, 179–181, 182, 183, 225–226, 230, 237, 250, 251
salvation, xvi, 4, 5, 9–238, 10, 11, 12, 13, 15, 56, 108, 150, 159, 164, 165, 173, 174, 175, 176, 177, 177–178, 178, 179, 181, 182–183, 186, 187, 188, 190, 191n12, 211n4, 243–244, 253n6
Second Vatican Council, xii, xx, 5, 90, 106, 150, 151, 156, 158, 159, 161, 177, 178, 183
sin, xi, xx, xxviii, 11–12, 23, 24, 37n4, 38n6, 54, 144, 179, 180, 216, 218, 223, 224, 237, 239, 242, 244; structure of, xxviii, 215–216, 218, 223, 224, 225, 226, 231, 244, 245, 246–247
Singer, Peter, 44
Skylstad, Bishop William S., xxiii, xxix
Sobrino, Jon, 145
solidarity, principle of, x–xi, xii–xiv, xxv, xxvii, xxx, 35, 36, 43, 83, 94, 95, 96, 104, 105, 108, 109–110, 110, 175, 189, 217, 218. *See also* future generations; poor; poverty
Stang, Dorothy, 145
Stern, Daniel N., 22
stewardship, xii, 116, 150, 155, 163, 174, 175, 208–209, 217. *See also* creation, stewards of
St. Francis Pledge to Care for Creation and the Poor, 16
Studiousness, 78
subsidiarity, principle of, xv, 162, 218
suburb, xxviii, 215–216, 218, 218–220, 221–222, 223, 224–225, 226, 226–227, 228–229, 230, 231, 232n12
Sun Come Up, xxix, 87, 259
sustainability, principle of, xxiv, xxx, xxxi, 67, 111, 120n13, 200, 204, 207, 208, 212n13, 227, 251, 252

technological solutions,. *See also* technology 225
technology,. *See also* technological solutions xiv, xvi, 96, 110, 118, 189, 200, 212n16, 220
Teilhard de Chardin, Pierre, xxvii, 174, 183–184, 185, 188, 190n7–191n8
theosis, 181, 189

transcendent humanism, 30–31, 33
Transcendent Presence, 203
Transformation, 6, 91, 104, 112–113, 115, 145, 180, 181, 182, 183, 184, 210, 238
trinitarian theology, 11, 16, 52, 244. *See also* Trinity
Trinity, 8, 9, 10, 11, 15, 16n5, 52, 156, 182. *See also* God the Father; Holy Spirit; Jesus Christ; trinitarian theology

Unabali, Bishop Bernard, xxiii, xxviii–xxix, 257, 263
United Nations Framework Convention on Climate Change (UNFCCC), 198, 209, 212n16
utilitarian. *See* utilitarianism
utilitarianism, xxvii, 44, 44–45, 45, 52, 55–56, 156

Vatican II. *See* Second Vatican Council
vernacular theology, 5, 6
von Balthasar, Hans Urs, 144, 178

water, xiv, xv, xxix, xxxi, 10, 37n1, 53, 69, 83, 87, 88–89, 90, 96, 96–97, 98, 99, 100n1, 100n6, 100n8, 113, 127, 128, 129, 133, 135, 136, 137, 139, 141, 145, 154, 156, 157, 158, 160, 161–163, 163, 169n28, 190n2, 198, 200, 205–206, 207, 211n8–212n9, 223, 225–226, 231, 239, 240, 241, 243, 244, 253n4, 253n8, 259, 260, 263; non-renewable resource, 100; partially renewable 97
White Jr., Lynn, 71
Weigel, George, 94, 95
Wilson, E.O., 132
wisdom, xii, xxv–xxvii, 3, 4, 5, 6, 7, 8, 9, 14, 15, 16, 56, 93, 108, 127, 128–129, 129, 129–131, 131–132, 133, 134, 135, 137, 139, 140–142, 143–144, 164, 203, 237, 243. *See also* Wisdom tradition
Wisdom tradition, xxvii, 4, 15, 128, 130, 134, 140, 242
Wittgenstein, Ludwig, 103

Yup'ik Village of Newtok, xxix, 263

Zizioulas, John (John of Pergamon), 181
zoning, 223, 230

About the Contributors

Mary Ashley is a doctoral student in Ethics and Social Theory at the Graduate Theological Union in Berkeley, California. She holds master's degrees in Social Welfare from the University of California-Los Angeles and in Theology from the Jesuit School of Theology. Her research centers on issues of large-scale environmental degradation including greenhouse gases and industrial animal agriculture. She was awarded a Newhall Fellowship to teach an introductory Christian ethics course on the relation between humans and animals. She currently enjoys life in Oakland, California and the opportunity to participate in the East Bay's strongly personalist environmental culture of BART, bikes, and backyard chickens.

Michael Baur is Associate Professor of Philosophy and Adjunct Professor of Law at Fordham University in New York City. He holds a Ph.D. in philosophy from the University of Toronto and a J.D. from Harvard Law School. He currently serves as the Director of the Natural Law Colloquium at Fordham University, Secretary of the Hegel Society of America, and series editor of the Cambridge Hegel Translation for Cambridge University Press. He has published on a variety of thinkers and topics, including "Law and Natural Law" in the *Oxford Handbook of Aquinas* (Oxford, 2012), "From Kant's Highest Good to Hegel's Absolute Knowing" in the *Blackwell Companion to Hegel* (2011), and "The Authority to Interpret, the Purpose of Universities, and the Giving of Awards, Honors, and Platform" in *Journal of Catholic Legal Studies* (2011).

John T. Brinkman, MM, is an historian of religions whose work contributes to the history of thought in East and Southeast Asia. He has focused his recent research on ecology and religion and is engaged in the inter-religious environmental dialogue as indicated by his participation in the United Nations Framework Convention on Climate Change (UNFCCC) from the 1997 COP 3 in Kyoto, Japan to the 2012 COP 18 in Doha, Qatar. His UNFCCC articles are published by the Oriens Institute in Tokyo and The Office of Global Concerns in Washington DC. He authored *Simplicity: A Distinctive Quality of Japanese Spirituality* (Peter Lang 1996) and co-edited the UNEP volume *Earth and Faith*. Fr. Brinkman serves on the Federation of Asian Bishops' Conferences (FABC) Climate Change Desk, contributed to the Asian Bishops declaration "Church Response to the Challenge of Climate Change in Asia: Toward a New Creation," and is in the process of developing the second FABC second climate change seminar to be held 22-25 October 2013.

David Cloutier is Associate Professor of Theology at Mount St. Mary's University in Emmitsburg, Maryland where he teaches courses in moral theology. He is the author of *Love, Reason, and God's Story: An Introduction to Catholic Sexual Ethics* (2008), and editor of the collection *Leaving and Coming Home: New Wineskins for Catholic Sexual Ethics* (2010). Recent articles include "Working with the Grammar of Creation: Benedict XVI, Wendell Berry, and the Unity of the Catholic Moral Vision" (*Communio*) and "The Problem of Luxury in the Christian Life" (*Journal of the Society of Christian Ethics*). He is currently working on a book focusing on the moral problem of luxury, and serves as a director at the Common Market, the consumer food cooperative of Frederick, Maryland.

Anselma Dolcich-Ashley holds a bachelor's and a master's degree in Biology from Georgetown University and a doctoral degree in Moral Theology from the University of Notre Dame (2011) for which she wrote a dissertation on the Catholic sexual abuse crisis in the United States. She is the author of "'Pastors and the Other Faithful': The Sexual Abuse Scandal and the Emergence of Authority Relationships in the Church" in *Visions of Hope: Emerging Theologians and the Future of the Church* (Orbis). A post-doctoral teaching fellow at Notre Dame, she also is engaged in local-food initiatives and other outdoor endeavors.

Elizabeth Groppe is Associate Professor of theology at Xavier University and a member of the Climate Change Task Force of the Archdiocese of Cincinnati. She is the author of *Yves Congar's Theology of the Holy Spirit* (Oxford 2004), *Eating and Drinking* (Fortress, 2011), and numerous journal articles. Her areas of research include Trinitarian theology, theology and ecology, and Christian-Jewish dialogue.

Scott G. Hefelfinger is a doctoral candidate in Moral Theology at the University of Notre Dame. He received a B.A. in Music Composition from the University of California at Berkeley, as well as an S.T.M. and S.T.L. in Systematic Theology from the International Theological Institute in Austria. His current research centers around Thomas Aquinas' conception of nature and cosmology and how his Aristotelian inheritance contributes to moral theology. Recent publications and translations have appeared in *Logos: A Journal of Catholic Theology* and *Letter & Spirit*. He is also completing a book-length translation project pertaining to the poetic theology of Thomas Aquinas. He and his Austrian wife live in South Bend with their two children.

Monsignor Kevin W. Irwin is a priest of the Archdiocese of New York and served as the Dean of the School of Theology and Religious Studies at The Catholic University of America. He holds the Walter J. Schmitz, Chair of Liturgical Studies. Msgr. Irwin is the author of fourteen books on liturgy and sacraments, including his most recent *101 Questions and Answers on the Mass* (Paulist 2012). He is a member of the North

American Academy of Liturgy, Catholic Academy of Liturgy, Society for Catholic Liturgy, and Catholic Theological Society of America.

The Most Rev. Donald Kettler is the 4th Bishop of the Diocese of Fairbanks, Alaska to which he was appointed by Pope John Paul II in 2002. After graduating from St. John's University and Seminary in Collegeville, Minnesota, he was ordained to the priesthood in 1970, served as Associate Pastor of parishes in Aberdeen and Sioux Falls from 1970 until 1979, coordinated work for the Sioux Falls diocesan offices for several years, and received a licentiate degree in Canon Law (J.C.L.) from The Catholic University of America. He was subsequently named the Judicial Vicar for the Diocese of Sioux Falls and served as a weekly TV Mass celebrant, a member of several diocesan councils and boards, and as rector of St. Joseph Cathedral in Sioux Falls.

Christiana Z. Peppard is Assistant Professor of Theology, Science and Ethics at Fordham University. Her publications address the value of fresh water in an era of economic globalization, Catholic social teaching, naturalism and theological ethics, and poetry as an ethical methodology. She received her Ph.D. from Yale University in 2011. She is the author of the forthcoming *Waters for Life* from Orbis Books and is co-editing a volume with Andrea Viccini, SJ on achieving sustainability justly for Orbis. She speaks nationally and internationally on topics at the intersection of science, theology, and ethics. In 2013 she received the Catherine Mowry LaCugna Award from the Catholic Theological Society of America.

Jame Schaefer focuses on interfacing theology, the natural sciences, and technology with special attention to religious foundations for ecological ethics. She received her Ph.D. from Marquette University where she currently serves as Associate Professor of Systematic Theology and Ethics, directs the Interdisciplinary Minor in Environmental Ethics, and advises Students for an Environmentally Active Campus. Her major recent publications include *Theological Foundations for Environmental Ethics: Reconstructing Patristic and Medieval Concepts* (Georgetown University Press 2009), *Confronting the Climate Crisis: Catholic Theological Perspectives* (Marquette University Press 2011), and the inaugural "Animals" entry in the *New Catholic Encyclopedia* (2013). Prior to entering academia, she held leadership posts in environmental advocacy groups and several energy and environment policy positions by appointment of county, state, federal, and bi-national governments.

The Most Rev. Bernard Unabali was appointed Bishop of Bougainville, Papua New Guinea by Pope Benedict XVI in 2009 and installed the next year as the first native bishop of the Diocese. After attending local schools in Bougainville, he studied for the priesthood at St. Peter Chanel Seminary at Ulapia, spent a spirituality year at Erave, was sent to Bomana Major Seminary in Port Moresby to study philosophy and theology, was ordained a priest for the Diocese of Bougainville in 1985, and earned a degree in Missiology at the Pontifical Urban University. He has served as a parish priest, the diocesan head of youth pastoral, formator at the major seminary, teacher of Missiology, vicar general of Bougainville, diocesan vocations director, spiritual director of Bougainville Catholic Women's Association, director and pastoral coordinator of Mabiri Ministri Skul, and director of Central Bougainville Pastoral Team.

Jeremiah Vallery is a candidate for a doctor of philosophy degree in Systematic Theology at Duquesne University in Pittsburgh, Pennsylvania. He is writing his dissertation on Benedict XVI's cosmic soteriology and its applicability to environmentalism. He has an M.A. and B.A. from Ave Maria University in Ave Maria, Florida. His research interests include the theology of Benedict XVI, eschatology, ecological ethics, and biblical hermeneutics. He is currently an Adjunct Instructor of Theology at the University of St. Thomas in Houston, Texas.

Brother Keith Douglass Warner, OFM is a Franciscan Friar in the St. Barbara Province and a practical social ethicist in the Franciscan tradition. He earned an M.A. in Spirituality from the Franciscan School of Theology at Graduate Theological Union and a Ph.D. in Environmental Studies at the University of California-Santa Cruz. In 2003 he began teaching at Santa Clara University where he serves as the Director of Education and Action Research at the Center for Science, Technology, & Society and directs the Global Social Benefit Fellowship. An active participant in the retrieval of the Franciscan intellectual tradition, he researches the emergence of environmental ethics within scientific and religious institutions.

Matthew Philipp Whelan is a doctoral candidate in the Graduate Program in Religious Studies at Duke University where he is dissertating on Catholic teaching pertaining to land and agriculture with a focus on the issue of agrarian reform. He holds degrees from the University of Virginia, Centro Agronómico Tropical de Investigación y Enseñanza, and Duke University. His articles have been published in *Nova et Vetera, Crosscurrents*, the *CTS Annual Volume, The Other Journal, Biodiversity and Conservation*, and *Agroforestería en las Américas*. He currently resides in Waco, Texas with his wife, Natalie, along with their two daughters, Chora and Edith.

Tobias Winright, is Associate Professor of Theological Ethics at Saint Louis University where he teaches courses in fundamental moral theology and social ethics. He has co-authored (with Mark Allman) *After the Smoke Clears: Just War and Post War Justice* (Orbis, 2010), edited *Green Discipleship: Catholic Theological Ethics and the Environment* (Anselm Academic, 2011), and co-edited (with Margaret Pfeil) *Violence, Transformation, and the Sacred* (Orbis, 2012). He is also Co-editor (with Mark Allman) of the *Journal of the Society of Christian Ethics* and Book Reviews Editor of the journal, *Political Theology*. His Ph.D. is from the University of Notre Dame where he wrote his dissertation entitled "The Challenge of Policing: An Analysis in Christian Social Ethics," and he also holds an M.Div. from

Duke Divinity School. He taught at Simpson College in Indianola, Iowa, from 1998 to 2003 and at Walsh University in North Canton, Ohio from 2003 to 2005.